Peptide Synthesis and Applications

Second Edition

Edited by

Knud J. Jensen

Department of Chemistry, Faculty of Sciences, University of Copenhagen, Frederiksberg, Denmark

Pernille Tofteng Shelton

IGM, Faculty of Life Sciences, University of Copenhagen, Zealand Pharma, Glostrup, Denmark

Søren L. Pedersen

IGM, Faculty of Life Sciences, University of Copenhagen, Gubra, Hørsholm, Denmark

Humana Press

Editors
Knud J. Jensen
Department of Chemistry
Faculty of Sciences
University of Copenhagen
Frederiksberg, Denmark

Pernille Tofteng Shelton
IGM, Faculty of Life Sciences
University of Copenhagen
Zealand Pharma, Glostrup, Denmark

Søren L. Pedersen
IGM, Faculty of Life Sciences
University of Copenhagen
Gubra, Hørsholm, Denmark

ISSN 1064-3745 ISSN 1940-6029 (electronic)
ISBN 978-1-4939-6004-0 ISBN 978-1-62703-544-6 (eBook)
DOI 10.1007/978-1-62703-544-6
Springer New York Heidelberg Dordrecht London

Printed on acid-free paper

Humana Press is a brand of Springer
Springer is part of Springer Science+Business Media (www.springer.com)

METHODS IN MOLECULAR BIOLOGY™

Series Editor
John M. Walker
School of Life Sciences
University of Hertfordshire
Hatfield, Hertfordshire, AL10 9AB, UK

For further volumes:
http://www.springer.com/series/7651

Preface

Protocols for Solid-Phase Peptide Synthesis

Peptides are used ubiquitously for studies in biology, biochemistry, chemical biology, peptide-based medicinal chemistry, and indeed many other areas of research. There is a solid number of marketed peptide drugs, and the prospects for the development of new peptide drugs are very encouraging. Most of these peptides are prepared by chemical synthesis, where solid-phase peptide synthesis is the predominant method for preparation of peptides on a laboratory scale and increasingly also on an industrial scale.

Practical methodologies for peptide synthesis are the focus of this book. Thus not all reported methods could be described in length. The aim of this book is to provide laboratory protocols for both the specialist and the nonspecialist. The basic protocols provided here for solid-phase peptide synthesis are intended as a practical introduction to peptide synthesis, while the chapters on posttranslational and other modifications hopefully will also appeal to experienced peptide chemists.

The first chapter provides an introduction to the basic concepts in peptide synthesis and provides a starting (reference) point for the nonspecialist. The subsequent chapters provide protocols based on experience from the contributors' laboratories. It commences with basic protocols for the synthesis of linear, unmodified peptides. The following chapters describe protocols for the synthesis of C-terminally modified peptides, particularly peptide thioesters and aldehydes. This is combined with methods for native chemical ligation and expressed protein ligation. Next, chapters on cyclic peptides and posttranslationally modified peptides describe their synthesis, including in phospho-, glyco-, and lipopeptides. Methods for the assembly of peptidomimetic peptoids are described. Technology is also an important aspect of this book, hence an overview of instruments is provided, which is followed by methods for the relatively new area of microwave heating in peptide synthesis. Finally, the assembly of peptide and glycopeptide microarrays is included in the chapters on glycopeptide synthesis.

We have aimed at presenting a broad range of synthetic methods and different approaches to the synthesis of peptides. To achieve this we asked peptide scientists from around the world to contribute protocols based on the chemistries they use in their own laboratories. We thank all the authors for contributing excellent chapters to this book.

Frederiksberg, Denmark — *Knud J. Jensen*
Zealand Pharma, Glostrup, Denmark — *Pernille Tofteng Shelton*
Gubra, Hørsholm, Denmark — *Søren L. Pedersen*

Contents

Contributors

MUHARREM AKCAN • *Institute for Molecular Bioscience, The University of Queensland, Brisbane, QLD, Australia*

CHRISTOPHER J. ARMISHAW • *Torrey Pines Institute for Molecular Studies, Port St. Lucie, FL, USA*

LARS BAUMANN • *Institute of Biochemistry, Universität Leipzig, Leipzig, Germany*

ANNETTE G. BECK-SICKINGER • *Institute of Biochemistry, Universität Leipzig, Leipzig, Germany*

OLA BLIXT • *Department of Chemistry, Faculty of Sciences, University of Copenhagen, Copenhagen, Denmark*

EMILIANO CLÓ • *Novo Nordisk A/S, Måløv, Denmark*

DAVID J. CRAIK • *Institute for Molecular Bioscience, The University of Queensland, Brisbane, QLD, Australia*

PAUL R. HANSEN • *Department of Drug Design and Pharmacology, Faculty of Health and Medical Sciences, University of Copenhagen, Copenhagen, Denmark*

KIM B. HØJLYS-LARSEN • *Symphogen A/S, Lyngby, Denmark*

MOHAMMED AKHTER HOSSAIN • *Florey Neuroscience Institutes, School of Chemistry, The University of Melbourne, Parkville, VIC, Australia*

KNUD J. JENSEN • *Department of Chemistry, Faculty of Sciences, University of Copenhagen, Frederiksberg, Denmark*

FENG LIN • *Florey Neuroscience Institutes, The University of Melbourne, Parkville, VIC, Australia*

LEILA MALIK • *Department of Chemistry, Faculty of Sciences, University of Copenhagen, Frederiksberg, Denmark*

JENS K. MUNK • *Department of Drug Design and Pharmacology, Faculty of Health and Medical Sciences, University of Copenhagen, Copenhagen, Denmark*

SØREN L. PEDERSEN • *IGM, Faculty of Life Sciences, University of Copenhagen, Gubra, Hørsholm, Denmark*

SØREN W. PEDERSEN • *Department of Drug Design and Pharmacology, University of Copenhagen, Copenhagen, Denmark*

RENÉE ROODBEEN • *Department of Chemistry, Faculty of Sciences, University of Copenhagen, Frederiksberg, Denmark*

FEDERICA ROSI • *Abt. Chemische Biologie, Max-Planck-Institut für Molekulare Physiologie, Dortmund, Germany*

FAZEL SHABANPOOR • *Florey Neuroscience Institutes, School of Chemistry, The University of Melbourne, Parkville, VIC, Australia*

MAX STEINHAGEN • *Institute of Biochemistry, Universität Leipzig, Leipzig, Germany*

KRISTIAN STRØMGAARD • *Department of Drug Design and Pharmacology, University of Copenhagen, Copenhagen, Denmark*

PERNILLE TOFTENG SHELTON • *IGM, Faculty of Life Sciences, University of Copenhagen, Zealand Pharma, Glostrup, Denmark*

GEMMA TRIOLA • *Abt. Chemische Biologie, Max-Planck-Institut für Molekulare Physiologie, Dortmund, Germany*
GRACE S. VANIER • *CEM Corporation, Matthews, NC, USA*
JOHN D. WADE • *Florey Neuroscience Institutes, School of Chemistry, The University of Melbourne, Parkville, VIC, Australia*

Chapter 1

Solid-Phase Peptide Synthesis: An Introduction

Knud J. Jensen

Abstract

This chapter provides an introduction to and overview of peptide chemistry with a focus on solid-phase peptide synthesis. The background, the most common reagents, and some mechanisms are presented. This chapter also points to the different chapters and puts them into perspective.

Key words Solid-phase peptide synthesis, SPPS, Fmoc, Boc, Linkers, Coupling reagents, HBTU, HATU, COMU, HOBt, HOAt, Oxyma, BAL

Abbreviations

Acm	Acetamidomethyl
BAL	Backbone Amide Linker
Boc	*Tert*-butyloxycarbonyl
BOP	Benzotriazol-1-yl-*N*-oxy-tris(dimethylamino)phosphonium hexafluorophosphate
Cbz	Benzyloxycarbonyl
COMU	(1-[(1-(Cyano-2-ethoxy-2-oxoethylideneaminooxy)-dimethylamino-morpho lino-methylene)] methanaminium hexafluorophosphate)
DIC	*N,N'*-Diisopropylcarbodiimide
DIEA	*N,N*-Diisopropylethylamine
DMF	N,N-Dimethylformamide
Fmoc	Fluoren-9-ylmethyloxycarbonyl
HATU	(N-[(Dimethylamino)-1H-1,2,3-triazole[4,5-b]pyridine-1-ylmethylene]-N-methyl-methanaminium hexafluorophosphate N-oxide)
HBTU	(N-[(1*H*-Benzotriazol-1-yl)(dimethylamino)methylene]-*N*-methylmethan-aminium hexafluorophosphate N-oxide)
HOBt	1-hydroxybenzotriazole
HOAt	3-Hydroxy-3H-1,2,3-triazolo[4,5-b]pyridine [1-hydroxy-7-azabenzotriazole
NHS	N-Hydroxysuccinimide
NMP	N-Methyl-2-pyrrolidinone
Pbf	2,2,4,6,7-Pentamethyl-dihydrobenzofuran-5-sulfonyl
Pfp	Pentafluorophenyl

Knud J. Jensen et al. (eds.), *Peptide Synthesis and Applications: Second Edition*, Methods in Molecular Biology, vol. 1047, DOI 10.1007/978-1-62703-544-6_1, © Springer Science+Business Media New York 2013

PyBOP	(1-benzotriazolyloxy-tris-pyrrolidinophosphonium hexafluorophosphate)
TFA	Trifluoroacetic acid
TFMSA	Trifluoromethanesulfonic acid
Trt	Trityl (triphenylmethyl)

1 Introduction

Synthetic peptides are ubiquitous in biology, biomedicine, drug discovery, and many other fields. Chemically synthesized peptides serve very diverse purposes, including as biopharmaceutical drugs, for epitope mapping, in peptide microarrays, and in vaccine development. While proteins generally are prepared by recombinant methods, chemical synthesis is the prevailing method for the preparation of peptides. This is due to the ease, predictability, and flexibility of chemical synthesis, which also allows the convenient incorporation of many non-proteinogenic modifications. Peptide synthesis has allowed the preparation of numerous peptides, from laboratory scale up to ton scale. However, there are also limitations—or current limitations—and having an understanding of both the possibilities and the limitations will enable the optimal use of peptides in research and development. This introductory chapter will provide an overview of some of the most common methods in solid-phase peptide synthesis (SPPS) but will start with solution synthesis of peptides to introduce some concepts.

This introduction would not be complete without mentioning a few key events in the history of peptide synthesis. In 1901 Emil Fischer reported the synthesis of a peptide, Leonidas Zervas developed the Z (Cbz) protecting group in 1932, and Vincent du Vigneaud was awarded the Nobel Prize in 1955 for the chemical synthesis of the cyclic peptide, oxytocin. In 1963 Robert Bruce Merrifield introduced the concept and the first implementation of solid-phase peptide synthesis for which he was awarded the 1984 Nobel Prize in chemistry [1, 2]. Solid-phase peptide synthesis has for decades been the primary source of synthetic peptides on a laboratory scale.

2 Solution Synthesis of Peptides and Chemical Ligation

Amino acids carry at least two functional groups, as the name indicates. When two amino acids have to be coupled together by an amide bond, one of the *N*α-amine and one of the carboxyl groups have to be protected (Fig. 1). This is needed to ensure formation of the correct amide bond, which is often called a "peptide bond." For bifunctional amino acids such as Gly, Ala, and Leu, only one protecting group is required for each amino acid. In trifunctional amino acids side-chain protecting groups are required to protect

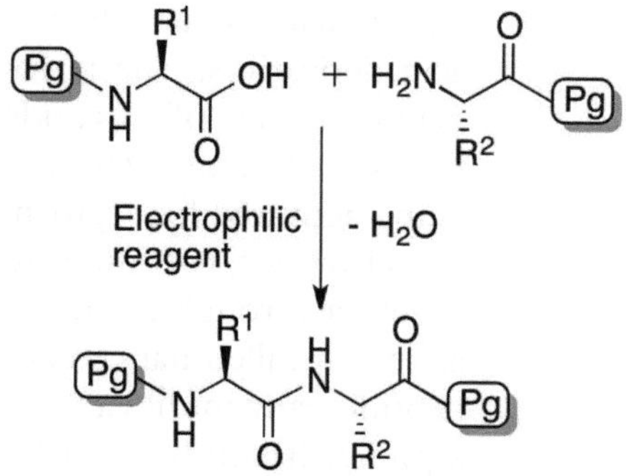

Fig. 1 Solution synthesis of a dipeptide (Pg, protecting group for the N^{α}-amino and moieties). Pg could, for example, be the Cbz and benzyl groups. Side-chain protecting groups may also be required

functional groups such as carboxyls (Asp, Glu), amines (Lys), thiols (Cys), hydroxyls (Ser, Thr, Tyr) imidazoles (His), and sometimes indoles (Trp) to prevent undesired reactions at these functional groups. These side-chain protecting groups need to be removed at the end of the synthesis.

Some of the protecting groups, which are used in solution synthesis, are benzyl derivatives. They are conveniently removable by hydrogenolysis, i.e., by treatment with hydrogen gas in the presence of a suitable catalyst. One example is the classical carboxybenzyl (Cbz) protecting group for amines, which was reported for peptide synthesis by Bergmann and Zervas in 1932 [3]. The Cbz group was the first in a long series of protecting groups where the amine forms part of a carbamate. Thus, the *N*α-amine of one amino acid could be protected by Cbz, while the carboxylic acid of the other amino acid could be protected as a benzyl ester. After coupling of the two amino acids by formation of the intended amide bond, the two protecting groups can be removed in one step. Amide bond formation requires that the hydroxyl of the carboxylic acid is converted to a better leaving group, which will be described in Subheading 6.

Not only can two amino acids be coupled together in solution, two *peptides* can also be coupled together in solution. This is often referred to as a segment condensation. A segment condensation based on direct amide formation requires that all the side-chains are protected on each peptide, such that one peptide has a free C-terminal carboxylic acid and the other a free N-terminal amino group. They are then coupled together by the formation of an amide bond. This means that the two protected segments can be prepared and purified separately before being coupled together. While this can be an attractive approach in some cases, it can need extensive optimization, as protected peptides can suffer from low solubility. Furthermore, carbamate *N*α-protecting groups such as

Cbz actually reduce the risk of epimerization in an activated amino acid derivatives. The nitrogen in the C-terminal amino acid in a peptide is part of an amide and not a carbamate, which increases the risk of epimerization at the C-terminal amino acid which is being activated for segment couplings.

However, two unprotected peptide segments can be coupled together through the use of a highly chemoselective reaction such as the so-called native chemical ligation (NCL) [4]. In NCL one peptide segment has a C-terminal thioester moiety, while the other segment has to have an N-terminal Cys residue, and the reaction proceeds via transthioesterification to provide a "native" amide bond; *see* Chapter 7 for NCL and Chapter 8 for the synthesis of thioesters. However, unprotected peptides can also be coupled through the formation of a nonnative bond, such as by highly chemoselective oxime [5, 6] or triazole formation. The former requires access to peptide aldehydes, the synthesis of which is described in Chapter 9.

3 Solid-Phase Peptide Synthesis

Merrifield's introduction of a functionalized solid support that allows for anchoring of an amino acid revolutionized the field of peptide science and inaugurated the SPPS methodology. Since the initial reports in the 1960s, all aspects of SPPS have been developed further and refined, thus extending the reach of synthetic peptide chemistry tremendously. Solid-phase synthesis has evolved into a highly efficient set of techniques for the preparation of numerous peptides and even small proteins. It has been crucial in the development of combinatorial and high-throughput chemistry and provides molecules for chemical biology, medicinal chemistry, and many other areas of research. While the term "solid-phase synthesis" is commonly used, maybe a more precise terminology would be "matrix assisted synthesis," as the resins most commonly used are anything but solids [7, 8]. The main characteristics of solid-phase synthesis are that (1) the first building block is attached (anchored) to a matrix, which can be filtered; (2) repeated cycles of chemical transformations (especially deprotection and coupling) are performed, also by automation; and (3) most often the final product is released from the matrix and deprotected in the same step. However, by proper choice of protecting groups and linker, the peptide can also either be deprotected while it remains attached to the support or be released from the support in the fully protected form (Fig. 2).

SPPS is defined by the set of *N*α-protecting groups, side-chain protecting groups, coupling reagents, linkers (handles), as well as the solid supports which can be used. Suitably *N*α- and side-chain protected amino acids are coupled sequentially in the *N*←*C*

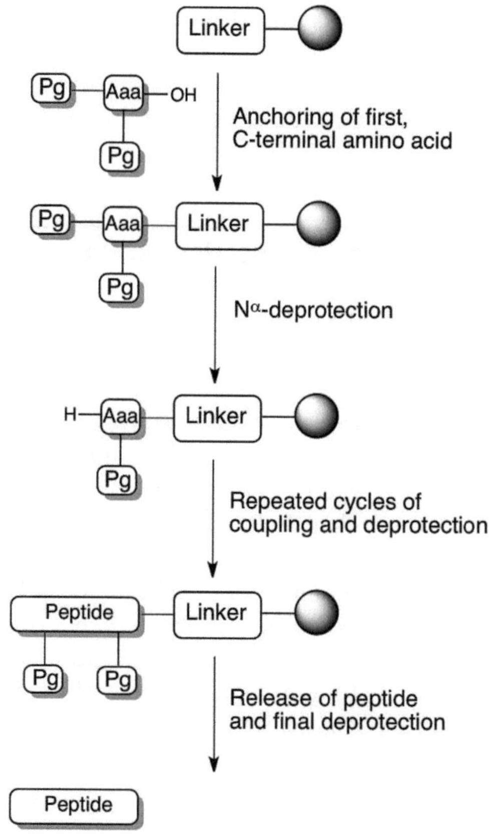

Fig. 2 Principles of solid-phase peptide synthesis (SPPS)

direction to a growing peptide chain anchored to the resin. Typically, the C-terminal amino acid is first anchored at the carboxy terminus to the solid support via a cleavable handle. Then the *N*α-protecting group can be removed without affecting the side-chain protecting groups; thus, the polypeptide chain is prepared for the next coupling cycle. SPPS reactions are driven to completion by the use of soluble reagents in excess, which can be removed by filtration and washing. Following the completion of the desired sequence of amino acids, the peptide is released from the solid support, and simultaneously the semipermanent side-chain protecting groups are typically removed concomitantly (Fig. 2, final step). The amino acid protecting groups, coupling reagents, and resins have been refined over the last three decades, and they are now very efficient in routine syntheses. The field of peptide synthesis continues to evolve to allow for the synthesis of even longer peptides and proteins, incorporation of posttranslational or unnatural modifications, use of microwave heating, and more.

At this point, it is important to remember that amino acids are defined as they appear in proteins. Proteins are depicted with the

N-terminal to the left and the C-terminal to the right. We thus recommend always drawing amino acids with the nitrogen left and the carboxylate right. When depicting SPPS reactions where the peptide is anchored by its C-terminal anchoring, hence, it is mandatory to place the resin to the *right*.

4 N^{α}-Amino Protecting Groups: Boc vs. Fmoc

The two most generally used protecting groups in SPPS are the fluoren-9-ylmethyloxycarbonyl (Fmoc) [9, 10] (Fig. 3) and the *tert*-butoxycarbonyl (Boc), with each $N\alpha$-protecting group defining an overall strategy for SPPS. The chemical conditions for removal of these transient protecting groups, i.e., base vs. acid, each define a "chemical window" of opportunities for the other chemical steps in the overall SPPS strategy. Therefore, the solid-phase strategy is defined by the choice of the $N\alpha$-protecting group for the amino acid building blocks. This $N\alpha$-protecting group will be removed after each coupling, and the chemical conditions required for removal thus define what the linker (often also referred to as a "handle") and the side-chain protecting groups have to be compatible with, hence, which conditions they are stable to. Conversely, the linkage to the solid support defines the chemistry possible in the following steps and the conditions for repetitive

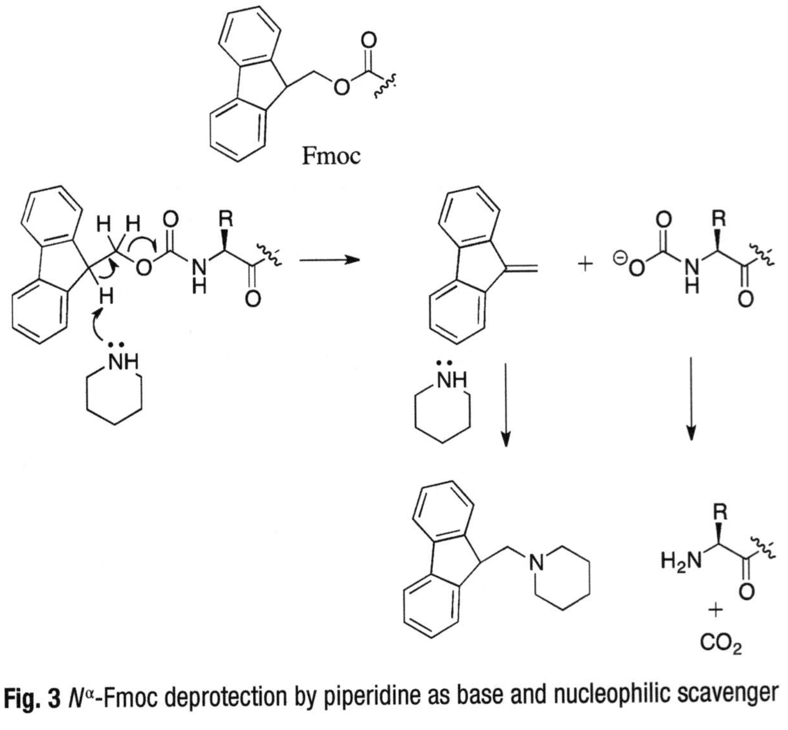

Fig. 3 N^{α}-Fmoc deprotection by piperidine as base and nucleophilic scavenger

removal co-determine the choice of semipermanent side-chain protecting groups and handle. In the following, Fmoc-based solid-phase peptide synthesis will be referred to as Fmoc-SPPS, while Boc-based solid-phase peptide synthesis will be referred to as Boc-SPPS. The standard protocols for Fmoc-SPPS are described in Chapters 2 and 3, while Chapter 4 describes some protocols for Boc-SPPS.

The Boc strategy, initially introduced by Merrifield, requires trifluoroacetic acid (TFA) for repetitive removal of the Boc groups while often relying on hydrofluoric acid (HF) for release of the final peptide from the support [11, 12]. Thus, the Boc strategy relies on differences in acid-lability of the *N*α- and side-chain protecting groups, hence, on graduated acid-lability. Due to the use of corrosive and toxic HF and the requirement for a specialized apparatus to handle HF in Boc-SPPS, the Fmoc strategy is often preferred over the Boc strategy for routine synthesis. The Fmoc group can be removed under mild conditions with secondary amines, typically a 1:4 solution of piperidine in DMF (Fig. 3) [13, 14].

Boc-SPPS normally relies on graduated acid-lability between the Boc (removed with TFA) and the linkage to the support (typically cleaved with HF or, alternatively, trifluoromethanesulfonic acid, TFMSA). In contrast, Fmoc-SPPS often displays an orthogonality [15] between the conditions for removal of the Fmoc protecting group (e.g., piperidine) and the conditions required for release from the support (often TFA). In the following, we will focus on Fmoc-SPPS.

5 Side-Chain Protecting Groups

The semipermanent side-chain protecting groups for Fmoc-SPPS have been developed extensively during the past decades [16]. For some trifunctional amino acids such as Cys, Asp, Glu, and Lys, side-chain protection is essential for successful peptide synthesis; however, generally all other trifunctional amino acids are also semipermanently side-chain protected. The currently used protecting groups are *tert*-butyl (*t*-Bu) ester for Glu and Asp; *t*-Bu ether for Ser, Thr, and Tyr; 2,2,4,6,7-pentamethyl-dihydrobenzofuran-5-sulfonyl (Pbf) [17] for Arg; and trityl (Trt) for Cys, Asn, Gln, and His (Fig. 4). The phenyl rings in Trt-derivatives can easily be modified with electron-withdrawing or electron-donating groups, which can be used to fine-tune their properties as protecting group. For example, the monomethyl and monomethoxy-trityl moieties are acid-labile protecting groups for the *N*ε-amine in Lys [18].

In addition, there are protecting groups that are typically not removed under the conditions required for the final release of the peptide from the support. They include the acetamidomethyl (Acm) protecting group for Cys, which can be removed selectively with heavy metal salts such as thallium trifluoroacetate or,

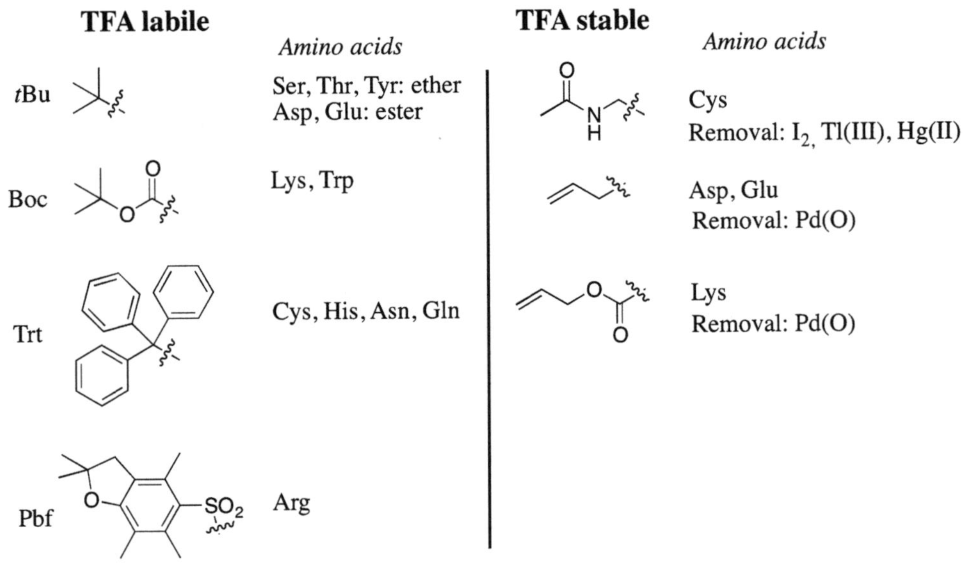

Fig. 4 Some common side-chain protecting groups used in Fmoc-SPPS. The relevant amino acids are mentioned below. The protecting groups in the *left column* can be removed with concentrated TFA, while the protecting groups in the *right column* are stable to TFA and are removed by other conditions

environmentally more benign, by iodine. Cys(Acm) can be used both in Boc- and Fmoc-SPPS. The conditions for Acm removal, which are orthogonal to the conditions for removal of most other protecting groups, enable the use of Acm for the directed and sequential installment of disulfide bridges in peptides with multiple disulfides. There are also side-chain protecting groups for Lys, which allow the chemoselective deprotection of the *N*ε-amine while leaving other side-chain protecting groups on the peptides. These Lys protecting groups include Alloc [19], MMT [20], as well as Dde [21] and ivDde. Similarly, selectively removable protecting groups for Glu include allyl and 2-phenyl-*iso*-propyl (Ph*i*Pr) [22].

6 Coupling Reagents

Activation of the carboxylic acid moiety of the amino acid is required to be able to react with the *N*α-amino group of the growing peptide chain. The first step is the reaction with an electrophile, in some cases in the presence of a base. Carbodiimide-based coupling reagents, such as DCC or DIC (*N*,*N*′-diisopropylcarbodiimide), have been used for decades (Figs. 6 and 7). Potential side reactions with carbodiimide-based reagents include the *O*-to-*N* rearrangement of the *O*-acylisourea intermediate and "overactivation" by formation of the symmetrical anhydride, which can lead to epimerization (Fig. 5). These side reactions can be prevented by the addition of auxiliary nucleophiles such as 1-hydroxybenzotriazole

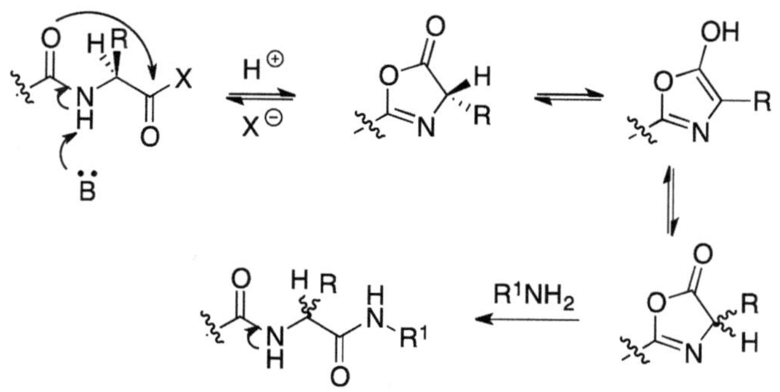

Fig. 5 Oxazolone mechanism for racemization

(HOBt) [23], or 1-hydroxy-7-azabenzotriazole (HOAt) [24], which form the corresponding activated esters. A relative newcomer is ethyl 2-cyano-2-(hydroxyimino)acetate (Oxyma) [25]. Auxiliary nucleophiles, such as HOBt, ensure that the optical integrity of the stereogenic center at the C-terminal of the activated amino acid residue is maintained throughout the coupling step. In automated syntheses, the coupling reagent DIC is strongly preferred over DCC, as DIC is a liquid and as the resulting urea is soluble.

Numerous so-called in situ coupling reagents have been developed to reduce coupling time and minimize epimerization. The most important are HBTU (*N*-[(1*H*-benzotriazol-1-yl)(dimethylamino)methylene]-*N*-methylmethanaminium hexafluorophosphate *N*-oxide) [26], HATU (*N*-[(dimethylamino)-1*H*-1,2,3-triazole[4,5-b]pyridine-1-ylmethylene]-*N*-methylmethanaminium hexafluorophosphate *N*-oxide) [27, 28], PyBOP (1-benzotriazolyloxy-tris-pyrrolidinophosphonium hexafluorophosphate) [29], and the novel COMU (1-[(1-(Cyano-2-ethoxy-2-oxoethylideneaminooxy)-dimethylamino-morpholino-methylene)] methanaminium hexafluorophosphate) [30] reagents (Fig. 6). Although initial reports showed the structures of HBTU and HATU as uronium salts, it has since then been shown that both compounds crystallize as aminium salts (guanidinium *N*-oxides) [31, 32]. A recent study from our group has demonstrated that while COMU is often a very efficient coupling reagent, both commercially available COMU and the activated Oxyma esters generated by it are more prone to hydrolysis than other in situ reagents and activated esters, respectively [33]. Solutions of COMU should either be freshly prepared or be stored to avoid exposure to humid air, e.g., under inert gas in an automated synthesizer.

Although peptides normally are assembled by in situ activation, i.e., by activation of the carboxylic acid immediately prior to coupling or (Figs. 7, 8, and 9) in the presence of the free amine, preactivated amino acid building blocks are also a viable possibility. Thus, pentafluorophenyl (Pfp) [34] esters of some Fmoc-amino acids are commercially available or can be prepared easily. For these couplings, an auxiliary nucleophile such as HOBt is added, which

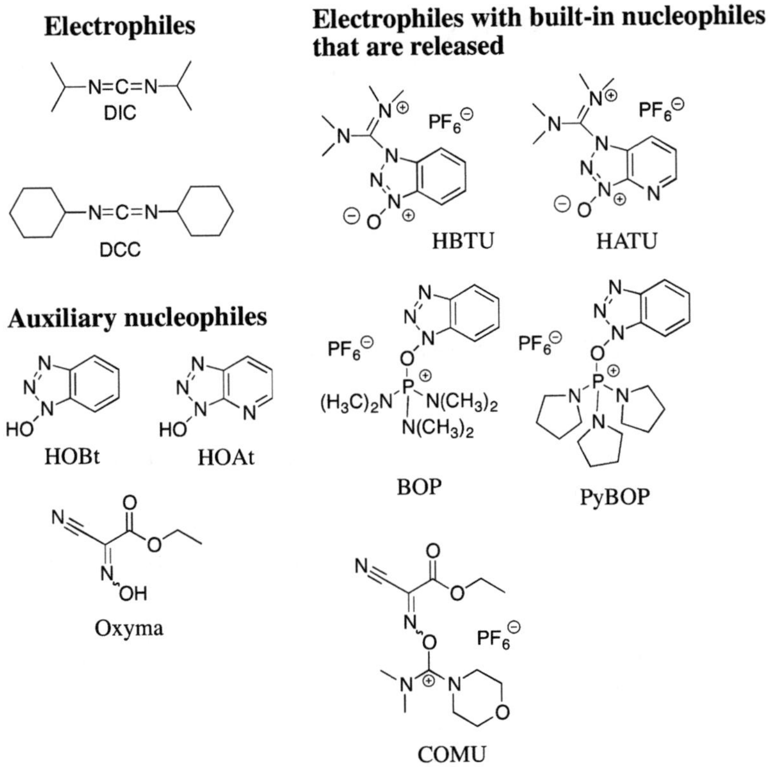

Fig. 6 Some coupling reagents: electrophiles, nucleophiles, and combinations

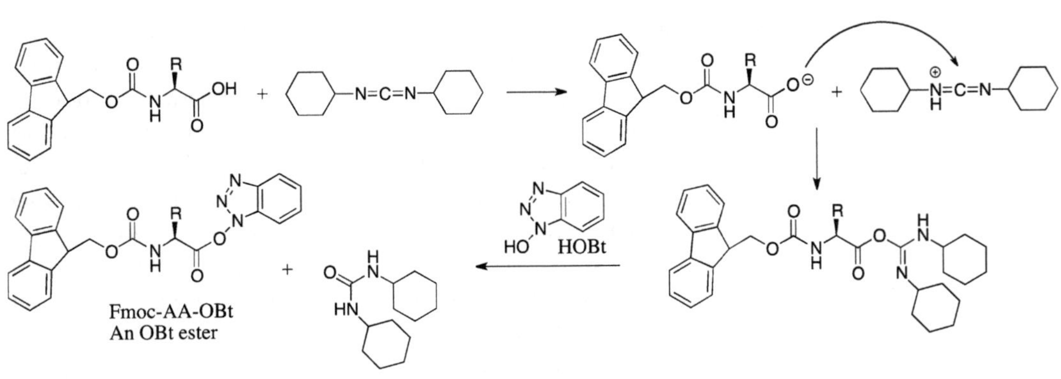

Fig. 7 Coupling chemistry: formation of activated ester using DIC

then generates the transient OBt ester. The *N*-hydroxysuccinimide (NHS) esters are mainly used in amide bond formations in aqueous solution and not on solid-phase.

While conductive heating occasionally has been applied to peptide synthesis, precise microwave irradiation to heat the reaction mixture during coupling and $N\alpha$-deprotection has become increasingly

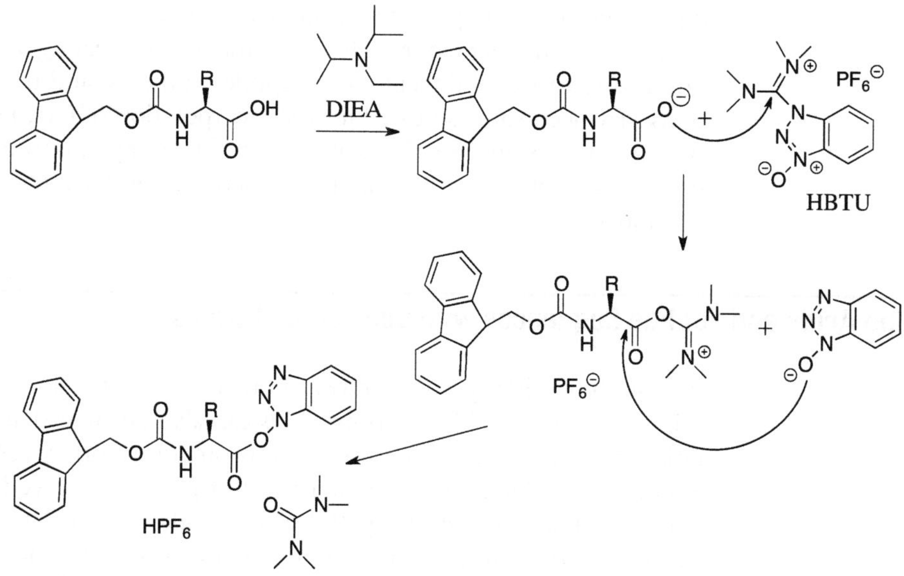

Fig. 8 Coupling chemistry: formation of activated ester using HBTU

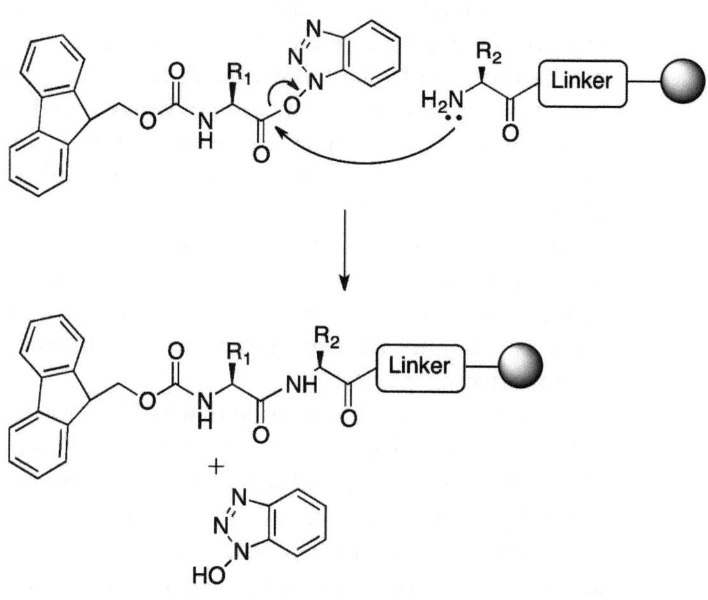

Fig. 9 Coupling chemistry: reaction of activated ester with peptidyl-resin for peptide chain elongation

popular. It has often provided dramatic reductions in synthesis times, accompanied by an increase in the crude peptide purity. Microwave heating has been proven particularly relevant for sequences that form β-sheet type structures and for sterically difficult couplings. While some reports state that the microwaves interact directly with dipole moments in peptides, systematic studies indicate that the effect is of a thermal nature. However, microwave heating as such is not a panacea for all difficulties in peptide syntheses, and the conditions may need to be adjusted for the incorporation of Cys, His, and Asp in peptides and for the synthesis of, for example, phosphopeptides, glycopeptides, and *N*-methylated peptides. *See* Chapters 10, 15, 16, and 17.

7 Amino Acids and Peptide Sequences with Special Challenges

Although peptide syntheses most often are successful, some amino acids are prone to side reactions. This can also be viewed as a limitation in the current chemistry for protection and coupling of amino acids. For example, when the Pbf group is cleaved of the Arg side-chain, the resulting Pbf electrophile can alkylate nearby Trp indole rings in a reaction which is not reversible under these conditions. This can often be minimized by using a Trp building block that is Boc protected at the indole nitrogen [35]. The Asn residue is prone to dehydration and aspartimide formation, although side-chain Trt protection solves some of the problem. The thiol moiety in Cys can be protected by a number of protecting groups during Fmoc-SPPS. While the Trt protecting groups can be removed by acidolysis, the Acm group is stable to acid and base but can be removed by salts of heavy metals, such as Hg and Tl, or by I_2. This provides a level of orthogonality, which can be utilized in the synthesis of peptides with multiple disulfide bridges. However, the incorporation of Cys in peptides is also likely to cause some level of racemization [36, 37]. His can also be epimerized during Fmoc-SPPS; however, the Trt side-chain protecting group is often well suited to prevent this. β-Branched amino acids such as Val and Ile can sometimes be difficult to incorporate quantitatively.

It seems reasonable to hope that SPPS in the foreseeable future will be able to reliably provide proteins by direct linear synthesis. However, low purities and sometimes even failure in achieving the desired peptide sequence are still frequently occurring problems, especially as the peptide becomes longer. The main reasons for this are believed to be steric hindrance and intra- and intermolecular packing. Amino acids which are prone to form β-sheets often lead to aggregation during peptide strand elongation, most likely due to their hydrogen bonding and hydrophobic properties. These problems often lead to premature terminations or deletions of the elongating peptide sequence, which can be tedious to purify afterwards. Intermolecular aggregation often leads to poor solvation of the peptidyl-polymer, but it is less pronounced when resins with a low loading are being used. *See* Chapter 2.

8 Resins

The most common resins for SPPS are based on polystyrene (PS), typically with 1 % cross-linking. A starting point for the production of many other resins is provided by chloromethyl-polystyrene (Merrifield resin), made either by "chloromethylation" of polystyrene or, better, by copolymerization to directly incorporate the chloromethyl moiety. Another important base resin is aminomethyl-polystyrene. Although chloromethyl-polystyrene can be used for Boc-SPPS, in by far most cases a dedicated linker (handle) is inserted between the base resin and the first amino acid.

While polystyrene is an inexpensive resin that has been used widely, especially in Boc-SPPS and in Fmoc-SPPS of shorter sequences, other resins provide certain advantages. An important class of resins is constructed from a polystyrene core onto which PEG chains have been attached. TentaGel (Rapp Polymere, Germany) carries amino groups at the end of the PEG chains. It is important to realize that these polystyrene microparticles have been functionalized with PEG inside out, thus not only on the surface of the resin bead. These supports often provide higher purity of synthesized peptide, especially with longer sequences and with peptides that are prone to aggregate.

Another class of resins consists mainly of PEG and contains no polystyrene. They rely on cross-linking of PEG chains. This class includes the poly(ethylene glycol)-poly-(*N*,*N*-dimethylacrylamide) copolymer (PEGA) supports, developed by Meldal [38], and the newer ChemMatrix [39] resins. While PEGA is unique in being permeable to proteins up to 35–70 kDa, which makes it well suited for biochemical studies of peptides immobilized to the support, the ChemMatrix resins have gained in importance for standard Fmoc-SPPS. *See* Chapter 2.

9 Linkers (Handles)

9.1 Linker Overview

Linkers or handles are bifunctional molecules that on one side allow anchoring to the support and on the other have the characteristics of a protecting group, which enables anchoring of the growing peptide chain during SPPS and release of the final product under well-defined conditions.

Many resins are commercially available with a suitable linker attached to them. However, most base resins can easily be functionalized with linkers. Typically, this is achieved by acylation of a resin containing a primary amino group. Nucleophilic displacement of the chloride in Merrifield resin is also a well-established method, however, this often then generates a benzylic ether linkage which may be somewhat labile.

Anchoring of the first amino acid normally occurs through the C-terminal and the linker has to be chosen according to whether the peptide will have a C-terminal carboxylic acid or an amide [40–45]. Here we will focus on linkers for Fmoc-SPPS. For the synthesis of peptides with a C-terminal carboxylic acid, 4-alkoxybenzyl alcohol-type (Wang) linkers are an obvious choice. The first amino acid is coupled to Wang-type handles by esterification, and attention to potential racemization in the formation of the ester bond has to be paid. Release of the peptide acid is achieved by treatment with conc. TFA. Substituted benzyl alcohol linkers, e.g., dialkoxybenzyl structures, provide higher acid-lability. Hence, the release of peptides can be achieved with lower concentrations of TFA (e.g., TFA-dichloromethane mixtures). Other linkers for the synthesis of peptide acids include trityl-based handles, e.g., the chloro-trityl chloride linker developed by Barlos. They have in general a higher acid-lability and the first amino acid is anchored racemization-free by nucleophilic displacement of the chloride. *See* Chapter 2.

For syntheses of peptides as their C-terminal amides in Fmoc-SPPS, the most common linker is a benzhydryl-type handle, the Rink amide linker. Most commonly used resins are available with a Rink amide linker preinstalled. The PAL [46, 47] ("peptide amide linker") handle, which has a trisalkoxybenzyl structure, is also very suitable for Fmoc-SPPS of peptide amides.

Many biologically active peptides or peptide building blocks are either cyclic or C-terminally modified, meaning that they have a C-terminal functionality other than a carboxylic acid or an amide. Their synthesis would be difficult or impossible through C-terminal anchoring. There are specialized linkers, which can provide this, for example, for the synthesis of peptide thioesters; *see* Chapter 8. Alternatively, a general strategy to obtain peptides with a C-terminus other than carboxylic acid or an amide uses anchoring of the peptide through a "vacant" backbone amide nitrogen. This so-called backbone amide linker (BAL) [48, 49] methodology is now widely used; *see* Chapter 9. In this strategy, the first amino acid is anchored by reductive amination, followed by acylation of the newly formed secondary amine. Thus, the growing peptide chain is anchored *not* through the C-terminal carboxyl but through a backbone amide nitrogen giving access to, in principle, any C-terminal modification. In the first implementation of this general strategy, amino acid derivatives were attached by convenient and reliable reductive amination to support-bound 5-(formyl-dimethoxyphenoxy)valeric acid, forming a trialkoxybenzylamine linkage. A defining feature of the BAL strategy is that the handle precursor is an aromatic aldehyde. BAL strategies have been used for the synthesis of peptide aldehydes [50], peptide thioesters [51], cyclic peptides, etc. The BAL strategy has mainly been used in Fmoc-SPPS, but it had also been adapted to Boc-SPPS [52].

Table 1
Cleavage conditions

	Leaving group					
Linker	Sulfonamide	Urea	Carbamate	Amide	Aniline	Amine
Dialkoxy-benzyl BAL	TFA-DCM 5:95, 5 min	TFA-TES 97.5:2.5, 15 min	TFA-DCM 1:19	TFA-DCM 3:7, 10 min	–	–
Trialkoxy-benzyl BAL	TFA-$CHCl_3$-H_2O 50:50:1, 1 h		TFA-$CHCl_3$-H_2O 50:50:1, 1 h	TFA-DCM 5:95–1:99, 1–2 h	TFA-H_2O 19:1, 2 h	TFA-DCM 95:5, 16 h
Indole BAL	TFA-DCM 1:99, 3 min	TFA-DCM 1:99, <2 min	TFA-DCM 1:99, 1 min	TFA-DCM 1:99, 8 h		

Another strategy for C-terminal modification and cyclization is side-chain anchoring. Here, the first amino acid is anchored through the side-chain functional group. This could be Asp/Asn, Glu/Gln, or indeed most other trifunctional amino acids.

9.2 Acidolytic Release and Deprotection of Peptides

All handles mentioned so far release the peptide from the support by acidolysis. The acid used for release of peptides in Fmoc-SPPS is normally TFA, as most resins swell well in TFA, peptides are often soluble in TFA, and TFA is volatile and easily evaporated. Concomitant to acidolytic release from the solid support, acid-labile side-chain protecting groups are also removed. This can generate a diverse set of carbocations, which are good electrophiles capable of reacting with nucleophilic side functionalities and thus causing side reactions. As mentioned, the cation derived from the Pbf protecting group can alkylate the indole ring in Trp. Thus, nucleophilic "scavengers" are added to the TFA. The most common is simply 5 % of water, which normally is sufficient for *tert-butyl* protecting groups. Silanes such as triethyl silane (TES) or trisisopropyl silane (TIS) are added to quench mainly trityl cations, while thiols can scavenge other cations. *See* Chapter 3.

The conditions for the release of a peptide from a linker depends not only on the linker structure but also on the nature of the leaving group, i.e., the functional group it releases. This has been summarized for BAL-type handles in Table 1, where the acid-lability of three linkers is also compared. In summary, the leaving group capability decreases in the order sulfonamide, carbamate ~ urea, secondary (substituted) amide > primary amide > amine.

9.3 Linkers for the Synthesis of Peptide Thioesters

Several handles release the peptides by nucleophilic displacement and the C-terminal functional group can be introduced in this step. In the 4-hydroxymethylbenzoic acid (HMBA) linker, the peptide is anchored through an ester bond. The peptide is released by nucleophiles, such as hydroxide or another competent nucleophile. HMBA is maybe best used in combination with PEGA resins. Another handle is a so-called safety-catch linker based on a sulfonamide, which is described in Chapter 8 [53, 54]. Here the first amino acid is anchored by acylation of a sulfonamide. The resultant linkage is stable during Fmoc-SPPS; however, the linkage is "activated" by *N*-alkylation which makes it susceptible to nucleophilic displacement. This is especially useful for the synthesis of peptide thioesters, which are created by release with thiols. The peptide is typically released into a solution with excess of thiol in DMF or NMP, which can be difficult to remove. Our group has reported a simple approach was developed where the linker is an C-terminal glutamic acid moiety [55]. The side-chain carboxylate is protected by a selectively removable protecting group, such as the Ph*i*Pr, and upon its removal, the carboxylate is activated to form the five-membered pyroglutamyl imide. Treatment with a thiolate will cleave the imide linkage releasing the peptide thioester into solution. Protocols for the synthesis of peptide thioesters are described in Chapter 8.

10 Other Synthetic Methods

The conformational properties, folding, and aggregational propensity of a peptide sequence depend to a large extend on the presence of the N–H's in the amides. When the hydrogen in an amide NH in the backbone is substituted for an alkyl moiety, it not only removes a potential hydrogen bond donor but it also changes the amide bond *cis–trans* equilibrium, which affects the neighboring sequence. The secondary amino acid Pro is generally considered an α-helix breaker. Secondary amino acid surrogates, which also mimic the effect of Pro, can potentially disrupt secondary structure formation. They are attractive in SPPS, if the modification which transiently made them secondary amino acid can be removed after peptide chain assembly. This has been used in two synthetic methods.

Weygand, Sheppard, and others introduced backbone amide protecting groups, such as DMB (2,4-dimethoxy benzyl) and HMB (hydroxy-methoxybenzyl), to help facilitate the synthesis of so-called difficult sequences [56–59]. A DMB substituted Gly is incorporated in the sequence together with the preceding amino acid in a dipeptide building block. It is incorporated with intervals in the sequence during peptide chain assembly. Especially Gly-Gly and Asp-Gly to prevent aspartimide formation are attractive. In contrast, HMB-protected amino acids are incorporated as the monomers. A preferred building block is Fmoc-(Hmb)Gly-OH. Both HMB and

DMB are benzylic with electron-donating groups and can thus be removed by treatment with TFA. However, they have some limitations, which include the higher cost.

Mutter and coworkers introduced pseudoprolines, which are Ser or Thr oxazolidines (N and O of the Ser or Thr residue) [60]. They are incorporated as the dipeptides, e.g., Gly-Ser. The oxazolidine ring can be opened with TFA-containing cocktails to unprotect the peptides.

11 Disulfide Bridge Formation

Disulfide bonds provide covalent constraints on the linear sequence of a peptide or protein and can also link two sequences together. Both aspects can be illustrated with the three disulfide bonds in insulin, of which one is intrastrand while the two other are interstrand. The formation of disulfide bridges has been studied extensively [61]. There are several unique aspects of disulfide bonds, including that they arise from the capability of sulfur to act either as nucleophile and electrophile (umpolung). The formation of a disulfide is formally an oxidation, while the cleavage is a reduction. The most simple and method for disulfide bond formation is "air oxidation" where the oxygen is the oxidant. It can be achieved by bubbling air through the reaction at requires basic pH. In contrast, Tam and coworkers reported that DMSO can promote disulfide bond formation in the extended pH range 3–8 [62]. Also, various redox buffers and solid supported reagents can be used. *See* Chapter 5.

12 Posttranslational and Other Modifications Introduced in SPPS

Posttranslational modifications of peptides and proteins are ubiquitous in nature. They include phosphorylation, *O*- and *N*-glycosylation, farnesylation and palmitoylation, and γ-carboxylation of glutamic acid [63]. Incorporation of some of these modified amino acids in a sequence can be achieved by solid-phase peptide synthesis. For some of these modifications, suitably protected amino acid building blocks are commercially available and can be incorporated into the peptide sequence using standard methods. This is certainly the case with phosphopeptides, where Fmoc-protected phosphorylated Ser and Thr building blocks, which carry a single *O*-benzyl protecting group on the phosphate, can be used in standard Fmoc-SPPS, however, with longer deprotection times. *See* Chapter 13.

Some Fmoc-protected *O*- or *N*-glycosylated amino acids, i.e., Ser, Thr and Asn, are also commercially available and can be used in SPPS. Typically, these building blocks contain mono-, di-, or trisaccharide glycans. The glycans are most commonly protected with *O*-acetyl or *O*-benzoyl ester moieties, which can be removed upon completion of the synthesis by treatment with, for example, a dilute solution of methoxide in methanol. While numerous glycopeptides have been prepared in this manner, it adds another level of complexity to the synthesis design and can cause new side reactions such as β-elimination of the glycan leaving a dehydroalanine residue in the peptide sequence [64]. *See* Chapter 14.

The synthesis of peptides with appended farnesyl, geranyl–geranyl, and palmitoyl groups has been achieved [65], but it has remained a task for specialists. However, Chapter 12 describes protocols for the synthesis of a range of lipopeptides. Other lipids can also be introduced, for example by simple acylation of a Lys N^{ε}-amine can be achieved while the peptide remains on the solid support or after it has been released from the support, provided there is only one Lys in the sequence.

13 Instruments for Automated SPPS

SPPS is very amenable to automation and the development of commercially available automated peptide synthesizers has come a long way allowing a high degree of predictability and reproducibility in the assembly of peptides. These synthesizers range from semiautomated systems, with automated washing and Fmoc removal steps, over fully automated synthesizers, which will assemble the whole sequence without intervention, to synthesizers that can prepare large numbers of peptides in parallel. Precise microwave heating has emerged as a new tool in peptide synthesis [66, 67]. Chapter 15 provides an overview of instrumentation, while Chapters 16 and 17 describe peptide synthesis with microwave heating on Biotage and CEM instruments, respectively.

14 Conclusions and Future Directions

An understanding of the possibilities and current limitations in solid-phase peptide synthesis is a good starting point for designing peptides for biochemical, biomedical, and biophysical studies. The chemical tools available to SPPS define which synthetic peptides are available and thus what peptide-based research and development can be carried out. Many peptides can be prepared with ease and predictability by solid-phase peptide synthesis. Peptides up to 15–20 amino acids in length are normally routinely available from companies specializing in on-demand synthesis. Longer peptides, peptides with unusual amino acids, and peptides carrying posttranslational modifications pose challenges. This book describes the protocols for Fmoc-SPPS of a wide variety of peptides and includes a chapter on Boc-SPPS. Several instrument manufacturers supply automated peptide synthesizers, which can be used not only by the specialist but also by the nonspecialist with a chemical interest.

References

1. Merrifield RB (1963) Solid-phase peptide synthesis. synthesis of a tetrapeptide. J Am Chem Soc 85:2149–2154
2. Merrifield RB (1985) Solid-phase peptide synthesis (nobel lecture). Science 232:341–347
3. Bergmann M, Zervas L (1932) Über ein allgemeines verfahren der peptid-synthese. Berichte 65:1192–1201
4. Dawson PE, Muir TW, Clark-Lewis I, Kent SBH (1994) Synthesis of proteins by native chemical ligation. Science 266:776–779
5. Rose K (1994) Facile synthesis of homogeneous artificial proteins. J Am Chem Soc 116:30–33
6. Brask J, Jensen KJ (2000) Carbopeptides: chemoselective ligation of peptide aldehydes to an aminooxy-functionalized d-galactose template. J Pept Sci 6:290–299
7. Hudson D (1999) Matrix assisted synthetic transformations: a mosaic of diverse contributions. I. The pattern emerges. J Comb Chem 1:333–360
8. Hudson D (1999) Matrix assisted synthetic transformations: a mosaic of diverse contributions. I. The pattern is completed. J Comb Chem 1:403–457
9. Carpino LA, Han GY (1970) The 9-Fluorenylmethoxycarbonyl function, a new base-sensitive amino-protecting group. J Am Chem Soc 92:5748–5749
10. Carpino LA, Han GY (1972) The 9-fluorenylmethoxycarbonyl function amino-protecting group. J Org Chem 37:3404–3409
11. Alewood P, Alewood D, Miranda L, Love S, Meutermans W, Wilson D (1997) Rapid in situ neutralization protocols for Boc and Fmoc solid-phase chemistries. In: Fields GB (ed) Methods in enzymology, vol 289. Academic, New York, pp 14–28
12. Stewart JM (1997) Cleavage methods following Boc-based solid-phase peptide synthesis. In: Fields GB (ed) Methods in enzymology, vol 289. Academic, New York, pp 29–43
13. Atherton E, Fox H, Harkiss D, Logan CJ, Sheppard RC, Williams BJ (1978) Mild procedure for solid-phase peptide synthesis—use of the fluorenylmethoxycarbonyl amino acids. Chem Commun 537–539
14. Atherton EH, Fox H, Harkiss D, Sheppard RC (1978) Application of polyamide resins to polypeptide synthesis—improved synthesis of beta-endorphin using fluorenylmethoxycarbonyl amino acids. Chem Commun 539–540
15. Barany G, Albericio F (1985) A 3-dimensional orthogonal protection scheme for solid-phase peptide synthesis under mild conditions. J Am Chem Soc 107:4936–4942
16. Isodro-Llobet A, Alvarez M, Albericio F (2009) Amino acid protecting groups. Chem Rev 109:2455–2504
17. Carpino LA, Shroff H, Triolo SA, Mansour E-SM, Wenschuh H, Albericio F (1993) The 2,2,4,6,7-pentamethyldihydrobenzofuran-5-sulfonyl group (Pbf) as arginine side-chain protectant. Tetrahedron Lett 34:7829–7832
18. Aletras A, Barlos K, Gatos D, Koutsogianni S (1995) Preparation of the very acid-sensitive Fmoc-Lys(Mtt)-OH—application in the synthesis of side-chain to side-chain cyclic peptides and oligolysine cores suitable for solid-phase assembly of MAPS and TASPs. Int J Pept Prot Res 45:488–496
19. Kunz H, Unverzagt C (1984) The allyloxycarbonyl (Aloc) moiety—conversion of an unsuitable into a valuable amino protecting group for peptide synthesis. Angew Chem Int Ed 23:436–437
20. Matysiak S, Böldicke T, Tegge W, Frank R (1998) Evaluation of monomethoxytrityl and dimethoxytrityl as orthogonal amino protecting groups in Fmoc solid-phase peptide synthesis. Tetrahedron Lett 39:1733–1734
21. Hillman JD, Orugunty RS, Smith JL (2007) U.S. Pat. Appl. 2007037963 A1, Chem Abstr 146:252110
22. Yue CW, Thierry J, Potier P (1993) 2-Phenyl isopropyl esters as carboxyl terminus protecting groups in the fast synthesis of peptide fragments. Tetrahedron Lett 34:323–326
23. König W, Geiger R (1970) Eine neue methode zur synthese von peptiden: aktivierung der carboxylgruppe mit dicyclohexylcarbodiimid unter zusatz von 1-hydroxy-benzotriazolen. Chem Ber-Recl 103:788–798
24. Carpino LA (1993) 1-Hydroxy-7-azabenzotriazole. An efficient peptide coupling additive. J Am Chem Soc 115:4397–4398
25. Subirós-Funosas R, Prohens R, Barbas R, El-Faham A, Albericio F (2009) Oxyma: an efficient additive for peptide synthesis to replace the benzotriazole-based HOBt and HOAt with a lower risk of explosion. Chem Eur J 15:9394–9403
26. Dourtoglou V, Gross B, Lambropoulou V, Zioudrou C (1984) O-Benzotriazolyl-*N*, *N*, *N'*, *N'*-tetramethyluronium hexafluorophosphate as coupling reagent for the synthesis of peptides of biological interest. Synthesis 7:572–574
27. Carpino LA (1993) J Am Chem Soc 115: 4397–4398
28. Carpino LA, El-Faham A, Minor CA, Albericio F (1994) Advantageous applications of azabenzotriazole (triazolopyridine)-based

coupling reagents to solid-phase peptide synthesis. J Chem Soc Chem Commun 201–203
29. Coste J, Lenguyen D, Castro B (1990) PyBOP—a new peptide coupling reagent devoid of toxic by-product. Tetrahedron Lett 31:205–208
30. El-Faham A, Funosas RS, Prohens R, Albericio F (2009) COMU: A safer and more effective replacement for benzotriazole-based uronium coupling reagents. Chem Eur J 15:9404–9416
31. Abdelmoty I, Albericio F, Carpino LA, Foxman BM, Kates SA (1994) Lett Pept Sci 1:57–67
32. Carpino LA, Imazumi H, El-Faham A, Ferrer FJ, Zhang C, Lee Y, Foxman BM, Henklein P, Hanay C, Mügge C, Wenschuh H, Klose J, Beyermann M, Bienert M (2002) The uronium/guanidinium peptide coupling reagents: finally the true uronium salts. Angew Chem Int Ed Engl 41:441–445
33. Tofteng AP, Pedersen SL, Staerk D, Jensen KJ (2012) Effect of residual water and microwave heating on half-lifes of reagents and reactive intermediates in peptide synthesis. Chem Eur J 18:9024–9031
34. Kisfaludy L, Schön I (1983) Preparation and applications of pentafluorophenyl esters of 9-fluorenylmethyloxycarbonyl amino acids for peptide. Synthesis 325–327
35. Fields CG, Fields GB (1993) Minimization of tryptophan alkylation following 9-fluorenylmethoxycarbonyl solid-phase peptide synthesis. Tetrahedron Lett 34:6661–6664
36. Han Y, Albericio F, Barany G (1997) Occurrence and minimization of cysteine racemization during stepwise solid-phase peptide synthesis. J Org Chem 62:4307–4312
37. Kaiser T, Nicholson GJ, Kohlbau HJ, Voelter W (1996) Racemization studies of Fmoc-Cys(Trt)-OH during stepwise Fmoc-solid-phase peptide synthesis. Tetrahedron Lett 37: 1187–1190
38. Meldal M (1992) PEGA: a flow stable polyethylene glycol dimethyl acrylamide copolymer for solid-phase synthesis. Tetrahedron Lett 33:3077–3080
39. Garcia-Martin F, Quintanar-Audelo M, Garcia-Ramos Y, Cruz LJ, Gravel C, Furic R, Cruz S, Tulla-Puche J, Albericio F (2006) ChemMatrix, a poly(ethylene glycol)-based support for the solid-phase synthesis of complex peptides. J Comb Chem 8:213–220
40. Fields GB, Tian Z, Barany G (2002) Principles and practice of solid-phase peptide synthesis. In: Grant GA (ed) Synthetic peptides, a user's guide, 2nd edn. W. H. Freeman and Company, New York, pp 93–219
41. Fields GB (ed) (1997) Methods in enzymology (solid-phase peptide synthesis), vol 289. Academic, San Diego
42. Chan WC, White PD (2004) Fmoc solid-phase peptide synthesis. Oxford University Press, Oxford
43. James IW (1999) Linkers for solid-phase organic synthesis. Tetrahedron 55:4855–4946
44. Guillier F, Orain D, Bradley M (2000) Linkers and cleavage strategies in solid-phase organic synthesis and combinatorial chemistry. Chem Rev 100:2091–2157
45. Albericio F, Giralt E (2004) Handles and supports. In: Goodman M, Felix A, Moroder L, Toniolo C (eds) Hoyben-Weyl, E22a, synthesis of peptides and peptidomimetics. Thieme, Stuttgart, pp 685–725
46. Albericio F, Barany G (1987) An acid-labile anchoring linkage for solid-phase synthesis of C-terminal peptide amides under mild conditions. Int J Pept Protein Res 30:206–216
47. Albericio F, Kneib-Cordonier N, Biancalana S, Gera L, Masada RI, Hudson D, Barany G (1990) Preparation and application of the 5-(4-(9-fluorenylmethyloxycarbonyl) aminomethyl-3,5-dimethoxyphenoxy)valeric acid (PAL) handle for the solid-phase synthesis of c-terminal peptide amides under mild conditions. J Org Chem 55:3730–3743
48. Boas U, Brask J, Jensen KJ (2009) The backbone amide linker for solid-phase synthesis. Chem Rev 109:2092–2118
49. Jensen KJ, Alsina J, Songster MF, Vágner J, Albericio F, Barany G (1998) Backbone amide linker (BAL) strategy for solid-phase synthesis of C-terminal modified and cyclic peptides. J Am Chem Soc 120:5441–5452
50. Guillaumie F, Kappel JC, Kelly NM, Barany G, Jensen KJ (2000) Solid-phase synthesis of C-terminal peptide aldehydes from amino acetals anchored to a backbone amide linker (BAL) handle. Tetrahedron Lett 41:6131–6135
51. Brask J, Albericio F, Jensen KJ (2003) Fmoc solid-phase synthesis of peptide thioesters by masking as trithioortho esters. Org Lett 5:2951–2953
52. Bourne GT, Meutermans WDF, Alewood PF, McGeary RP, Scanlon M, Watson AA, Smythe ML (1999) A backbone linker for BOC-based peptide synthesis and on-resin cyclization: synthesis of stylostatin. J Org Chem 64: 3095–3101
53. Ingenito R, Bianchi E, Fattori D, Pessi A (1999) Solid-phase synthesis of peptide C-terminal thioesters by Fmoc/t-Bu chemistry. J Am Chem Soc 121:11369–11374
54. Backes BJ, Ellman JA (1999) An alkanesulfonamide 'safety-catch' linker for solid-phase synthesis. J Org Chem 64:2322–2330
55. Tofteng AP, Sørensen KK, Conde-Frieboes KW, Hoeg-Jensen T, Jensen KJ (2009) Fmoc solid-phase synthesis of C-terminal peptide

thioesters via formation of a backbone pyroglutamyl imide moiety. Angew Chem Int Ed 48:7411–7414
56. Weygand F, Steglich W, Bjarnason J (1966) Leicht Abspaltbare Schutzgruppen Für Säureamidfunktionen, Tetrahedron Lett 7:3483–3487
57. Narita M, Fukunaga T, Wakabayashi A, Ishikawa K, Nakano H (1984) Syntheses and properties of tertiary peptide bond-containing polypeptides. Int J Pept Protein Res 23: 306–314
58. Blaakmeer J, Tijsse-Klasen T, Tesser GI (1991) Enhancement of solubility by temporary dimethoxybenzylsubstitution of peptide bonds. Int J Pept Protein Res 37:556–564
59. Quibell M, Packman LC, Johnson T (1995) Synthesis of the 3-repeat region of human tau-2 by the solid-phase assembly of backbone amide-protected segments. J Am Chem Soc 117:11656–11668
60. Wöhr T, Wahl F, Nefzi A, Rohwedder B, Sato T, Sun X, Mutter M (1996) Pseudo-prolines as a solubilizing, structure-disrupting protection technique in peptide synthesis. J Am Chem Soc 118:9218–9227
61. Annis I, Hargittai B, Barany G (1997) Disulfide bond formation in peptides. In: Fields GB (ed) Methods in enzymology, vol 289. Academic, New York, pp 198–220
62. Tam JP, Wu C-R, Liu W, Zhang J-W (1991) Disulfide bond formation in peptides by dimethyl sulfoxide. Scope and applications. J Am Chem Soc 113:6657–6662
63. Walsh CT (2006) Posttranslational modifications of proteins. Roberts and Company Publishers, Englewood, Colorado
64. Jansson AM, Hilaire PMS, Meldal M (2004) Synthesis of glycopeptides. In: Goodmann M et al (eds) Synthesis of peptides and peptidomimetics, vol 22b. Thieme, Stuttgart, pp 235–322
65. Brunsveld L, Kuhlmann J, Waldmann H (2006) Synthesis of palmitoylated ras-peptides and -proteins. Methods 40:151–165
66. Pedersen SL, Tofteng AP, Malik L, Jensen KJ (2012) Microwave heating in solid-phase peptide synthesis. Chem Soc Rev 41:1826–1844
67. Malik L, Tofteng AP, Pedersen SL, Sørensen KK, Jensen KJ (2010) Automated 'X-Y' robot for peptide synthesis with microwave heating: application to difficult peptide sequences and protein domains. J Pept Sci 16:506–512

Chapter 2

Linkers, Resins, and General Procedures for Solid-Phase Peptide Synthesis

Pernille Tofteng Shelton and Knud J. Jensen

Abstract

This chapter describes the basic protocols for solid-phase peptide synthesis using the Fmoc group as the N^{α}-protecting group (Fmoc-SPPS). The chapter introduces resins and their handling, choice of linkers, and the most common methods for peptide chain assembly. The proper choice of resins and linkers for solid-phase synthesis is a key parameter for successful peptide synthesis. This chapter provides an overview of the most common and useful resins and linkers for the synthesis of peptides with *C*-terminal amides, carboxylic acids, and more. The chapter finishes with robust protocols for general solid-phase peptide synthesis, i.e., the standard operations.

Key words Solid-phase peptide synthesis, SPPS, TentaGel, ChemMatrix, PEGA, Polystyrene, HBTU, HATU, DIC, COMU, HOBt, HOAt, Oxyma

1 Introduction

In the past decades, numerous resins and linkers have been developed and commercialized, which have enabled a wide range of applications. The resin determines the physical properties, e.g., swelling of the peptidylresin and may also limit the conditions and chemistries under which the resins are stable [1, 2]. The linker determines not only the conditions, which can be used during peptide chain assembly, but also the conditions required for release of the peptide and, finally, also the nature of the *C*-terminal functionality.

The proper choice of resin is important for the efficient and economical preparation of a peptide. The first resin to be used by Merrifield in the 1960s was polystyrene and this efficient type of resin is still widely used today [3]. Over the years, a limited number of resins, which contain polyethylene glycol (PEG) linked to polystyrene, have been developed. Also, resins made solely from PEG and a cross-linker have become commercially available. However, companies which specialize in reagents for solid-phase peptide synthesis tend to use their own names for resins and linkers as well as

Knud J. Jensen et al. (eds.), *Peptide Synthesis and Applications: Second Edition*, Methods in Molecular Biology, vol. 1047, DOI 10.1007/978-1-62703-544-6_2, © Springer Science+Business Media New York 2013

Table 1
Overview of the three subgroups of resins for Fmoc-based SPPS and their swelling properties in selected solvents [26]

Resin subgroups	Commercial name	Initial loading (mmol/g)	Approximate swelling (mL/g)						
			DCM	DMF (NMP)	Ether	Water	TFA	THF	MeOH
PS (1 % DVB)	Aminomethylated PS	0.4–1.5	7	4	4	N.A	2	9	1.6
PEG-PS	Amino TentaGel (TG)[a]	0.15–0.3	6	5	2	3.6	N.A.	5.0	3.6
PEG based	Amino PEGA[b]	0.2–0.4	13	11	N.A	16	N.A.	13	13
	ChemMatrix (CM)	0.4–0.6	11	8	N.A	11	14	N.A.	9

[a]The swelling volumes are for standard grade TG resins
[b] The swelling properties listed are for 0.2 mmol/g resins

for premade linker-resins. This can make it difficult to survey the commercially available linkers and resins.

Linkers in SPPS provide a reversible linkage between the peptide chain and the solid support (resin). Furthermore, in most cases, the linker provides protection and blockage of the *C*-terminal α-carboxyl group during synthesis. An exception here is the use of linkers which do not attach to the α-carboxylic acid but leave the *C*-terminal free for modifications. This can be achieved by side-chain anchoring or by a backbone amide linker (Chapter 9). Furthermore, the choice of linker determines which chemistry can be used during peptide chain assembly, as well as release of the peptide in the final step (Chapter 3). The strategies for Boc- and Fmoc-SPPS require different linkers, due to different conditions for repetitive removal of the N^{α}-protecting groups [4]. For further information on linkers for SPPS using the Boc-strategy, *see* Chapter 4, as the following chapter mainly will focus on linkers for Fmoc-based solid-phase chemistry.

1.1 Resins

The varieties of resins for SPPS may seem bewildering. However, there are only three different subgroups of resins depending on what type of material they contain (Table 1). The most common classes of resins are the classic polystyrene (PS) resins, the PS-functionalized polyethylene glycol (PEG) resins, and pure cross-linked PEG resins. Conventional PS resins have in numerous cases proven successful in the synthesis of short- to medium-length peptides. However, PEG-based resins often outperform the PS-based resins in the synthesis of medium-length and long peptides as well as peptides which contain "difficult sequences." This has to be balanced with the fact that PEG-containing resins in general also are more costly. The length of the PEG as well as its percentage of the total resin, the amount of cross-linking, as well as possible batch-to-batch variation will influence how the resin performs during Fmoc-SPPS. Thus, when synthesizing medium to long peptide sequences, choosing the optimal resin is likely to be a very important factor.

The polystyrene resin is a polymer cross-linked with 1 % of divinylbenzene (DVB) and with a loading of 0.2–1.2 mmol/g. This type of support is compatible with DMF and DCM but not compatible with water (*see* Table 1). The most well-established PEG-functionalized PS resin today is TentaGel (TG) reported by Bayer and Rapp [5, 6]. TG resins are prepared by grafting of PEG (50–70 %) to low cross-linked polystyrene by an ether linkage. The TG resins have excellent swelling properties in most solvents compatible with PEG, such as DCM, DMF, NMP, water, ethanol, and tetrahydrofuran (THF) (*see* Table 1). When changing the solvent from an organic media to a water-based media, it is recommended to use a solvent gradient by going from DCM to THF or ethanol and then to water. This will maintain the optimal swelling properties of the TG resins. There are other PEG-PS-based resins developed and commercialized under a variety of brand names. The TG resins have evolved over 20 years and have been used in many scientific studies, making it a well-tested and reliable solid support. This type of support is highly suited for the synthesis of longer peptides.

The pure cross-linked PEG resins contain no polystyrene. This group includes the poly(ethylene glycol)-poly-(*N,N-dimethylacrylamide*) copolymer (PEGA) developed by Meldal [7] and the more recent ChemMatrix (CM) resin which was reported by Albericio and co-workers in 2006 [8]. A unique advantage of PEGA resins is that they swell very well in water and that peptide substrates anchored to PEGA thus can be used in interaction studies with proteins up to 35–70 kDa. PEGA resins are supplied swollen in ethanol, due to the very sticky nature of the resin beads. The beads are easily damaged when shrunk or dried and are therefore best handled in a swollen state. The newer CM resins are fully PEG based and contain primary ether bonds and are reported to be chemically robust. CM resins have the advantage compared to the PEGA-type resins that they are easily handled in a dry state and therefore in regard to handling are more comparable to the PS- or PEG-PS-based resins. The CM resins have found usage in the synthesis of longer peptides and even small proteins due to its excellent swelling properties, which is beneficial for diffusion and for accessibility of reactive sites. They are used with the same procedures as for polystyrene and TG; however, they swell more in most solvent, in particular TFA (*see* Table 1). It should be mentioned that the CM resins are still relatively new compared to the TG family of resins.

Most of the abovementioned resins are commercially available either as amino-functionalized resins or with a variety of different linkers attached. The amino-functionalized resins allow for anchoring of specific linkers by a standard amide bond.

1.2 Linkers for Fmoc-Based SPPS

A large variety of linkers suitable for Fmoc-SPPS are available [9, 10], and many resins can be purchased with the linker preinstalled. A focused table with the linkers divided into subclasses is included in this chapter (*see* Table 2), and Chapters 1 and 3 give further guidelines regarding cleavage conditions. Most linkers release the peptide upon TFA treatment as the *C*-terminal acids and amides, the classic examples being the Wang and Rink-amide linkers, respectively [11]. The Rink-amide linker and other aminomethyl-based linkers can be installed on the resin by coupling of the Fmoc-protected linker to the resin using standard couplings procedures. Another class of linkers are trityl based, where a classic example is the 2-chlorotrityl chloride resin which is used in the production of peptide acids [12]. Often 2-chlorotrityl resins are attached to polystyrene resins by direct on-resin synthesis. However, premade trityl linkers can also be coupled to a variety of resins, typically amino-functionalized base resins. Specialized linkers which release the peptide as esters, secondary amines, or thioesters have also been developed, e.g., the safety-catch linker and aryl hydrazide linkers [13, 14]. The safety-catch linker is cleaved by alkylation of the sulfonamide which enables release of the modified peptides upon treatment with different nucleophiles (Chapter 8). The aryl hydrazide linker is cleaved by oxidation to an acyldiazene that enables release with suitable nucleophiles. Another class of handles which provides *C*-terminal modified peptides is the backbone amide linker (BAL) (Chapter 9) [15]. Here the peptide is anchored through a backbone amide, typically of the *C*-terminal residue. These linkers leave the *C*-terminal free to be modified, and peptide esters, aldehydes, as well as thioesters have been synthesized by this method [16, 17].

2 Materials

1. Selection of commercially available resins:
   (a) Aminomethylated polystyrene (PS) resins:
      - Sold as either high loading (HL, 0.5–1.5 mmol/g) or low loading (LL, 0.3–0.5 mmol/g) from many distributors, however, in varying qualities.

   (b) Amino PS-PEG (TentaGel TG) resins:
      - TentaGel® resins sold as standard grade (S) loading approx. 0.25 or research grade (R) loading approx. 0.15 mmol/g from Rapp Polymere.

Table 2
Overview linkers for Fmoc-based SPPS

Linker types	Name	Final *C*-terminal functionality	Linker structure	Described in this book
Aminomethyl				
	Rink-amide linker	Peptide amides	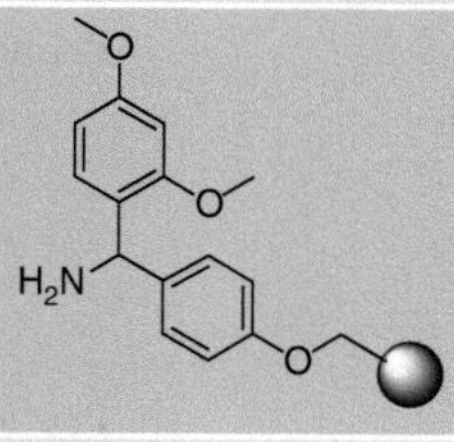	Chapter 2
	Sieber linker	Peptide amides	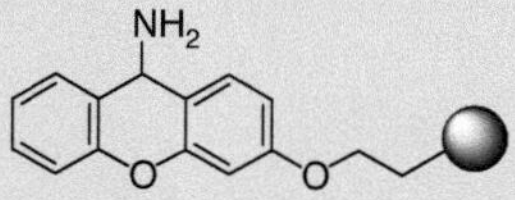	
	PAL linker	Peptide amides	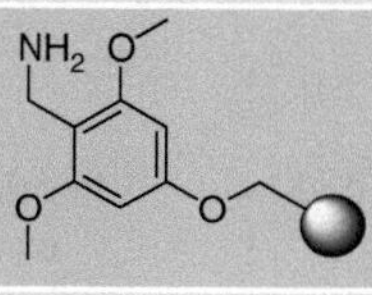	
Hydroxymethyl				
Wang-type resins:	Wang/PHB linker	Peptide acids	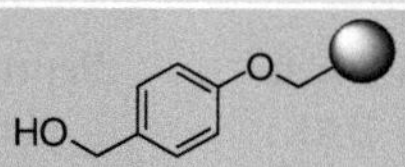	Chapter 2
	HMPA linker	Peptide acids	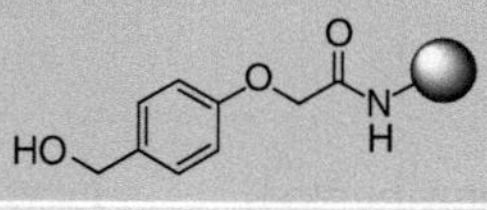	Chapter 2
	HMBA linker	Protected peptide acids, amides, alcohols, hydrazides	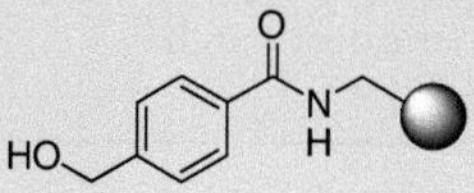	Chapter 3
Variants	Rink acid linker	Peptide acids	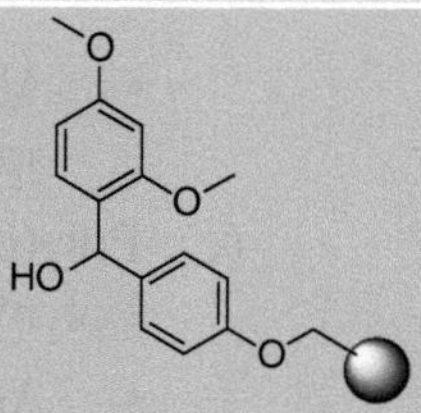	
	SASRIN linker	Peptide acids	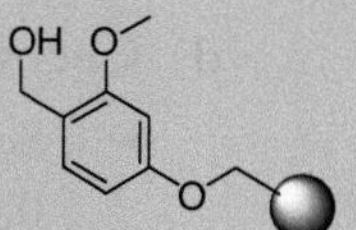	

(continued)

Table 2
(continued)

Linker types	Name	Final *C*-terminal functionality	Linker structure	Described in this book
2-Chlorotrityl				
		Peptide acids	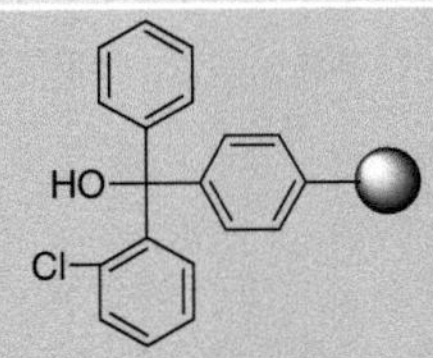	Chapter 2
		Peptide amines	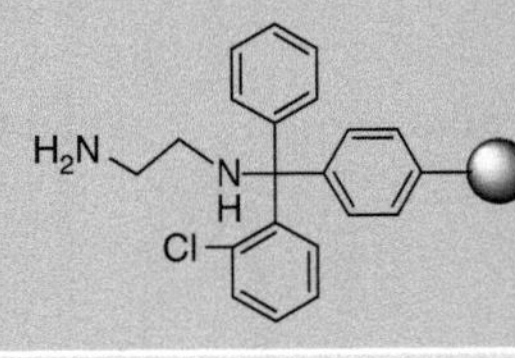	
Others				
	Aryl hydrazide linker	Peptide amines or esters	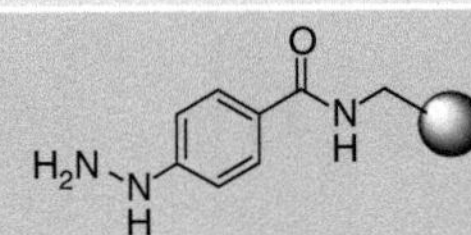	
BAL	*ortho*-PALdehyde linker (o-BAL)	Peptide acids, aldehydes, thioesters, among others	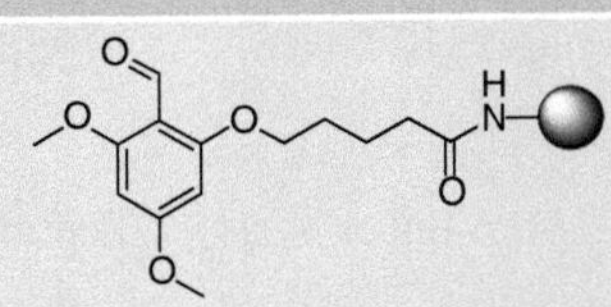	Chapter 9
Safety catch	4-sulfamylbutyryl/ Kenner safety-catch linker	Peptide thioesters	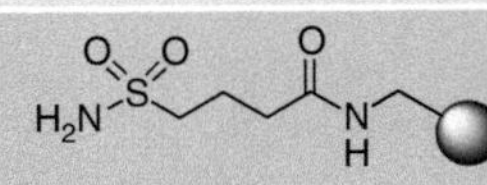	Chapter 8

- Alternatively, there is NovaSyn® TG with a loading of approx. 0.2–0.3 mmol/g from Novabiochem®.

(c) Poly(ethylene glycol)-poly(acryl amide)copolymer (PEGA, Polymer Laboratories, now Agilent) resins:

- Available with loadings of 0.2–0.4 mmol/g.

(d) ChemMatrix® (CM) resins:

- Sold from PCAS BioMatrix Inc with a loading of approx. 0.46 mmol/g. Furthermore, this resin is sold from distributors, such as Biotage AB, and from Novabiochem®, where it is sold under the trade name NovaPEG.

2. Solvents:

(a) Dichloromethane (DCM).

(b) *N*,*N*-Dimethylformamide (DMF).

   (c) *N*-Methyl-2-pyrrolidinone (NMP).
   (d) Methanol (MeOH).
   (e) Tetrahydrofuran (THF).
   (f) Trifluoroacetic acid (TFA).
   (g) Piperidine.
   (h) Pyridine.
   (i) Acetic anhydride.
3. Coupling reagents:
   (a) Diisopropylcarbodiimide (DIC).
   (b) 1-[(1-(Cyano-2-ethoxy-2-oxoethylideneaminooxy)-dimethylamino-morpholinomethylene)]methanaminium hexafluorophosphate (COMU).
   (c) *N*-[(Dimethylamino)-1H-1,2,3-triazolo[4,5-*b*]pyridin-1-ylmethylene]-N-methylmethanaminium hexafluorophosphate *N*-oxide (HATU).
   (d) *N*-[(1*H*-Benzotriazol-1-yl)(dimethylamino)methylene]-*N*-methylmethanaminium hexafluorophosphate *N*-oxide (HBTU).
   (e) 1-Hydroxybenzotriazole (HOBt).
   (f) Ethyl(2-cyano-2-(hydroxyimino)acetate) (Oxyma).
   (g) 1-(Mesitylene-2-sulfonyl)-3-nitro-*1H*-1,2,4-triazole (MSNT).
4. Other reagents:
   (a) *N,N*-Diisopropylethylamine (DIEA).
   (b) 4-(*N,N*-Dimethylamino)pyridine (DMAP).
   (c) 1-methylImidazole (MeIm).
5. Equipment:
   (a) Polypropylene syringes equipped with a polyethylene filter.
   (b) A shaking table or equivalent.
   (c) A "Teflon" block design for multiple filtrations or an equivalent setup for manual SPPS.
6. Building blocks:
   (a) Fmoc-protected amino acids.
   (b) Fmoc-protected pentafluorophenyl esters, if needed.
   (c) Fmoc-protection pseudoproline dipeptides, if needed.
   (d) Fmoc-protected *N*-Hmb amino acids.
   (e) Fmoc-protected *N*-Dmb amino acids.
7. Kaiser test: ninhydrin, phenol, ethanol, pyridine, potassium cyanide (KCN).

3 Methods

3.1 Resin Swelling, Washing, and Drying

Many resins benefit from an initial swelling procedure in order to increase the final peptide yield. The standard procedures described here (*see* Subheading 3.1.1) are suited for aminomethylated resins and preloaded hydroxy-functionalized resins. A more elaborate swelling procedure is recommended for the ChemMatrix resins in order to obtain higher final yields (*see* Subheading 3.1.2) [18]. Furthermore, standard resin washing and drying are described in the following section.

3.1.1 Standard Resins Swelling

1. Place the resin in a syringe equipped with a polypropylene filter.
2. Add DCM until the resin is completely covered. The volume of solvent depends on the type of resin used.
3. Empty the syringe by applying vacuum and repeat the DCM treatment.
4. Cover the resin in DMF and leave for 15–30 min.
5. Remove the DMF by applying vacuum. The resin is now ready for synthesis. N.B. As many resins are purchased in the Fmoc-protected form, a N^{α}-deprotection should be performed first!

3.1.2 ChemMatrix Resin Swelling

1. Place the resin in a syringe equipped with a polystyrene filter.
2. Add MeOH until the resin is covered and leave for 1 min.
3. Remove MeOH by applying vacuum and repeat treatment with MeOH.
4. In a similar manner, wash with DMF (2 × 1 min), DCM (3 × 1 min) TFA-DCM (1:99) (3 × 1 min) DIEA-DCM (1:19) (3 × 1 min), DCM (3 × 1 min), DMF (3 × 1 min).
5. The resin is now ready for synthesis. N.B. As many resins are purchased in the Fmoc-protected form, a N^{α}-deprotection should be performed first!

3.1.3 Washing of the Resin

1. Add DMF (approx. 5 mL/0.5 mmol) to the peptidyl-resin.
2. Leave for 1 min before removing the solvent by vacuum filtration.
3. Repeat three to five times depending on previously performed chemistry.

3.1.4 Drying of Standard PS, TG, or CM Resins

Some resins such as PEGA require a more specific washing procedure. It is therefore always recommended to check the supplier's recommendation. The method described below can be used for standard PS, TG, or CM resins loaded with a variety of linkers.

1. Perform a thorough resin wash with DMF (*see* Subheading 3.1.3)
2. Wash in a similar manner with ethanol (3 × 1 min) and diethyl ether (2 × 1 min).
3. Apply vacuum for 10 min.
4. (a) If the next step is a peptide cleavage step (*see* Chapter 3), leave the resin for 1–2 h at room temperature.
   (b) If the next step is a resin loading test (*see* Subheading 3.7), place the resin in a desiccator under vacuum for 12–18 h prior to the loading test.

3.2 Loading of Amino-Functionalized Resins

For amino-functionalized resins, anchoring of the first residue to the solid support is performed using standard amide coupling reactions. This linkage will upon cleavage yield a *C*-terminal amide, and if there is no specific requirement for the *C*-terminal functionality, this functionality is the preferred. The most common amino-functionalized linker is the Rink-amide linker, however PAL handle is also very reliable and may have a higher acid-lability (*see* Table 2) [19, 20]. Standard coupling procedures are described in Subheading 3.5.

3.3 Loading of Hydroxy-Functionalized Resins

For anchoring of the first amino acid onto hydroxymethyl-based resin, which upon cleavage provides a *C*-terminal carboxylic acids, it is recommended to use a protocol without the use of tertiary bases, such as DIEA. This type of protocol is designed to minimize the degree of self-acylation, hence, double incorporation, and racemization of the first residue. The most common hydroxymethyl-based resins are Wang-type linkers (*see* Table 2). Two different protocols are described below for loading of hydroxymethyl-based resins: (1) the symmetrical anhydride method and (2) the MSNT/MeIm method. The MSNT/MeIm method is recommended for difficult situations, which includes the attachment of amino acids that are prone to epimerization. For the synthesis of *C*-terminal acids where the first residue is either Cys or Pro, it is recommended to use the trityl-based resins. Many of the resins with a hydroxymethyl linker can be obtained with the first amino acid already preloaded.

3.3.1 The Symmetrical Anhydride Method

1. Place the hydroxy-functionalized resin in a dry flask and add dry DMF until the resin is completely covered (*see* **Note 1**).
2. Let the resin swell for 30 min before applying vacuum to remove the DMF.
3. Place the desired Fmoc-protected amino acid (10 equiv. relative to the resin loading) in a dry round-bottomed flask containing a magnetic stirrer.
4. Add dry DCM (approx. 3 mL/mmol amino acid derivative) to dissolve the Fmoc-protected amino acids. A few drops of DMF may be needed to aid complete dissolution.

5. Prepare a solution of DIC (5 equiv. relative to the resin loading) in dry DCM (approx. 1 mL/mmol DIC) and add to the flask containing the dissolved Fmoc-protected amino acids.
6. Stir the mixture for 15 min at 0 °C, keeping the reaction mixture free of moisture with a calcium chloride drying tube. If any precipitation is observed, add more DMF (dropwise) and leave stirring for an additional 10 min (*see* **Note 2**).
7. Remove the DCM by evaporation using a rotary evaporator.
8. Redissolve in a minimum volume of DMF and add to the swelled hydroxy-functionalized resin prepared in **steps 1** and **2**. The resin should be completely covered.
9. Prepare a solution of DMAP (0.1 equiv. relative to resin loading) in DMF (approx. 1 mL/mmol) and add to the resin mixture.
10. Apply a stopper to the flask and leave for 1 h at room temperature with occasional swirling (*see* **Note 3**).
11. Remove excess reagent by filtration and wash the resin with DMF (×5).
12. Perform a loading test as described in Subheading 3.8.
13. If the loading is less than 70 % of the theoretical, the procedure described above should be repeated.

3.3.2 The MSNT/MeIm Method

1. Place the hydroxy-functionalized resin in a dry reaction vessel and swell the resin in DCM
2. Remove DCM by applying vacuum and add fresh DCM until the resin is completely covered.
3. Flush the vessel with nitrogen and seal with a septum.
4. Place the desired Fmoc-protected amino acid (5 equiv. relative to the resin loading) in a dry round-bottomed flask containing a magnetic stirrer.
5. Add dry DCM (approx. 3 mL/mmol amino acid derivative) to dissolve the Fmoc-protected amino acids. A few drops of THF may be needed to aid complete dissolution.
6. Add MeIm (3.75 equiv. relative to the resin loading) followed by MSNT (5 equiv. relative to the resin loading). Flush the flask with nitrogen and seal with a septum. The mixture is stirred until the MSNT has dissolved.
7. Using a syringe, transfer the amino acids solution to the vessel containing the swelled hydroxy-functionalized resin prepared in **steps 1–3**. Leave the resin mixture to react for 1 h at room temperature applying occasional swirling (*see* **Note 3**).
8. Remove the septum and remove excess reagent by filtration and wash the resin with DCM (×3).

9. Perform a loading test as described in Subheading 3.8.
10. If the loading is less than 70 % of the theoretical, the procedure described above should be repeated.

3.4 Loading of Chlorotrityl Resins

The last protocol which will be described in the following is loading of the trityl-based linkers to yield *C*-terminal carboxylic acids upon cleavage. This resin is a good alternative to hydroxymethyl-based resins due to the absence of epimerization during loading of the first amino acids. It is recommended in particular for *C*-terminal His, Cys, Pro, Met, and Trp which are highly prone to give unwanted side reaction when using the symmetrical anhydride method for loading of hydroxymethyl-based resins. This resin can in a similar manner as described below be used for synthesis of a large variety of *C*-terminal functionalities, however, it may require a trityl linker with different substituents to tune its properties. Peptide release from this type of resin is described in Chapter 3.

1. In a Falcon tube, dissolve the Fmoc-protected amino acids (1.2 equiv. relative to the resin loading) and DIEA (5 equiv. relative to the resin loading) in dry DCM (10 mL/g resin) (*see* **Note 4**).
2. If necessary, add a small amount of DMF to aid dissolution of the mixture.
3. Add the mixture to the resin and stir for approx. 2 h at room temperature.
4. Wash the resin with DCM/MeOH/DIEA (17:2:1) (3 × 1 min), DCM (3 × 1 min), DMF (2 × 1 min), and DCM (2 × 1 min).
5. Perform a loading test as described in Subheading 3.8.

3.5 Standard Coupling Procedures

The following section describes three general protocols for standard amide bond formation in SPPS: (1) the activation using aminium or phosphonium salts, (2) the DIC/HOBt method, and the use of (3) preformed active esters [21]. The first two procedures are the most common whereas the latter is used in more specialized situations.

Instead of weighing out the individual reagents prior to each cycle, a stock solution of different coupling reagent and Fmoc-protected amino acids can be made. The shelf life for these solutions depends on the general storage and preparation conditions which includes the water content in the DMF used for making the solutions, the average temperature by which the solution is kept, the amount of time the solutions is open, and for how long the solutions is kept open (*see* Table 3). The majority of by-products present in the solutions of amino acids upon standing are due to loss of Fmoc or other protecting groups. The least stable protecting group is the trityl moiety and trifunctional amino acids with a

Table 3
Guidelines for storage of different Fmoc-protected amino acids in DMF/NMP

Fmoc-protected amino acids	Storage time at room temperature (DMF)	Storage time at 4 °C (DMF)
Bifunctional Fmoc-protected amino acids	<2 weeks	<4 weeks
Trifunctional Fmoc-protected amino acids, except trityl protected	<2 weeks	<4 weeks
Trityl-protected Fmoc-protected amino acids	<5 days	<10 days
Fmoc-Trp(Boc)-OH	<2 days	<5 days

Table 4
Guidelines for the storage of different coupling reagents in DMF (NMP)

Coupling reagent and other reagents	Storage time in open containers (DMF)	Storage time in closed containers (DMF)
HOBt and Oxyma	<2 weeks	–
HBTU and DIC	<1 week	<2 weeks
HATU	<1 days	<1 week
COMU	<4 h	<5 days

trityl protection group should not be kept in solution for over a week at room temperature. The other relatively unstable amino acid is Trp(Boc) which also degrades rapidly and should not be stored in solution for more than 2 days at room temperature. Otherwise, Fmoc-Aaa-OH solutions in DMF can last at least 2 weeks at room temperature and up to 4 weeks at 4 °C. The shelf life for all the solutions is increased if they are kept under nitrogen which is the case for many automated peptide synthesizer.

Some laboratories routinely add HOBt to stock solutions of Fmoc-protected amino acids. However, it appears that the presence of HOBt in the stock solutions accelerates the decomposition of the stock solutions and breakdown products can be found after only 1 week (Table 3).

The stability of different coupling reagents are also of great importance since a partially degraded stock solution of coupling reagents will have a negative impact on average amino acid incorporation (*see* Table 4). It has been reported recently that the stability of commercial COMU in open vials in comparison to HATU and HBTU is very low [22]. The key issue here is the water content of the DMF used for preparing the stock solution of the coupling reagent. It is in particular recommended to use fresh (thus relatively anhydrous) or anhydrous DMF for preparing stock solutions of COMU. This will increase the storage time in open or closed containers. Furthermore, inert gas covers or closed systems which provide a cover of dry nitrogen are applied in many automated systems and will also aid the stability of the coupling reagent in solution. Furthermore, as is the case for solutions of Fmoc-protected amino acids, lowering the temperature will also increase stability.

Another method for improving the synthesis of difficult peptides is to incorporate either backbone-protected Dmb or Hmb derivatives or pseudoproline dipeptides (described in Chapter 1). It is recommended to incorporate Dmb or Hmb derivatives or pseudoproline dipeptides at every sixth residue if possible or before a region of hydrophobic residues. Moreover, there should be a spacing of at least two residues between Dmb- or Hmb-protected derivatives, pseudoproline dipeptides, and Pro residues. These derivatives are purchased Fmoc protected and can be used with standard coupling procedures as described below. These specialized amino acid derivatives can, however, add significantly to the cost of synthesis. Also, the removal of the Dmb and Hmb protecting groups might be slow.

3.5.1 HBTU/HATU/COMU Manual Coupling

The standard coupling reagents as described in Chapter 1 are generally applied using the same protocol. As a general rule, the coupling reagents can be ranked by their coupling efficiency as HBTU<HATU/COMU.

1. Place the resin in a syringe equipped with a polypropylene filter. If this is the beginning of a synthesis, go to Subheading 3.1 for resin swelling (*see* **Note 5**).
2. In a separate flask or Falcon tube, mix the Fmoc-protected amino acid (4 equiv. relative to the resin loading) and HATU (3.8 equiv.) and add DMF (for a scale of 0.1 mmol, approx. 2.0 mL DMF is used for an overall concentration of 0.2 M in regard to the amino acids).
3. When all is dissolved, add DIEA (7.8 equiv.).
4. Immediately after addition of the base, add the mixture to the syringe containing the resin.
5. Mix well and react while shaking for 45 min at room temperature. Sometimes longer coupling times or double couplings are required (*see* guidelines in Table 5).
6. Remove the excess reagents by filtration.
7. Wash the resin with DMF (×4).
8. Optionally, a Kaiser test can be performed to investigate whether the coupling needs to be repeated. However, the Kaiser test is not well suited for peptides with an *N*-terminal Pro (*see* Subheading 3.9).

Table 5
Guidelines for standard coupling conditions at room temperature

Conditions	Coupling time
Standard	45 min
Predicted difficult coupling	90 min
After incorporation of the first 20 amino acids	2×45 min
After steric hindered amino acids, e.g., Aib	2×45 min

Furthermore, the use of microwave heating has been successful for fast and high-yielding couplings. Robust protocols are described in Chapters 16 and 17

3.5.2 DIC/HOBt Manual Synthesis

The DIC/HOBt coupling procedure differs from the latter protocol by being performed in the absence of base. This can be beneficial in many situations and minimize the risk of racemization. HOAt is a more powerful additive than HOBt, but due to their potentially explosive properties, the commercial availability of these reagents has become restricted. An alternative additive to HOBt is Oxyma Pure which is now commercially available. Oxyma, which is the nucleophile implicit in the very effective coupling reagent COMU, has in many cases been reported to be superior to HOBt but not always to HOAt.

1. Place the resin in a syringe equipped with a polypropylene filter. If this is the beginning of a synthesis, go to Subheading 3.1 for resin swelling (*see* **Note 5**).
2. In a separate flask or Falcon tube, mix the Fmoc-protected amino acid (5 equiv. relative to the resin loading) and HOBt (5 equiv.) and add the minimum amount of DMF to dissolve.
3. Add DIC (5 equiv.) and stir the mixture for 20 min.
4. Mix well and react while shaking for 45 min at room temperature. Sometimes longer coupling times or double couplings are required (*see* Table 5).
5. Remove the excess reagents by filtration.
6. Wash the resin with DMF (×4).
7. Optionally, a Kaiser test can be performed to investigate whether the coupling needs to be repeated (*see* Subheading 3.9).

3.5.3 Use of Preformed Active Esters

The majority of activated esters such as OBt esters are not stable in storage, and they are therefore prepared fresh for each step as described above. However, the pentafluorophenyl esters (OPfp) are stable preformed esters and commercially available for most common proteinogenic amino acids. They are efficient acylation agents; however, due to their higher price, they are not used except in more specific situations.

1. Place the resin in a syringe equipped with a polypropylene filter. If this is the beginning of a synthesis, go to Subheading 3.1 for resin swelling (*see* **Note 5**).
2. In a separate flask or Falcon tube, mix the Fmoc-protected amino acid pentafluorophenyl ester (5 equiv. relative to the resin loading) and HOBt (5 equiv.) and add the minimum amount of DMF to dissolve.
3. Mix well and react while shaking for 45 min at room temperature. Sometimes longer coupling time or double couplings are required (*see* Table 5).
4. Remove the excess reagents by filtration.
5. Wash the resin with DMF (×4).
6. Optionally, a Kaiser test can be performed to investigate whether the coupling needs to be repeated (*see* Subheading 3.9).

3.6 Capping or Acylation of the N-Terminal

In some cases, it can be beneficial to "cap" unreacted amino groups after the completion of a coupling reaction; three classical methods using acetic anhydride are explained below. However, this capping step is in most cases not a necessity, and it has been shown that the classic acetic anhydride method is not suited for automated peptide syntheses due to the low stability of the acetic anhydride solution. Alternative protocols have been developed using Z(2Cl)-OSu in NMP:DCM:DIEA [23]. The three methods described below can also be used for a *N*-terminal acylation of the final peptidyl-resin.

Method 1:

1. After completion of a coupling reaction, the resin is washed thoroughly with DMF.
2. Add 25 % acetic anhydride in DMF. The total volume should exceed the height of the resin threefold.
3. Leave for 2 h while shaking.
4. Wash resin with DMF (×4) and continue on to a Fmoc removal (Subheading 3.7) or for *N*-terminal acylation continue to a peptide cleavage reaction (Chapter 3).

Method 2:

1. After the completion of a coupling reaction, the resin is washed thoroughly with DMF.
2. Add 25 % acetic anhydride in DMF. The total volume should exceed the height of the resin threefold.
3. Leave for 5 min while shaking.
4. Add 1.5 equiv. of DIEA and leave for 30 min under shaking.
5. Wash resin with DMF (×4) and continue on to a N^{α}*-deprotection* (*see* Subheading 3.7) or for *N*-terminal acylation continue to a peptide cleavage reaction (Chapter 3).

Method 3:

1. Make a fresh solution of 20 % acetic anhydride, 20 % pyridine in DMF.
2. Add this mixture to the resin after completion of the coupling step (or for *N*-terminal acylation after completion of the Fmoc removal). The total volume should exceed the height of the resin threefold.
3. Leave the resin shaking for 2 h.
4. Wash resin with DMF (×4) and continue on to a ***N***$^{\alpha}$***-deprotection*** (*see* Subheading 3.7) or for *N*-terminal acylation continue to a peptide cleavage reaction (*see* Chapter 3).

3.7 Standard Fmoc Removal

After each coupling step, the Fmoc group is removed with a solution of 20 % piperidine in DMF. Most commercially available amino-functionalized resins or preloaded hydroxy-functionalized resins require an initial *N*$^{\alpha}$-Fmoc removal before synthesis. For longer peptide sequences, an increase in deprotection time can be beneficial further into the synthesis (after 25–30 cycles). Furthermore, for certain combinations of Asp-Xxx (in particular Asp-Gly) in the sequence, a side reaction is aspartimide formation. This side reaction can be minimized by adding 0.1 M formic acid or 1 M HOBt (1 M Oxyma) to the piperidine solutions applied in the Fmoc removal steps [24]. However, only the use of Asp(OtBu)-(Dmb) Gly-OH dipeptides completely prevent aspartimide formation.

1. Add 20 % piperidine in DMF (approx. 5 mL/0.5 mmol) to the resin (placed in a syringe equipped with a polypropylene filter).
2. Leave to react for 2 min before removal of solvent by vacuum filtration.
3. Repeat the treatment with 20 % piperidine in DMF and leave standing for 15 min.
4. Remove solvent by vacuum filtration and wash the resin with DMF (×5)
5. N.B. For longer peptide sequences, it can be beneficial to repeat **steps 3–4**.

3.8 Fmoc Quantification/Loading Test

In many cases, it is necessary to quantify the loading of the resin in order to determine the available amino groups.

1. Take out a small sample of resin (enough for 3×5 mg dry resin) and dry it down as described in Subheading 3.1.4.
2. Measure out three resin sample of approx. 5 mg each in an Eppendorf tube.
3. Add 1 mL of 20 % piperidine in DMF to each tube and leave for 10 min under shaking.
4. Leave the sample for 5 min without shaking (*see* **Note 6**).

5. Extract 100 µl from the top liquid (without resin) and dilute in a UV cell to 1 mL with 20 % piperidine in DMF (*see* **Note 7**).
6. Create a reference by filling 1 mL piperidine in DMF in a silica UV cuvette.
7. Measure the absorbance at 290 nm (A_{290}) where $\varepsilon_{290} = 5{,}800$ M^{-1} cm^{-1}.
8. Calculate the loading for each sample according to the following equation:

$$\text{Loading}\,(\text{mmol}/\text{g}) = \frac{(V \times A_{290\text{nm}})}{(\varepsilon \times d \times m_{\text{sample}})},$$

where V = volume in L, d is the cuvette size in cm, and $\varepsilon_{290} = 5{,}800$ M^{-1} cm^{-1}.

3.9 Kaiser Test

In order to perform the Kaiser test, three different solutions are prepared. These solutions are stable and can be used for months [25]. However, the Kaiser test is not well suited for peptides with *N*-terminal Pro residues.

Solution 1: Dissolve 5 g ninhydrin in 100 mL ethanol.

Solution 2: Dissolve 40 g phenol in 10 mL ethanol.

Solution 3: Add 2 mL of a 0.001 M aq. KCN solution to 98 mL pyridine

Procedure

1. Wash resin with DCM (×2).
2. Sample a few beads of resin in an Eppendorf tube.
3. Add one to two drops of each of the three solutions described above.
4. Mix well and heat to 120 °C for 3–5 min.
5. The presences of free resin-bound amines are indicated by blue resin beads.

4 Notes

1. This method is not suited for loading of His or Cys.
2. The total amount of DMF should not exceed 10 % of the total volume. Adding DMF will aid solubility of the mixture but will decrease reactivity.
3. It is not recommended to use magnetic stirring in combination with resins. Magnetic stirring can grind the resin and thereby ruin the resin hence the following synthesis. It is recommended to use a shaker. Automated peptide synthesizers usually use either vortex stirring or nitrogen bobbling.

4. It can be necessary to dry down the Fmoc-protected amino acids to increase the success of the reaction. Drying can be done by repeated evaporation in dioxane.
5. Remember to remove the Fmoc group from the purchased resin.
6. The resin will sink to bottom of the tube.
7. Here it is important to use either a quartz or fused silica cuvette due to the properties of DMF.

References

1. Hudson D (1999) Matrix assisted synthetic transformations: a mosaic of diverse contributions. I. The pattern emerges. J Comb Chem 1:333–360
2. Hudson D (1999) Matrix assisted synthetic transformations: a mosaic of diverse contributions. II. The pattern is completed. J Comb Chem 1:403–457
3. Merrifield RB (1963) Solid phase peptide synthesis. I. The synthesis of a tetrapeptide. J Am Chem Soc 85:2149–2154
4. Fields GB, Noble RL (1990) Solid-phase peptide-synthesis utilizing 9-fluorenylmethoxycarbonyl amino acids. Int J Pept Protein Res 35:161–214
5. Rapp W, Zhang L, Habich R, Bayer E (1989) in Peptides 1988 (Eds. Jung G and Bayer E), de Gruyter, Berlin, p. 199
6. Bayer E (1991) Towards the chemical synthesis of proteins. Angew Chem Int Ed 30: 113–129
7. Meldal M (1992) Pega: a flow stable polyethylene glycol dimethyl acrylamide copolymer for solid phase synthesis. Tetrahedron Lett 33:3077–3080
8. Garcia-Martin F, Quintanar-Audelo M, Garcia-Ramos Y, Cruz LJ, Gravel C, Furic R, Cruz S, Tulla-Puche J, Albericio F (2006) ChemMatrix, a poly(ethylene glycol)-based support for the solid-phase synthesis of complex peptides. J Comb Chem 8:213–220
9. Songster MF, Barany G (1997) Solid-phase peptide synthesis. In: Fields GB (ed) Methods in enzymology. Academic, San Diego, pp 126–174
10. Guillier F, Orain D, Bradley M (2000) Linkers and cleavage strategies in solid-phase organic synthesis and combinatorial chemistry. Chem Rev 100:2091–2158
11. Rink H (1987) Solid phase synthesis of protected peptide fragments using a trialkoxy-diphenyl-methylester resin. Tetrahedron Lett 28:3787–3790
12. Bray B (2003) Large-scale manufacture of peptide therapeutics by chemical synthesis. Nat Rev Drug Discov 2:587–593
13. Camarero J, Hackel BJ, de Yoreo JJ, Mitchell AR (2004) Fmoc-based synthesis of peptide α-thioesters using an aryl hydrazine support. J Org Chem 69:4145–4151
14. Woo Y-H, Mitchell AR, Camarero JA (2007) The use of aryl hydrazide linkers for the solid phase synthesis of chemically modified peptides. Int J Pept Res Ther 13:181–190
15. Jensen KJ, Alsina J, Songster M, Vagner J, Albericio F, Barany G (1998) Backbone amide linker (BAL) strategy for solid-phase synthesis of C-terminal-modified cyclic peptides. J Am Chem Soc 120:5441–5452
16. Boas U, Brask J, Christensen J, Jensen KJ (2002) The ortho backbone amide linker (o-BAL) is an easily prepared and highly acid-labile handle for solid-phase synthesis. J Comb Chem 4:223–228
17. Brask J, Albericio F, Jensen KJ (2003) Fmoc solid-phase synthesis of peptide thioesters by masking as trithioortho esters. Org Lett 5:2951–2953
18. García-Ramos Y, Parodís-Bas M, Tulla-Puche J, Albericio F (2010) ChemMatrix® for complex peptides and combinatorial chemistry. J Pept Sci 16:375–378
19. Albericio F, Barany G (1987) An acid-labile anchoring linkage for solid-phase synthesis of C-terminal peptide amides under mild conditions. Int J Pept Protein Res 30:206–216
20. Albericio F, Kneib-Cordonier N, Biancalana S, Gera L, Masada RI, Hudson D, Barany G (1990) Preparation and application of the 5-(4-(9-fluorenylmethyloxycarbonyl) aminomethyl-3,5-dimethoxyphenoxy)-valeric acid (PAL) handle for the solid-phase synthesis of C-terminal peptide amides under mild conditions. J Org Chem 55:3730–3743
21. Valeur E, Bradley M (2009) Amide bond formation: beyond the myth of coupling reagents. Chem Soc Rev 38:606–631
22. Tofteng AP, Pedersen SL, Staerk D, Jensen KJ (2012) Effect of residual water and microwave heating on the half-life of the reagents and reactive intermediates in peptide synthesis. Chem Eur J 18:9024–9031

23. Ball HL, Mascagni P (1995) N-(2-chlorobenzyloxycarbonyloxy)-succinimide as a terminating agent for solid-phase peptide synthesis: application to a one-step purification procedure. Lett Pept Sci 2:49–57
24. Palasek SA, Cox ZJ, Collins JM (2007) Limiting racemization and aspartimide formation in microwave-enhanced Fmoc solid phase peptide synthesis. J Pept Sci 13:143–148
25. Kaiser E, Colescott RL, Bossinger CD, Cook PI (1970) Color test for detection of free terminal amino groups in the solid-phase synthesis of peptides. Anal Biochem 34:595–598
26. Information on swelling is available from different distributors (December 2012) http://www.rapp-polymere.com, http://www.pcasbiomatrix.com, http://www.merckmillipore.com

Chapter 3

Peptide Release, Side-Chain Deprotection, Work-Up, and Isolation

Søren L. Pedersen and Knud J. Jensen

Abstract

After having successfully synthesized a peptide, it has to be released from the solid support, unless it is being used for on-resin display. The linker and, in some cases, the cleavage mixture determine the *C*-terminal functionality of the released peptide. In most cases, the peptide is released with concomitant removal of side-chain protecting groups. However, some combinations of linkers and side-chain protecting groups enable a two-stage procedure, either using orthogonal chemistry or graduated labilities. Herein, we describe protocols for the release of peptides from the most commonly used linker types providing a variety of different *C*-terminal functionalities, including acids, amides, amines, and aldehydes. Moreover, suggestions for determination of peptide Purities and for storage conditions are provided.

Key words Peptide cleavage, Side-chain deprotection, Rink amide linker, Wang linker, Hydrazine benzoyl linker, HMBA linker

1 Introduction

Release of the assembled peptide and removal of side-chain protecting groups are key steps in peptide synthesis. However, the treatment of a peptidyl-resin with a cleavage cocktail does not initiate just one simple reaction, but several simultaneous reactions that can compete and interfere with each other. Suitable scavengers need to be used to avoid unwanted irreversible side reactions that can modify the assembled peptide.

The aim of release and deprotection is to separate the peptide from the solid support as well as from the side-chain protecting groups. To avoid side reactions, the peptide should only be exposed to the cleavage mixture for the minimum time it takes to release and deprotect it. The peptide is then removed from the support by filtration and from the protecting groups by precipitation, centrifugation, and decantation. The methods described herein are for release of peptides assembled via Fmoc-based solid-phase peptide synthesis (Fmoc-SPPS).

Knud J. Jensen et al. (eds.), *Peptide Synthesis and Applications: Second Edition*, Methods in Molecular Biology, vol. 1047, DOI 10.1007/978-1-62703-544-6_3,

Most cleavage and deprotection strategies for Fmoc-SPPS are trifluoroacetic acid (TFA) based and vary in the percentage of acid, types of scavengers used, and reaction time. These methods will be described in detail. Next, some methods for the nucleophilic release of C-terminally modified peptides from HMBA and hydrazinobenzoyl linkers are also presented. Moreover, suggestions for the solubilization, quantification, and storage of peptides will be discussed in the chapter.

1.1 Peptide Release and Side-Chain Deprotection

Some linkers can result in a number of different *C*-terminal functionalities either by having different initial functional groups attached to the acid-labile linkers, such as for the Rink and trityl linkers, or by varying the nucleophilic reagent used to release peptides from the HMBA and hydrazino benzoyl linkers. Altering the initial functionality changes the acid-lability of the Rink linker, *see* Chapters 1 and 9. Acidolytic release of peptides from the solid support and the removal of side-chain protecting groups carry the risk of side reactions. To prevent the reaction of the debris from side-chain protecting groups as well as the reattachment of the peptide onto the linker, scavengers are often used to trap the cations. Water is a very common scavenger used in most cleavage cocktails. Other very useful scavengers include triethylsilane (TES), ethanedithiol (EDT), thiophenol, and the non-malodorous 2,2′-(ethylenedioxy)diethanethiol (also called 1,2-bis(2-mercaptoethoxy) ethane and 3,6-dioxa-1,8-octane-dithiol or just DODT).

The removal of side-chain protecting groups and release of the peptide from the support can either occur in one or two steps. While the one-step protocol in Fmoc-SPPS combines an acidolytic handle with acid-labile side-chain protecting groups for the convenient production of fully unprotected peptides, the two-step protocol enables regioselective chemistry on the peptide, for example, the synthesis of branched peptide structures.

1.1.1 Linkers

Boc-SPPS will be described in Chapter 4; in this chapter, we will solely focus on linkers suitable for Fmoc-SPPS. The sensitivity of the linker towards acid depends on (1) the stability of the carbocation formed upon cleavage; (2) the nature of the leaving group, e.g., whether it is a carboxylic acid or amide; and (3) the steric environment; *see* Chapters 1 and 9. The development of linkers with higher acid-lability has mainly focused on the two first factors. The addition of methoxy or alkoxy groups to benzyl or benzhydrylamine linkers has resulted in more acid-labile linkers (Wang and Rink linkers), thus they can be cleaved using relatively mild acidic conditions. This is typically achieved with solutions of TFA in dichloromethane (DCM) and in the presents of scavengers. There are also at least three different modes of anchoring: (1) classical *C*-terminal anchoring, (2) side-chain anchoring, and (3) backbone amide linkage (BAL).

Most acid-labile linkers have so far been constructed from benzyl, benzhydryl, or trityl cores. Wang and co-workers introduced the 4-alkoxybenzyl alcohol linker (Wang linker) for the synthesis of peptide acids that is presently the linker of choice for the synthesis of peptide acids by Fmoc-SPPS [1]. Since then, several variations of the Wang linker have been described, *see* Chapter 1. The attached esters are typically cleaved using 50–100 % of TFA in DCM for 30–120 min. Several different linkers are commercially available that can be cleaved using dilute acid—i.e., 3-methoxy-4-hydroxymethylphenoxyacetic acid (HMPA) [2], 4-[(2′,4′-dimethoxyphenyl) hydroxymethyl]phenoxy linker (Rink acid) [3], 2-methoxy-4-alkoxybenzyl alcohol (SASRIN) [4], *ortho* backbone amide linker (o-BAL) [5, 6], and 2-chlorotrityl linker [7, 8] (Table 1). The peptide ester bond formed from the 2-chlorotrityl linker is cleavable using 1 % TFA in DCM, and the Rink acid linker is cleavable using as low as 0.2 % TFA in DCM which can deliver fully side-chain protected peptide acids. The trityl linker is also the linker of choice to prepare *C*-terminal peptide amines which can also be released using only 1 % TFA in DCM. The acid-sensitivity of the trityl linker can be tuned by the substituents. Barlos and co-workers have shown that the acid-lability increased in the order 2-chlorotrityl, trityl, 4-methyltrityl, and 4-methoxytrityl [9]. Besides fully protected peptide acids, the *o*-BAL handle can also be used to prepare *C*-terminal peptide aldehydes, peptide thioesters, as well as cyclic peptides [5, 10, 11] (Table 1). The *o*-BAL applications will be described in Chapter 9; thus, the release conditions for this linker will not be further described in this chapter.

Several linkers provide *C*-terminal peptide amides. The most prominent of these, the 4-[(2′,4′-dimethoxyphenyl)aminomethyl] phenoxyacetamide methyl linker (Rink Amide) [3], is after peptide chain assembly cleavable using 10–95 % TFA in DCM. Trialkylsilane is recommended for scavenging carbocations arising from protecting groups and degradation products from the linker. Another linker type, the infrequently used Sieber xanthenyl linker, can be cleaved using only 1 % TFA in DCM resulting fully side-chain protected peptide amides [12].

Linkers cleavable by nucleophiles include the safety-catch linker [13, 14], hydrazino benzoyl linker [15, 16] and 4-hydroxymethylbenzoic acid (HMBA) [14, 17] (Table 1). All three linkers can provide fully protected peptides, and as the HMBA and hydrazinobenzoyl linkers are acid-stable, the peptide side-chains can be deprotected prior to release of the peptide. The HMBA linker is a versatile but less often used linker for Fmoc-SPPS. Varying the nucleophile employed in the cleavage cocktail; a wide range of *C*-terminal modified peptides can be obtained: primary amide (NH_3 in methanol) [18, 19], hydrazides (5 % NH_2NH_2 in DMF) [20], acids (0.1 M aqueous NaOH) [21], and methyl esters (5 % DIEA, 5 % MeOH in DMF at 50 °C) [22] (Fig. 1).

Table 1
Overview of *C*-terminal functional groups versus linkers and deprotection conditions

Linker	Final *C*-terminal functionality	Linker structure	Cleavage mixture
2-Chlorotrityl	Peptide acids	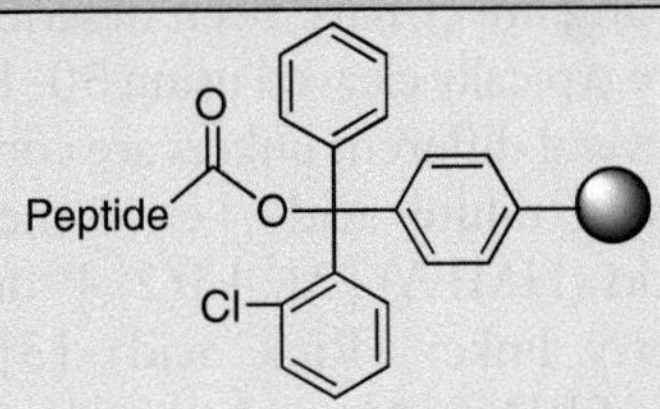	95 % TFA, 2.5 % TES, 2.5 % H_2O
	Protected peptide acids		1 % TFA, 3 % TES, 96 % DCM
	Peptide amines	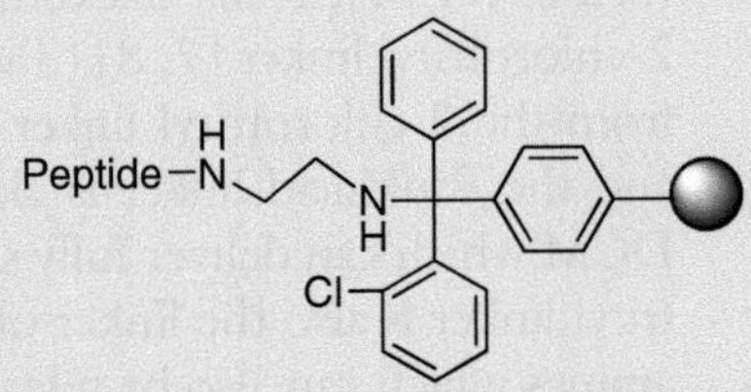	95 % TFA, 2.5 % TES, 2.5 % H_2O
HMBA	Protected peptide acids	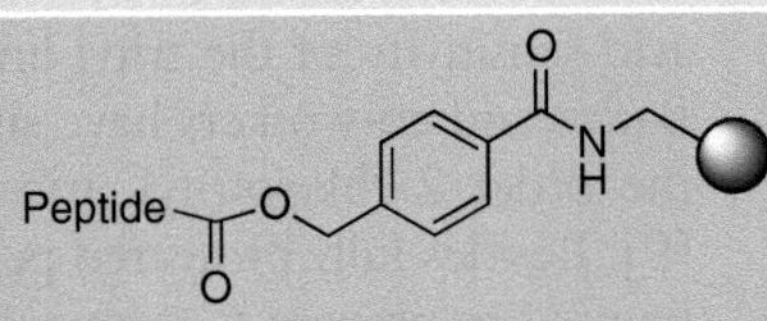	0.1 M NaOH
	Peptide amides		NH_3/MeOH
	Peptide alcohols		$NaBH_4$/EtOH
	Peptide hydrazides		5 % NH_2NH_2 in DMF
HMPA	Peptide acids	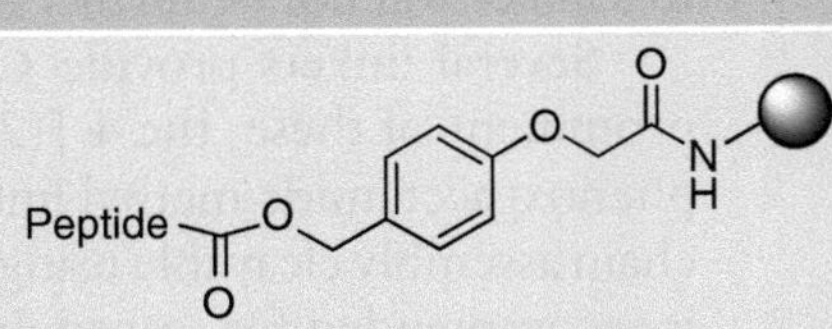	95 % TFA, 2.5 % TES, 2.5 % H_2O
Hydrazino benzoyl	Peptide amides	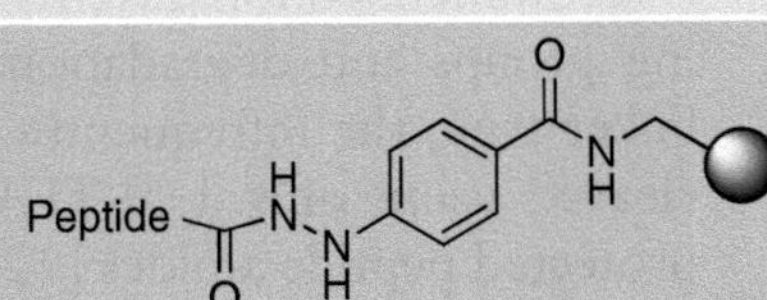	Cu(II) (cat.), Pyridine, RNH_2
	Peptide esters		Cu(II) (cat.), Pyridine, ROH

(continued)

Table 1 (continued)

Linker	Final *C*-terminal functionality	Linker structure	Cleavage mixture
o-BAL	Peptide acids Protected peptide acids	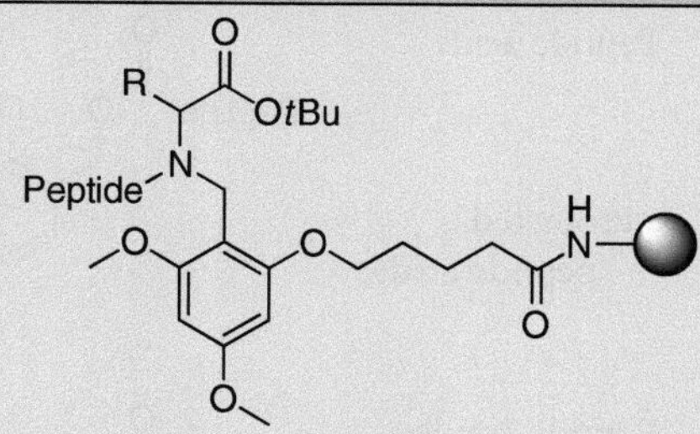	95 % TFA, 2.5 % TES, 2.5 % H_2O 1 % TFA, 3 % TES, 96 % DCM
	Peptide aldehydes (*C*-terminal glycinal)	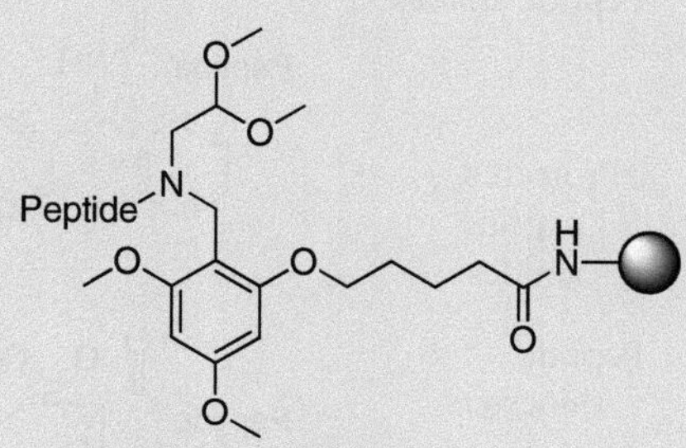	95 % TFA, 5 % H_2O
	Peptide thioesters (*C*-terminal Gly)	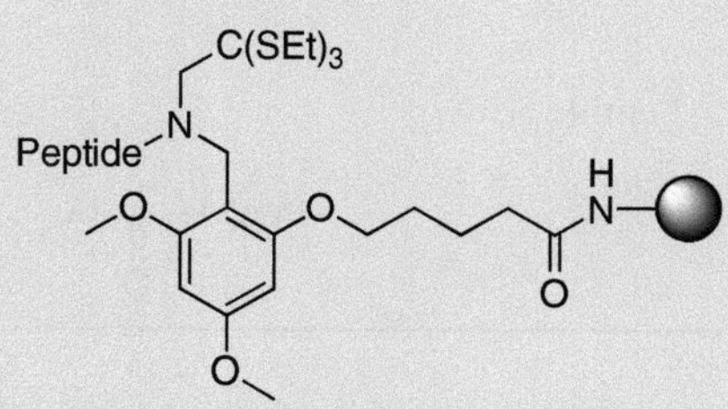	95 % TFA, 5 % H_2O
Rink acid	Peptide acids Protected peptide acids	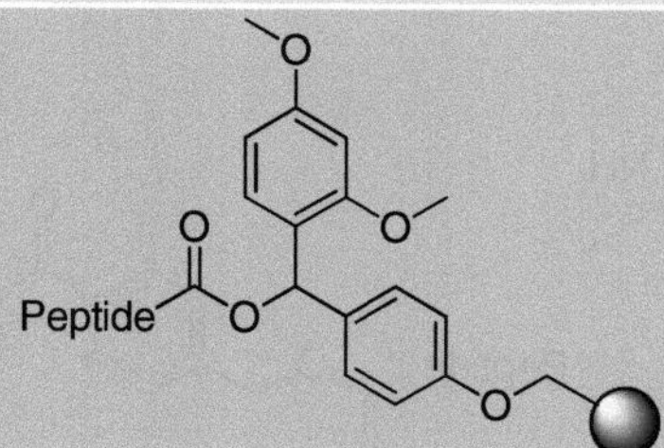	95 % TFA, 2.5 % TES, 2.5 % H_2O 1 % TFA, 3 % TES, 96 % DCM
Rink Amide	Peptide amides	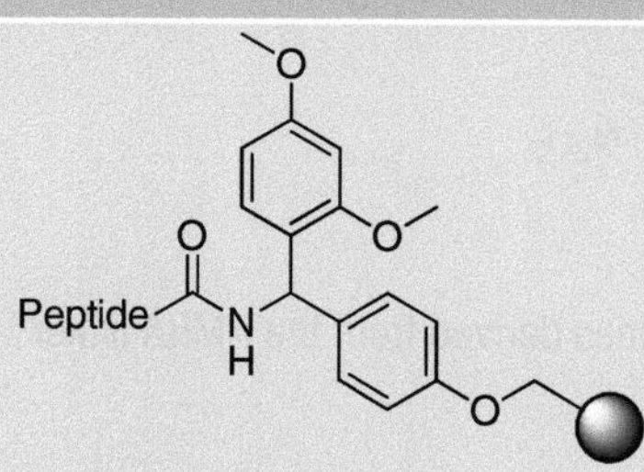	95 % TFA, 2.5 % TES, 2.5 % H_2O

(continued)

Table 1 (continued)

Linker	Final *C*-terminal functionality	Linker structure	Cleavage mixture
SASRIN	Peptide acids	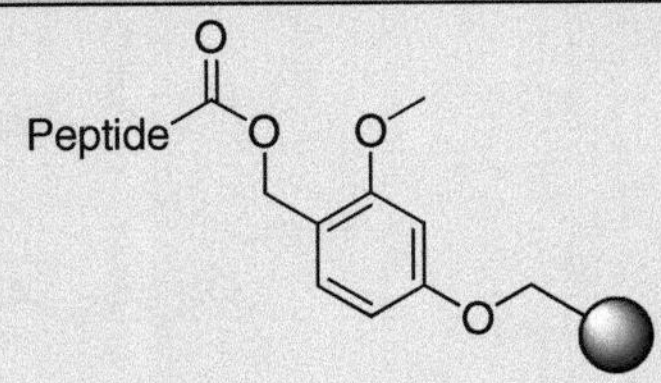	95 % TFA, 2.5 % TES, 2.5 % H_2O
	Protected peptide acids		1 % TFA, 3 % TES, 96 % DCM
Sieber linker	Peptide amides	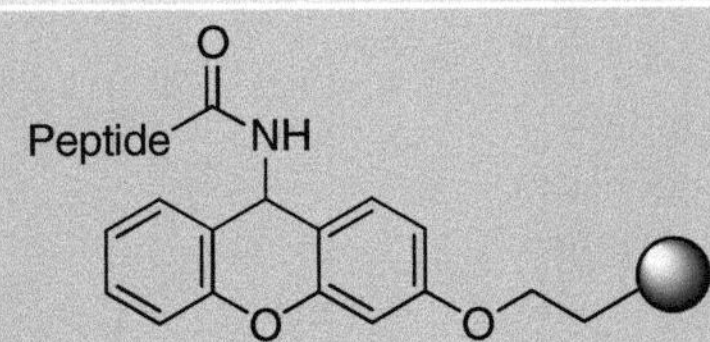	95 % TFA, 2.5 % TES, 2.5 % H_2O
	Protected peptide amides		1 % TFA, 3 % TES, 96 % DCM
4-Sulfamylbutyryl	Peptide thioesters	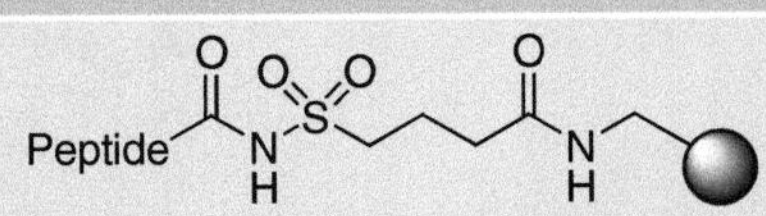	1. TMS-CHN_2 2. RSH, NaSPh (cat.)
Wang	Peptide acids	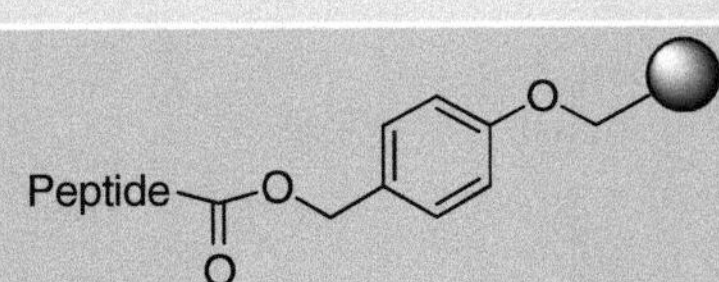	95 % TFA, 2.5 % TES, 2.5 % H_2O

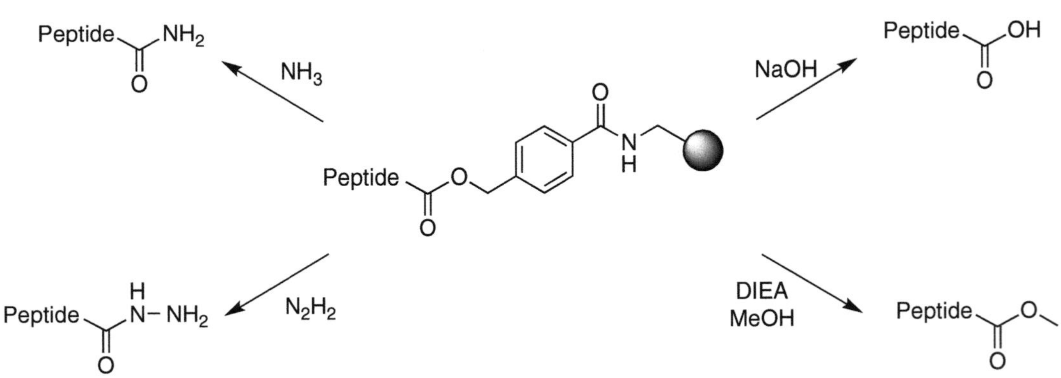

Fig. 1 *C*-Terminal functionalities derived from the HMBA linker

Peptides can be released from the hydrazinobenzoyl linker by Cu(II) acetate activation and a nucleophile such as an amine or MeOH in DCM [16, 23]. Application of the safety-catch linkers is discussed in Chapter 8; thus, it is not further described in the present chapter.

Table 2
N^{α}-Fmoc protected amino acids, their general protecting groups and molecular weight

Amino acids		Functionality protected	MW of protected amino acids (g/mol)	MW of unprotected amino acids (g/mol)	MW minus water (g/mol)
Fmoc-Ala-OH	A		311.3	89.1	71.1
Fmoc-Arg(Pbf)-OH	R	Guanidino	648.8	174.2	156.2
Fmoc-Asn(Trt)-OH	N	Amide	596.7	132.1	114.1
Fmoc-Asp(O*t*Bu)-OH	D	Carboxyl	411.5	123.1	115.1
Fmoc-Cys(Trt)-OH	C	Sulfhydryl	585.7	121.2	103.1
Fmoc-Gln(Trt)-OH	Q	Amide	610.7	146.2	128.1
Fmoc-Glu(O*t*Bu)-OH	E	Carboxyl	425.5	147.1	129.1
Fmoc-Gly-OH	G		297.3	75.1	57.1
Fmoc-His(Trt)-OH	H	Imidazole	619.7	155.2	137.1
Fmoc-Ile-OH	I		353.4	131.2	113.2
Fmoc-Leu-OH	L		353.4	131.2	113.2
Fmoc-Lys(Boc)-OH	K	Amino	468.5	146.2	128.2
Fmoc-Met-OH	M		371.5	149.2	131.2
Fmoc-Phe-OH	F		387.4	165.2	147.2
Fmoc-Pro-OH	P		337.4	115.1	97.2
Fmoc-Ser(*t*Bu)-OH	S	Hydroxyl	383.4	105.1	87.1
Fmoc-Thr(*t*Bu)-OH	T	Hydroxyl	397.5	119.1	101.1
Fmoc-Trp(Boc)-OH	W	Indole	526.6	204.2	186.2
Fmoc-Tyr(*t*Bu)-OH	Y	Phenol	459.6	181.2	163.2
Fmoc-Val-OH	V		339.4	117.2	99.1

Pbf 2,2,4,6,7-pentamethyl-dihydrobenzofuran-5-sulfonyl, *Trt* Trityl, *tBu* *t*-butyl ether, *Boc* *t*-butyloxycarbonyl

1.1.2 Side-Chain Deprotection of Amino Acids

Proteinogenic amino acids are either bi- or trifunctional, while some non-proteinogenic amino acids can contain up to four functional groups. Most trifunctional amino acids require a side-chain protecting group during solid-phase peptide synthesis. Upon TFA treatment of acid-labile side-chain protecting groups, carbocation ions are generated which in some cases potentially can react with other functional groups in the peptide. Thus, special precaution must be undertaken to minimize undesired side reactions during TFA cleavage. The protecting groups for 20 proteogenic amino acids and their masses are listed in Table 2.

Some amino acids can be problematic during TFA-mediated release and deprotection. Methionine can easily be oxidized to methionine sulfoxide both during release and synthesis. It can be reduced back to methionine either postsynthetically using dithiothreitol (DTT) and *N*-mercaptoacetamide [24] or during cleavage using 1,2-ethanedithiol (EDT) and trimethylsilyl bromide (TMS-Br) [25]. DODT (3,6-Dioxa-1,8-octane-dithiol) is a very good thiol alternative to EDT in the cleavage mixture as it is non-malodorous [26]. Nevertheless, to avoid methionine oxidation during cleavage avoid oxygen and perform the final concentration of the cleavage cocktail using nitrogen or argon flow. Peptides containing cysteine residues are prone to form disulfide bonds either intramolecularly, creating cyclic peptides, or intermolecularly, leading to oligomers. Disulfide bonds can often be reduced by DTT or other reducing agents. Often if only one cysteine residue is present in the peptide sequence, 2.5 % triethylsilane in TFA is adequate to keep the thiols in the reduced form. Trp-containing peptides can also be oxidized and readily be alkylated by protecting groups released from other amino acid side-chains if scavengers are not added to the cleavage mixture [14]. Modification of Trp can generally not be reverted. Minimal oxidation or alkylation of the indole ring is observed when using Boc as protecting group for Trp. His and Tyr can be alkylated in the same manner as Trp, if carbocations are not properly scavenged. *N*-terminal glutamine can also undergo cyclization to form a pyroGlu under acidic conditions.

1.2 Cleavage Cocktails and Isolation of Crude Peptide Product

The cleavage cocktail is chosen depending on the Depending on the amino acid composition, the side-chain protecting groups, and the nature of the cleavable linker [27]. The most common universal cleavage cocktail in past decades has been "Reagent K" (Table 3), but further developments have resulted in cleavage mixtures which are less toxic and malodorous. Thus, with *t*Bu and trityl amino acid side-chain protecting groups and the widely used Rink Amide and Wang linkers peptides can normally be released using either TFA/TES/H_2O 95/2.5/2.5. In peptides with multiple Cys(Trt), Met, or Arg(Pbf) residues, TFA/TES/EDT/H_2O (92.5/2.5/2.5/2.5) or TFA/DODT (95/5) can be useful. In special circumstances Met, Trp, and Arg residues can cause substantial problems during the cleavage, and sometimes cleavage cocktails such as reagent K or R can be an advantage. Consequently, it is recommended to perform a test cleavage (mini-cleavage) before cleaving the entire amount of resin.

The amount of cleavage cocktail used depends on the amount of resin and its ability to swell. The cleavage solution should cover the resin—approximately 5 mL per 0.5 g of resin—less for polystyrene (PS)-based resins and more for polyethylene glycol (PEG)-based resins. The reaction time is typically 2–4 h. Too long reaction times can cause side reactions, and too short

Table 3
Cleavage cocktails

Reagent	Recipe	Cleavage time [h]
B	TFA/phenol/H_2O/TES (88/5/5/2)	1–4
K	TFA/phenol/H_2O/thioanisole/EDT (82.5/5/5/5/2.5)	1–4
L	TFA/DTT/H_2O/TES (88/5/5/2)	1–4
P	TFA/phenol (95/5)	1–4
R	TFA/thioanisole/EDT/anisole (90/5/3/2)	1–4
T	TFA/TES (95/5)	1–4
	TFA/H_2O (95/5)	1–4
	TFA/TES/H_2O (95/2.5/2.5)	1–4
	TFA/TES/EDT/H_2O (92.5/2.5/2.5/2.5)	1–4
	TFA/DODT (95/%)	2–3

TFA trifluoroaceticacid, *TES* triethylsilane, *EDT* 1,2-ethanedithiol, *DTT* dithiothreitol, *DODT* 2,2′-(Ethylenedioxy) diethanethiol

cleavage time will result in non-quantitative deprotection and peptide release from the resin. Cleavage times can be reduced to minutes when using microwave heating [28–30]; however, cleavage associated side reactions will likely also be accelerated. The cleavage cocktails must always be prepared using high-quality TFA and fresh scavengers (*see* **Note 1**).

The cleavage reaction is terminated by removing the cleavage cocktail from the solid support, and subsequently, the cocktail can be concentrated, for example, under a stream of nitrogen in a well-ventilated fume hood, followed by a precipitation in ice-cold diethylether. Some side-chain protecting group debris is soluble in diethyl ether, and after centrifugation, the ether phase is decanted followed by another ether precipitation to remove the final protecting groups. It is recommended that the diethyl ether phase is collected and set aside until the target peptide has been identified in the crude precipitate. The peptide can preferable be dissolved in water-acetonitrile and lyophilized for storage until purification.

For several reasons, peptides may not precipitate in cold diethyl ether. Very short peptides (less than 5–6 amino acids), very hydrophobic peptide sequences, or fully side-chain protected peptides tend to be soluble in cold diethyl ether. Fully protected and very hydrophobic sequences may precipitate in cold H_2O. Sometimes the peptide can be worked up by dissolving the peptide in 10–20 % aqueous acetic acid followed by an extraction using chloroform to remove scavengers.

1.3 Characterization

1.3.1 Peptide Solubility

Many peptides have good solubility and can easily be dissolved in water; however, a little acetic acid may aid dissolution. If the peptide is not completely soluble in water/acetic acid, acetonitrile can be added to increase the solubility of the more hydrophobic peptides. Nevertheless, some crude peptide mixtures may be notoriously insoluble in mixtures of acetonitrile-H_2O. Peptides containing large stretches of hydrophobic residues can form compact aggregates such that the peptide minimizes hydrophobic interactions with water. To dissolve peptides with a propensity to form aggregates, larger amounts of acetic acid (up to 50 %) in acetonitrile as well as sonication may lead to solubilization. However, if the peptide is still not solubilized a minimum amount of DMF or DMSO can be added to the peptide-acetonitrile-acetic acid mixture. However, these are high-boiling solvents which may be difficult to remove afterwards by evaporation, and precautions have to be taken to avoid early elution of the peptide during HPLC. Peptides containing many basic residues can often be dissolved by addition of larger amounts of acetic acid, and acidic peptides can often be dissolved in basic buffer such as bicarbonate. If possible, peptides dissolved in basic buffers should be pH adjusted to neutral prior to HPLC injection. Larger peptides may be dissolved in chaotropic salts such as guanidine hydrochloride or urea which can assist in breaking secondary structures of the peptides.

1.3.2 LC-MS

Many peptide laboratories use acetonitrile and water containing 0.1 % formic acid as standard eluents for LC-MS characterization and quantification. Formic acid is used to acidify the eluents. TFA, which is commonly used in HPLC, can cause a suppression of the MS ionization in LC-MS. RP-HPLC should be performed in acidic solvents as most silica have a tendency to dissolve at pH above 7. Some newer column materials are more stable to higher pH, but the column performance is often better at acidic conditions. There are three common column types for peptide characterization, C4, C8, or C18. Generally, C18 columns are used for small- to medium-sized peptides that are fairly hydrophilic, and C4 and C8 columns are normally used for medium to large peptides sequences as these tend to elute at a higher percentage of acetonitrile. It is not only the alkyl chain lengths that have an effect on peptide purification, but column dimensions, particle, and pore size are also very important to obtain good separation. Particle size has an impact on resolution, and small particles tend to separate better, but the effect of particle size is most pronounced during isocratic separation. However, smaller particle size increases the backpressure generated over the column. The pores of RP-HPLC columns have the majority of their interactions with the analyte on the surface area of the particles. Due to the fact that most surface area is inside the pores, the pores must be large enough for the analyte to enter. Small peptides do not require large pores and can therefore enter the small pore size particles. Smaller pore size particles have higher surface area which increases the loading capacity. In contrast larger peptides require larger pore sizes to access the full surface area of the particles.

The backbone amides in peptides can be monitored at wavelength of approximately 215 nm, while peptide sequences containing Tyr, Phe, or Trp have strong absorbances originating from the aromatic rings at approximately 254 nm.

If DMF, DMSO or acetic acid is used to dissolve the peptide expect a large solvent peak in the beginning of the gradient. Normally peptides that need to be dissolved in DMF or DMSO are very hydrophobic and are therefore retained longer on the column; thus, they will normally not interfere with the DMF peak.

In our laboratory, we use a Dionex Ultimate 3000 LC system with a PDA detector. An A–B eluent system (A: Acetonitrile containing 0.1 % formic acid; B: H_2O containing 0.1 % formic acid) is applied on either a C4 analytical column (Phenomenex, Jupiter, 300 Å, 5 μm, 3.90×150 mm) or C18 analytical column (Phenomenex, Gemini, 110 Å, 3 μm, 4.60×50 mm) using standard B gradient elution from 5 to 100 % over 14 or 10 min, for C4 and C18, respectively. The identification of compounds is achieved using an ESI-MS (MSQ Plus Mass Spectrometer, Thermo).

1.3.3 Mass' of Byproducts

To characterize unwanted reactions, the mass of the most common by-products are listed in Table 4.

1.3.4 Suggestions for Storage

For long-time storage, it is recommended to keep the peptides attached to the solid support. Peptides that are side-chain deprotected may react inter- and intramolecularly as well as with contaminants. Moreover, peptides and proteins may contain residual solvents, such as TFA, which can lead to unwanted reactions for example hydrolysis of acid-labile peptide bonds such as Asp-Pro, and deamination of Gln or Asn. Thus, lyophilization is recommended prior to storage. Finally, long-time storage of peptides is recommended as powders or in slightly acidic frozen solutions at −80 °C.

2 Materials

1. Cleavage solution A: 95 % TFA, 2.5 % TES, 2.5 % H_2O.

   To 4.75 mL of trifluoroacetic acid (TFA), add 0.125 mL of triethylsilane (TES) and 0.125 mL H_2O and mix the solution (cocktail).

2. Cleavage solution B: 95 % TFA, 5 % DODT.

   To 4.75 mL of trifluoroacetic acid (TFA), add 0.25 mL of 2,2′-(ethylenedioxy)diethanethiol (DODT).

Table 4
List of modifications observed by mas spectrometry

-98	H_2PO_4 (loss of)	Artifact observed during MS of pSer and pThr peptides
-80	Dephosphorylation	Avoid prolonged storage of phosphopeptides in aqueous acid
-34	Dehydroalanine from Cys	
-32	Lanthionine formation from Cys	
-27	Oxidation of Arg to Glu	
-18	Dehydration	
-18	Pyroglutamic acid formed from Glu	
-18	Aspartimide formation from Asp	Use Fmoc-Asp(OtBu)-(Dmb)Gly-OH or Fmoc-Asp(OMpe)-OH
-17	Pyr formed from Gln	Avoid *N*-terminal Gln residues
-17	Succinimide formation from Asn	
-16	H-Phosphonate formation if a phosphopeptide	Use benzyl protected phosphoramidites and anhydrous t-butyl hydroperoxide for oxidation
-13	N-piperidinylalanine formation form *p*Ser or *p*Thr (elimination-addition)	Avoid using microwave heating for Fmoc removal
-2	Disulfide formation from Cys	
-1	Amide formation (C-terminus)	
1	Deamination of Asn or Gln to Asp and Glu	Avoid leaving peptides in aqueous media for prolonged periods
2	Reduction of Cys	
2	Reduction of indole double bond of Trp	Use Fmoc-Trp(Boc)-OH, and avoid use of triethyl silane (TES)
12	Cys to thioproline conversion (*C*-terminal)	Due to formaldehyde contamination of acetic acid
12	Formaldehyde adduct of Trp	
12	Formaldehyde imine on Lys	
14	Methyl ester	Ensure that peptide resin is free from methanol prior to cleavage
14	Hydrazide formation from free acid	
16	Oxidation of Met to sulfoxide	
16	Sulfenic acid from Cys	
17	Ammonium adduct	
22	Sodium adduct	

(continued)

Table 4 (continued)

28	Ethyl	
28	Formyl	Use good quality DMF and avoid heating peptides dissolved in formic aid
32	Oxidation of Met to sulfone	
38	Potassium adduct	
40	Uncleaved pseudoproline	Extend cleavage time and add EDT
41	Acetonitrile adduct of Lys	
42	Ac	
44	Disodium (observed in MS)	
44	CO_2 adduct of Trp	Dissolve in 1 % aqueous acetic acid and freeze dry
45	Nitro	
48	Cysteic acid formation by oxidation of Cys	
51	β-Piperidinylalanine formation from C-terminal Cys (elimination-addition)	Use trityl linker such as 2-chlorotrityl chloride
56	*t*Bu	Alkylation adduct of Cys, Met, Trp, Tyr
57	Gly insertion	Avoid Gly symmetrical anhydride
60	Sodium-potassium adduct	
67	Asp piperidine formation	Use Fmoc-Asp(*t*Bu)-(Dmb)Gly-OH or Fmoc-Asp(OMpe)-OH
71	Acm	
76	Mercaptoethanol	
80	Sulfonation (SO_3H) of Arg, Ser, Thr, from Pbf or Pmc	Recleave with TMS-Br/thioanisole/EDT/TFA
80	Sulfation of Tyr	
80	Phosphorylation	
88	S*t*Bu	
90	Bzl, uncleaved phosphoamino acid protecting group or alkylation adduct of Cys, Trp or Tyr	
91	Bzl alkylation of Met	
92	EDT disulfide adduct	Reduce with TCEP
96	TFA	Treat peptide with 0.1 M ammonium bicarbonate to cleave TFA esters
97	Proline insertion	Avoid double coupling of Pro

(continued)

Table 4
(continued)

98	Tetramethylguanidinium termination by-product on amine	Use phosphonium reagents instead of aminium
100	Boc	
104	MeBzl	
104	Bz	
117	Succinyl	
120	MeOBzl	
134	Z	
136	Hmb	
150	Dmb	
152	DTT disulfide adduct	
154	Tos	
166	Dnp	
172	Ethanedithiol/TFA cyclic adduct	Use Trp(Boc) instead of Trp and avoid prolonged cleavage times when EDT is used as scavenger
180	Tmob	
182	Mts	
188	Alkylation of Trp by Pbf chroman	
189	α-Cyano sinapinic acid adduct	
201	HMP (hydoxymethylphenyl)/TFA adduct	
204	S-Farnesyl	
206	*iv*Dde	
212	Mtr	
216	Mca	
222	Fmoc	
226	Mbh	
233	Dansyl	
238	Bpoc	
238	Palmitoylation	
242	Trt	
243	Rink Amide linker adduct	Use Rink Amide AM linker instead of Rink Amide

(continued)

Table 4 (continued)

251	Dabsyl	
252	Pbf	
256	Mtt	
266	Pmc	
272	Mmt	
299	Knorr linker adduct	Use Rink Amide AM instead of Rink Amide MBHA
312	Dmab	

3. Cleavage solution C: 96 % DCM, 3 % TES, 1 % TFA.

   To 4.8 mL of dichloromethane (DCM), add 0.15 mL of triethylsilane (TES) and 0.05 mL of trifluoroacetic acid (TFA) and mix the solution.

4. Cleavage solution D: 7 N ammonia in methanol (Sigma-Aldrich).

5. Cleavage solution E: 5 % hydrazine hydrate in DMF.

   To 95 mL DMF 5 mL of hydrazine hydrate is added.

6. Cleavage solution F: 0.1 M aqueous NaOH.

   Dissolve 0.4 g of sodium hydroxide (NaOH) in 100 mL of H_2O.

7. Cleavage solution G: 5 % DIEA, 5 % MeOH in DMF.

   5 mL of *N,N*-Diisopropylethylamine (DIEA), and 5 mL of methanol (MeOH) are added to 90 mL of dimethylformamide (DMF).

8. Cleavage solution H: $NaBH_4$ in 50 % aqueous EtOH.

   To 2.5 mL of water, add 2.5 mL of ethanol, and 126 mg of sodium borohydride ($NaBH_4$) is added and mixed.

9. 0.1 M Hydrogen chloride (HCl).

10. Dimethylformamide (DMF).

11. Dichloromethane (DCM).

12. H_2O.

13. 5 % *N,N*-Diisopropylethylamine (DIEA) in dichloromethane.

    5 mL of DIEA is mixed with 95 mL of dichloromethane (DCM).

14. 50 % aqueous of ethanol.

    50 mL of H_2O is mixed with 50 mL of ethanol.

3 Methods

Below are release and cleavage protocols for an array of different linker types and side-chain protecting groups. All the described cleavage protocols are for a 0.1 mmol scale.

3.1 Concentrated TFA Cleavage

Cleaving peptides from Rink Amide, HMPA, and Wang linkers can be divided into two categories: (1) peptide sequences containing none or few Arg(Pbf), Cys(Trt), or Met residues and (2) peptide sequences containing several Arg(Pbf), Cys(Trt), or Met residues. Either category requires different cleavage mixtures. Peptide sequences containing none or few Cys and Met residues and/or unprotected Trp are typically cleaved using cleavage solution A. If the test cleavage using cleavage solution A reveals problems with oxidation of Cys or Met residues or problems in the deprotection of multiple Arg residues, cleavage solution B usually helps as it contains thiols for scavenging:

1. Place the resin in a syringe equipped with a polypropylene filter.
2. Wash the resin with 3× DMF, 3× DCM. It is important that the last washes are with DCM and not DMF, as DMF cannot be fully removed from the resin by evaporation.
3. Dry resin under high vacuum for at least 4 h.
4. Add 5 mL of *cleavage solution A or B*:
   (a) *Solution A* for standard peptide sequences.
   (b) *Solution B* for peptide sequences prone to oxidize.
5. Leave to react for 2 h at room temperature.
6. Collect the TFA-peptide mixture by filtration.
7. Wash the resin twice with 3 mL of TFA and combine filtrate.
8. Concentrate the TFA-peptide mixture by nitrogen flow to approx. 1 mL.
9. Add 15 mL of cold diethyl ether.
10. Centrifuge the precipitate for 5 min at 3,000 × *g*.
11. Decant the ether and wash the precipitate twice with additional cold diethyl ether (10 mL).
12. Dissolve the residual solid in acetonitrile-water 1:5 (10 mL) and lyophilize.

3.2 Cleavage with Dilute TFA

The protocol for treatment with dilute TFA is for peptide sequences attached to acid-sensitive linkers such as Rink acid, 2-chlorotrityl, 2-methoxy-4-alkoxybenzyl alcohol (SASRIN), and Sieber resins. This protocol provides fully side-chain protected peptide sequences.

3.2.1 Release of Fully Protected Sequences

1. Place the resin in a syringe equipped with a polypropylene filter.
2. Wash 3× DMF, 3× DCM.
3. Dry resin under high vacuum for at least 4 h.
4. Add 5 mL of *cleavage solution C.*
5. Leave reaction for 30 min at room temperature.
6. Collect the TFA-DCM-peptide mixture by filtration.
7. Repeat **steps 4–6** three times.
8. Combine filtrate.
9. Concentrate the TFA-DCM-peptide mixture under a stream of nitrogen to approx. 0.5 mL.
10. Add 25 mL of cold water to precipitate the fully protected peptide.
11. Centrifuge the precipitate for 5 min at 3,000 ×*g*.
12. Decant the water.

3.3 Base-Mediated Cleavage

Base-mediated cleavage is for peptide attached to HMBA or hydrazine resin. These linkers can provide an array of different *C*-terminal functionalities depending on the nucleophile used in the cleavage mixture. Before cleaving the peptide from the HMBA and hydrazine linkers, the side-chains can be removed by treating the peptidyl-resin with cleavage solution A for 2 × 1 h at room temperature or the fully protected peptides can be accessed by avoiding TFA cleavage. The methods described below are to obtain unprotected peptides.

3.3.1 Peptide Carboxamides Using the HMBA Linker

1. Place the resin in a syringe equipped with a polypropylene filter.
2. Wash 3× DMF, 3× DCM.
3. Add 5 mL of *cleavage solution A.*
4. Leave to react for 2 × 1 h at room temperature.
5. Wash three times with DCM, three times with 5 % DIEA in DCM, and three times with DCM
6. Dry resin under high vacuum for at least 4 h (*see* **Note 2**).
7. Swell the resin in anhydrous DMF (5 mL) for 30 min.
8. Wash with anhydrous methanol.
9. Add 5 mL of *cleavage solution D.*
10. Collect the ammonia-methanol-peptide mixture by filtration.
11. Wash the resin twice with 3 mL of methanol and combine filtrate.
12. Remove the ammonia-methanol mixture by nitrogen flow.

13. Add 15 mL of cold diethyl ether to precipitate peptide.
14. Centrifuge the precipitate for 5 min at 3,000 × *g*.
15. Decant the ether and wash the precipitate twice with additional cold diethyl ether (10 mL).
16. Dissolve the residual solid in acetonitrile-water 1:5 (10 mL) and lyophilize.

3.3.2 Peptide Hydrazides Using the HMBA Linker

1. Place the resin in a syringe equipped with a polypropylene filter.
2. Wash 3× DMF, 3× DCM.
3. Add 5 mL of *cleavage solution A*.
4. Leave to react for 2 × 1 h at room temperature.
5. Wash three times with DCM, three times with 5 % DIEA in DCM, and three times with DCM
6. Dry resin under high vacuum for at least 4 h.
7. Swell the resin in anhydrous DMF (5 mL) for 30 min.
8. Add 5 mL of *cleavage solution E*.
9. Collect the hydrazine-DMF-peptide mixture by filtration.
10. Wash the resin twice with 3 mL of DMF and combine filtrate.
11. Evaporate the filtrate to dryness by a rotary evaporator.
12. Add 15 mL of cold diethyl ether to precipitate peptide.
13. Centrifuge the precipitate for 5 min at 3,000 × *g*.
14. Decant the ether and wash the precipitate twice with additional cold diethyl ether (10 mL).
15. Dissolve the residual solid in acetonitrile-water 1:5 (10 mL) and lyophilize.

3.3.3 Peptide Acids Using the HMBA Linker

This method works best on water-compatible supports such as PEGA:

1. Place the resin in a syringe equipped with a polypropylene filter.
2. Wash three times with DMF, three times with DCM.
3. Add 5 mL of *cleavage solution A*.
4. Leave for reaction for 2 × 1 h at room temperature.
5. Wash 3× DCM, 3× 5 % DIEA in DCM and 3× DCM.
6. Dry resin under high vacuum for at least 4 h.
7. Swell the resin in water (5 mL) for 30 min.
8. Add 10 mL of *cleavage solution F*.
9. Leave to react for 15 min at room temperature.
10. Filtrate the NaOH-water-peptide mixture into a flask standing in a cooled ice-water bath containing 5 mL of 0.1 M HCl.

11. Wash the resin twice with 3 mL of H_2O and combine filtrate.
12. Adjust pH to 7.0.
13. Lyophilize (*see* **Note 3**).

3.3.4 Peptide Methyl Esters Using the HMBA Linker

1. Place the resin in a syringe equipped with a polypropylene filter.
2. Wash 3× DMF, 3× DCM.
3. Add 5 mL of *cleavage solution A*.
4. Leave to react for 2 × 1 h at room temperature.
5. Wash three times with DCM, three times with 5 % DIEA in DCM, and three times with DCM.
6. Dry resin under high vacuum for at least 4 h.
7. Swell the resin in DMF (5 mL) for 30 min.
8. Add 15 mL of *cleavage solution G*.
9. Leave for reaction for 20 h at 50 °C.
10. Cool to ambient temperature.
11. Filtrate the MeOH-DMF-peptide mixture into a flask.
12. Wash the resin twice with 3 mL of MeOH and ones with TFA and combine filtrate.
13. Evaporate to dryness on a rotary evaporator.
14. Precipitate peptide by addition of cold diethyl ether (15 mL).
15. Centrifuge the precipitate for 5 min at 3,000 × *g*.
16. Decant the ether and wash the precipitate twice with additional cold diethyl ether (10 mL).
17. Dissolve the residual solid in acetonitrile-water 1:5 (10 mL) and lyophilize.

3.3.5 Peptide Alcohols Using the HMBA Linker

This protocol is only for water-compatible resins such as PEGA resin:

1. Place the resin in a syringe equipped with a polypropylene filter.
2. Wash 3× DMF, 3× DCM.
3. Add 5 mL of *cleavage solution A*.
4. Leave to react for 2 × 1 h at room temperature.
5. Wash three times with DCM, three times with 5 % DIEA in DCM, and three times with DCM.
6. Dry resin under high vacuum for at least 4 h.
7. Swell the resin with 5 mL 50 % aqueous ethanol for 30 min.
8. Add 5 mL *cleavage solution H*.
9. Leave for reaction for 4 h at room temperature.
10. Collect filtrate by filtration.

11. Wash resin with 5 mL of 50 % aqueous ethanol.
12. Adjust pH to 7.
13. Lyophilize.

4 Notes

1. Thiols are noxious. Preparation of the cleavage cocktails and the cleavage reaction should be performed in a well-ventilated fume hood. To avoid the smell of thiols, it is recommended to soak all glassware and disposables in 50 % bleach after use and prior to disposal.
2. If the peptidyl-resin is not thoroughly dried prior to this cleavage procedure, peptide acid may be obtained as a by-product.
3. Be aware that the peptide is lyophilized together with sodium chloride that can be removed either during preparative HPLC purification or by gel filtration.

References

1. Fields GB, Noble RL (1990) Solid-phase peptide-synthesis utilizing 9-fluorenylmethoxycarbonyl amino acids. Int J Pept Protein Res 35:161–214
2. Sheppard RC, Williams BJ (1982) Acid-labile resin linkage agents for use in solid-phase peptide-synthesis. Int J Pept Protein Res 20: 451–454
3. Rink H (1987) Solid-phase synthesis of protected peptide fragments using a trialkoxy-diphenyl-methylester resin. Tetrahedron Lett 28:3787–3790
4. Mergler M, Tanner R, Gosteli J, Grogg P (1988) Peptide synthesis by a combination of solid-phase and solution methods I: a new very acid-labile anchor group for the solid phase synthesis of fully protected fragments. Tetrahedron Lett 29:4005–4008
5. Boas U, Brask J, Jensen KJ (2009) Backbone amide linker in solid-phase synthesis. Chem Rev 109:2092–2118
6. Boas U, Brask J, Christensen JB, Jensen KJ (2002) The ortho backbone amide linker (o-BAL) is an easily prepared and highly acid-labile handle for solid-phase synthesis. J Comb Chem 4:223–228
7. Barlos K, Gatos D, Kapolos S, Papaphotiu G, Schäfer W, Wenqing Y (1989) Veresterung von partiell geschützten peptid-fragmenten mit harzen. Einsatz von 2-chlortritylchlorid zur synthese von Leu15-gastrin I. Tetrahedron Lett 30:3947–3950
8. Barlos K, Gatos D, Kutsogianni S, Papaphotiou G, Poulos C, Tsegenidis T (1991) Solid-phase synthesis of partially protected and free peptides containing disulfide bonds by simultaneous cysteine oxidation-release from 2-chlorotrityl resin. Int J Pept Protein Res 38:562–568
9. Eleftheriou S, Gatos D, Panagopoulos A, Stathopoulos S, Barlos K (1999) Attachment of histidine, histamine and urocanic acid to resins of the trityl-type. Tetrahedron Lett 40:2825–2828
10. Guillaumie F, Kappel JC, Kelly NM, Barany G, Jensen KJ (2000) Solid-phase synthesis of C-terminal peptide aldehydes from amino acetals anchored to a backbone amide linker (BAL) handle. Tetrahedron Lett 41:6131–6135
11. Brask J, Albericio F, Jensen KJ (2003) Fmoc solid-phase synthesis of peptide thioesters by masking as trithioortho esters. Org Lett 5:2951–2953
12. Sieber P (1987) A new acid-labile anchor group for the solid-phase synthesis of C-terminal peptide amides by the Fmoc method. Tetrahedron Lett 28:2107–2110
13. Mende F, Seitz O (2011) 9-Fluorenylmethoxycarbonyl-based solid-phase synthesis of peptide α-thioesters. Angew Chem Int Ed 50:1232–1240
14. Guillier F, Orain D, Bradley M (2000) Linkers and cleavage strategies in solid-phase organic synthesis and combinatorial chemistry. Chem Rev 100:2091–2158
15. Semenov AN, Gordeev KY (1995) A novel oxidation-labile linker for solid-phase peptide synthesis. Int J Pept Protein Res 45:303–304

16. Millington CR, Quarrell R, Lowe G (1998) Aryl hydrazides as linkers for solid phase synthesis which are cleavable under mild oxidative conditions. Tetrahedron Lett 39:7201–7204
17. Atherton E, Gait MJ, Sheppard RC, Williams BJ (1979) The polyamide method of solid phase peptide and oligonucleotide synthesis. Bioorg Chem 8:351–370
18. Story SC, Aldrich JV (1992) Preparation of protected peptide amides using the Fmoc protocol—comparison of resins for solid-phase synthesis. Int J Pept Protein Res 39:87–92
19. Jullian M, Hernandez A, Maurras A, Puget K, Amblard M, Martinez J, Subra G (2009) N-terminus FITC labeling of peptides on solid support: the truth behind the spacer. Tetrahedron Lett 50:260–263
20. Abd-Elgaliel WR, Gallazzi F, Lever SZ (2007) Total solid-phase synthesis of bombesin analogs with different functional groups at the C-terminus. J Pept Sci 13:487–492
21. Minkwitz R, Meldal M (2005) Application of a photolabile backbone amide linker for cleavage of internal amides in the synthesis towards melanocortin subtype-4 agonists. QSAR Comb Sci 24:343–353
22. Hutchins SM, Chapman KT (1996) Fischer indole synthesis on a solid support. Tetrahedron Lett 37:4869–4872
23. Brunsveld L, Kuhlmann J, Waldmann H (2006) Synthesis of palmitoylated Ras-peptides and -proteins. Methods 40:151–165
24. Beck W, Jung G (1994) Convenient reduction of S-oxides in synthetic peptides, lipopeptides and peptide libraries. Lett Pept Sci 1:31–37
25. Malik L, Tofteng AP, Pedersen SL, Sørensen KK, Jensen KJ (2010) Automated 'X-Y' robot for peptide synthesis with microwave heating: application to difficult peptide sequences and protein domains. J Pept Sci 16:506–512
26. Teixeira A, Benckhuijsen WE, de Koning PE, Valentijn ARPM, Drijfhout JW (2002) The use of DODT as a non-malodorous scavenger in fmoc-based peptide synthesis. Protein Pept Lett 9:379–385
27. Guy CA, Fields GB (1997) (Editor: Fields GB) Trifluoroacetic acid cleavage and deprotection of resin-bound peptides following synthesis by Fmoc chemistry. In: Solid-phase peptide synthesis, San Diego, 289, Elsevier Academic, pp 67–83
28. Banerjee J, Hanson AJ, Muhonen WW, Shabb JB, Mallik S (2010) Microwave-assisted synthesis of triple-helical, collagen-mimetic lipopeptides. Nat Protoc 5:39–50
29. Harris P, Williams G, Shepherd P, Brimble M (2008) The synthesis of phosphopeptides using microwave-assisted solid phase peptide synthesis. Int J Pept Res Ther 14:387–392
30. Brandt M, Gammeltoft S, Jensen K (2006) Microwave heating for solid-phase peptide synthesis: general evaluation and application to 15-mer phosphopeptides. Int J Pept Res Ther 12:349–357

Chapter 4

Synthesis of Peptides Using *Tert*-Butyloxycarbonyl (Boc) as the α-Amino Protection Group

Søren W. Pedersen, Christopher J. Armishaw, and Kristian Strømgaard

Abstract

The use of the *tert*-butyloxycarbonyl (Boc) as the *N*α-amino protecting group in peptide synthesis can be advantageous in several cases, such as synthesis of hydrophobic peptides and peptides containing ester and thioester moieties. The primary challenge of using Boc SPPS is the need for treatment of the resin-bound peptide with hazardous hydrogen fluoride (HF), which requires special equipment.

Key words *tert*-Butyloxycarbonyl, Boc, Hydrogen fluoride, Scavengers, In situ neutralization

1 Introduction

In contemporary solid-phase peptide synthesis (SPPS), there are two general strategies that are employed based on the nature of the *N*α-amino protecting group of the growing peptide chain and the corresponding side-chain protection groups (Table 1). In the first strategy, the 9-fluorenylmethyloxycarbonyl (Fmoc) group is used for *N*α-amino protecting, which is selectively removed via β-elimination under strongly basic conditions, most commonly using piperidine. The orthogonal side-chain protection groups are often based on highly acid labile *tert*-butyl groups, which are removed concomitantly to cleavage from the resin with trifluoroacetic acid (TFA). The Fmoc/*t*Bu strategy is the most widely applicable strategy for routine solid-phase peptide synthesis and is described further in Chapter 2.

The other often used strategy employs the *tert-butyloxycarbonyl* (Boc) group for the protection of the *N*α-amino moiety [1–3], where the stepwise removal of Boc is achieved by treatment with acid, most often TFA, which is a strong acid, but a relatively weak nucleophile. Side-chain protecting groups are often based on TFA stable benzyl (Bn) groups, which are cleaved with very strong acid, primarily anhydrous hydrogen fluoride

Knud J. Jensen et al. (eds.), *Peptide Synthesis and Applications: Second Edition*, Methods in Molecular Biology, vol. 1047, DOI 10.1007/978-1-62703-544-6_4, © Springer Science+Business Media New York 2013

Table 1
Advantages and disadvantages for Boc/Bn versus Fmoc/*t*Bu strategy

	Boc/Bn strategy	Fmoc/*t*Bu strategy
Requires special equipment	Yes	No
Cost of reagents	Lower	Higher
Solubility of peptides	Higher	Lower
Deprotection efficiency	High	May be lower
Purity of hydrophobic peptides	High	May be lower
Problems with aggregation	Less frequently	More frequently
Synthesis time	~20 min/amino acid	~20–60 min/amino acid
Final deprotection	HF	TFA
Safety	Potentially dangerous	Relatively safe
Orthogonal	No	Yes

(HF) [4]. Alternatively, trifluoromethanesulfonic acid (TFMSA) or trimethylsilyl trifluoromethanesulfonate (TMSOTf) can be used; however, these reagents have not found general applicability, as incomplete deprotection of side-chains and increased side reactions are observed. Thus, the Boc/Bn strategy is using the principle of graded acid stability, where differential acid strengths are used for the stepwise removal of the *N*α-Boc protecting group and side-chain deprotection and peptide cleavage. Although Merrifield's original method for the solid-phase synthesis of peptides in the 1960s employed the Boc/Bn protecting group strategy, this method still finds widespread use today [5].

The Fmoc/*t*Bu strategy is the preferred method for many laboratories undertaking routine SPPS, mainly due to the requirements of using the highly corrosive HF for cleavage of peptides in the Boc/Bn strategy; however, the latter strategy has a number of advantages (Table 1). Generally, peptides with base-sensitive functional groups are advantageously prepared by the Boc/Bn strategy; this includes depsi peptides (ester-containing peptides) and in particular peptides with C-terminal thioester (as described in Chapter 8), which are used, for example, in native chemical ligation, and for which no generally applicable Fmoc/*t*Bu strategies

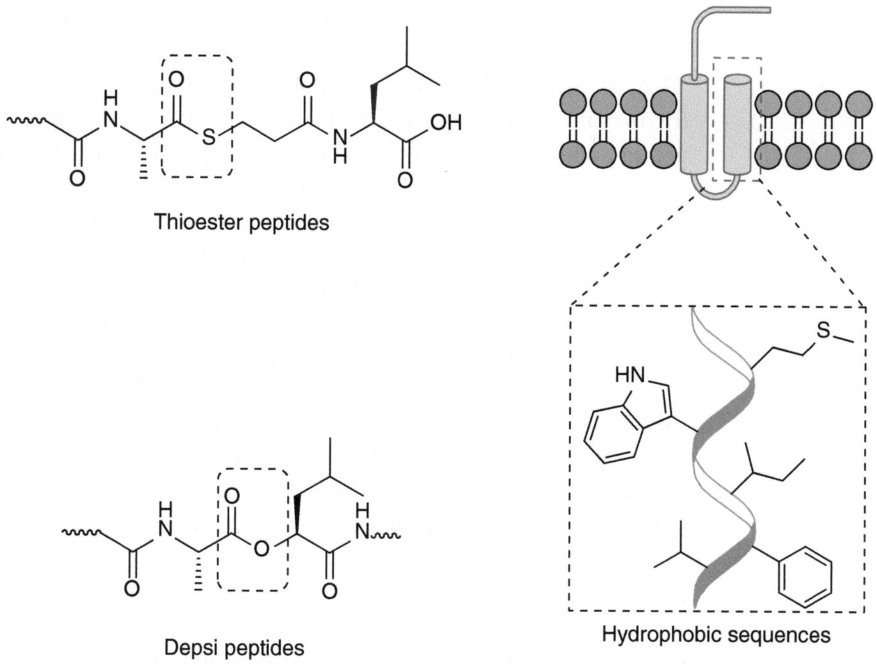

Fig. 1 Examples of peptides and peptides functionalities where Boc peptide synthesis can be advantageous

exist (Fig. 1). Finally, the Boc/Bn strategy tends to be preferred in the synthesis of hydrophobic peptide sequences such as those from membrane-anchored segments (Fig. 1). On the other hand, peptides with acid-sensitive functional groups, in particular glycosylated and phosphorylated peptides, are not achievable by the Boc/Bn strategy due to the strongly acidic conditions employed upon final cleavage.

In the Boc/Bn strategy for SPPS of peptides, special equipment is required to handle anhydrous HF to perform the cleavage of the desired peptide from the resin and concomitant deprotection of side-chain protection groups. This is in contrast to the Fmoc/*t*Bu strategy, where TFA is used and no special equipment is necessary. The cleavage with HF has the advantage, however, that the solubility of the cleaved peptides is generally higher than those generated from the Fmoc/*t*Bu strategy, as peptide fluoride salts have increased solubility compared to peptide TFA salts.

Similarly, problems with aggregation of peptides and formation of secondary structure during the synthesis are generally more of an issue with the Fmoc/*t*Bu strategy compared the Boc/Bn strategy. This is primarily because the removal of a Boc group with TFA yields a positively charged α-amino group, whereas the removal of an Fmoc group yields a neutral α-amino group. Thus, the positively charged α-amino group confines the formation of secondary structure on the resin.

In the original Boc/Bn strategy developed by Merrifield [5], the stepwise removal of the Boc protecting group is achieved with TFA, as previously described. The acidolytic deprotection leads to formation of an ammonium-TFA salt at the N-terminal α-amino group, which in contrast to the basic deprotection of Fmoc-SPPS can affect the efficiency of the following coupling. This is overcome by including a two-fold excess of activating base such as *N,N-diisopropylethylamine* (DIEA) in the coupling step thereby avoiding a separate neutralization step. The dual action mode of the base is termed in situ neutralization [6, 7]; the positively charged amino group is simultaneously neutralized and coupled to the incoming activated amino acid. Reactions are driven to completion by the use of excess (two- to fourfold) activated amino acid which typically is formed in situ by adding the amino acid together with coupling reagent and activator base and allows high coupling yields (>99 %) to be routinely achieved. After each deprotection and coupling step, the resin is washed thoroughly with *N,N-dimethylformamide* (DMF) to remove excess reagents.

Importantly, the amino acid side-chain protecting groups and the C-terminal linking functionality must be stable to repetitive TFA treatments used for the stepwise removal of the *N*α-Boc protecting group; the most frequently used protecting group are listed in Table 2, and for an extended discussion of selection of appropriate side-chain protection groups, *see* ref. 8. As such, final removal of the peptide from the solid support occurs simultaneously with side-chain deprotection with strong acid, primarily anhydrous HF via hydrolytic cleavage. Anhydrous HF is a highly volatile (b.p. 19 °C), nonoxidizing acid, which generally provides very clean cleavage of peptides from the resin. However, HF is also a highly toxic and corrosive substance that violently corrodes standard laboratory glassware and therefore needs to be handled in special equipment, such as a dedicated, corrosion resistant Teflon apparatus. During treatment with HF, a range of scavengers are used to prevent potentially deleterious side reactions such as alkylation by generated benzyl carbocations. For this purpose, scavengers such as *p*-cresol, *p*-thiocresol, and resorcinol are used in excess to mimic the side-chain protecting groups and prevent these side reactions.

1.1 Boc Strategy in SPPS

The general scheme that outlines the strategy of Boc synthesis is shown in Fig. 2, which is principally similar to the scheme for Fmoc synthesis (Chapter 2). Prior to the synthesis, the appropriate resins and linkers (Table 3), as well as appropriately protected building blocks (Table 2), should be selected. Generally, two resins, the chloromethyl (or Merrifield) [1] or phenylacetamidomethyl (PAM) [9] resin, are used to generate peptide carboxylates, whereas 4-methylbenzhydrylamine (MHBA) [10] resin is the principal resin for making peptide carboxamides. Boc synthesis is particularly attractive for the synthesis of peptide thioester.

For thioester resins, the TAMPAL resin is prepared by coupling commercially available S-Tritylmercaptopropionic acid to Boc-Leu-OCH2 PAM resin, followed by deprotection with TFA / TIPS / water (90:5:5) [11]. As previously indicated, a wealth of different, appropriately protected amino acid building block are commercially available; the most frequently used are listed in Table 2, and Albericio and colleagues have recently discussed this in great detail [8].

Once the appropriate resin/linker and set of building blocks have been selected, the first Boc protected amino acid is attached

Table 2
Frequently used side-chain protecting groups

Side-chain functionality	Protecting group	Side-chain functionality	Protecting group
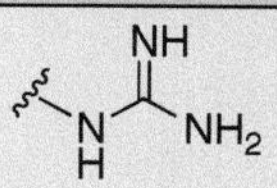 Arg	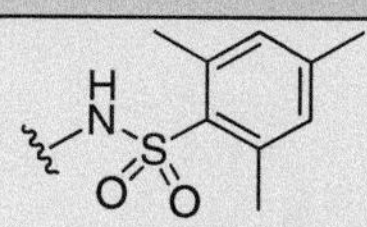 Mesityl-2-sulfonyl (Mts)		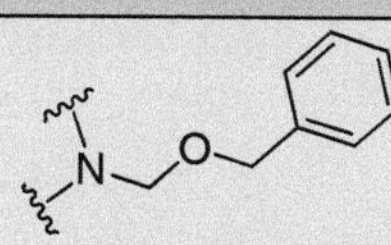 *N*-Benzyloxymethyl (Bom)
	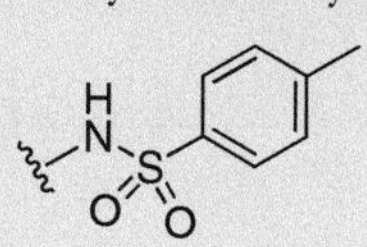 *p*-Toluenesulfonyl (Tosyl, Tos)		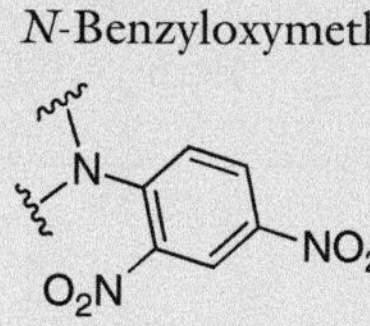 *N*-2,4-Dinitrophenyl (Dnp)
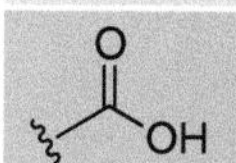 Asp/Glu	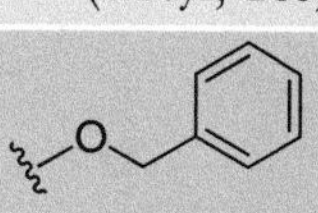 *O*-Benzyl (OBzl)	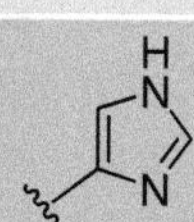 His	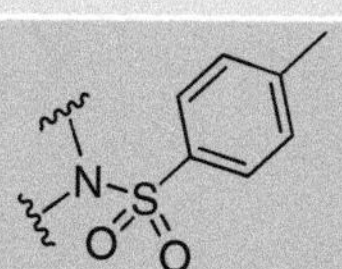 *p*-Toluenesulfonyl (Tosyl, Tos)
	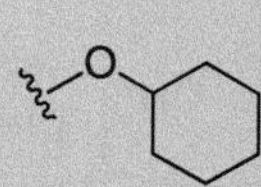 *O*-Cyclohexyl (OcHex)		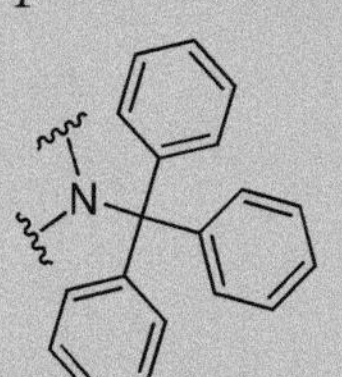 Trityl (Trt)
	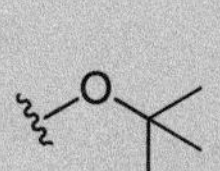 *O*-*tert* Butyl (OtBu)		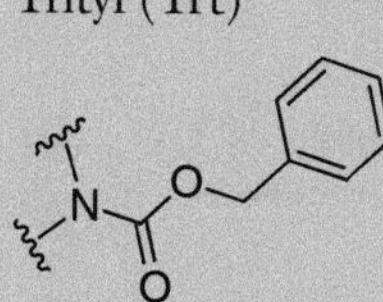 Benzyloxycarbonyl (Z)

(continued)

Table 2
(continued)

Side-chain functionality	Protecting group	Side-chain functionality	Protecting group
—SH Cys	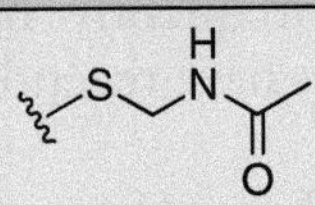 Acetamidomethyl (Acm)	—NH_2 Lys	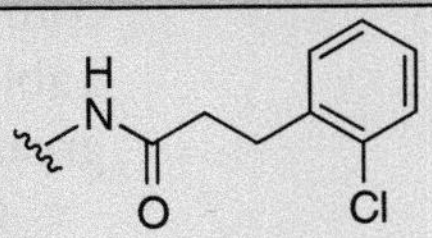 2-Chlorobenzyloxycarbonyl (ClZ)
	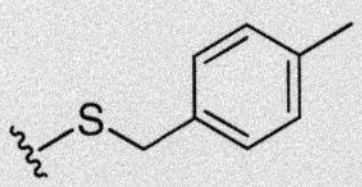 *p*-Methylbenzynyl (Meb)	—OH Ser/Thr	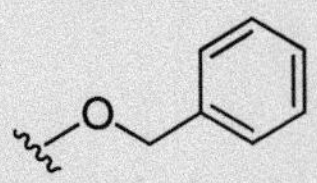 Benzyl (Bn)
	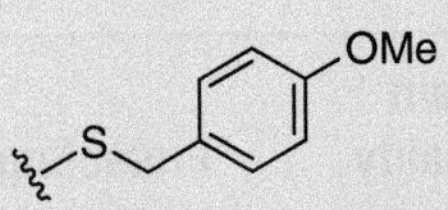 *p*-Methoxybenzyl (Mob)	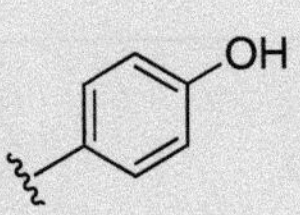 Tyr	v2-Bromobenzyloxycarbonyl (BrZ)
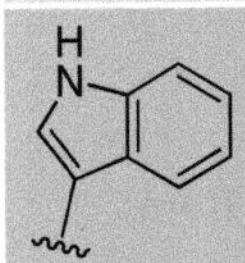 Trp	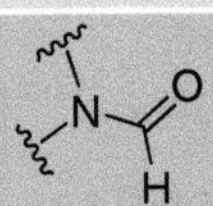 Formyl (CHO)		

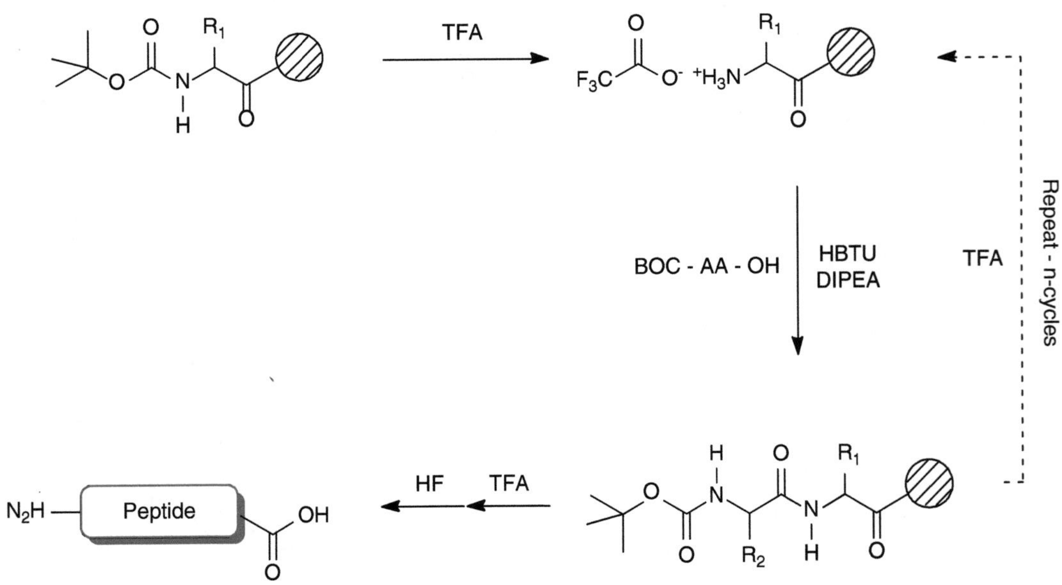

Fig. 2 Principles of Boc SPPS

Table 3
The most frequently used linkers/resin in Boc SPPS

C-Terminus	Resin	Comments	Linker structure
Amide	MBHA	• Efficient cleavage from resin • Stable towards acidic deprotection	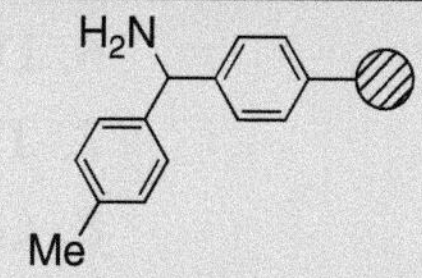
Carboxylic acid	Merrifield	• Efficient cleavage from resin • Unstable in neat TFA→Deprotect in TFA/DCM (1:1)	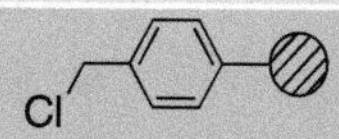
	PAM	• Highly table towards TFA • Many preloaded variants are commercial available	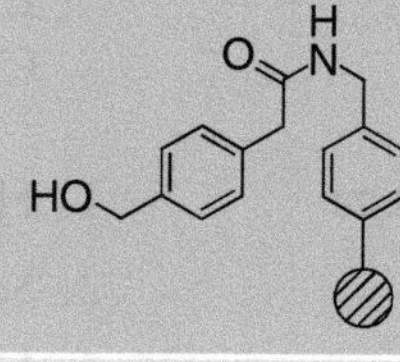
Thioester	TAMPAL	• Converts into a C-terminal thioester when treated with thiol-agent after cleavage from resin	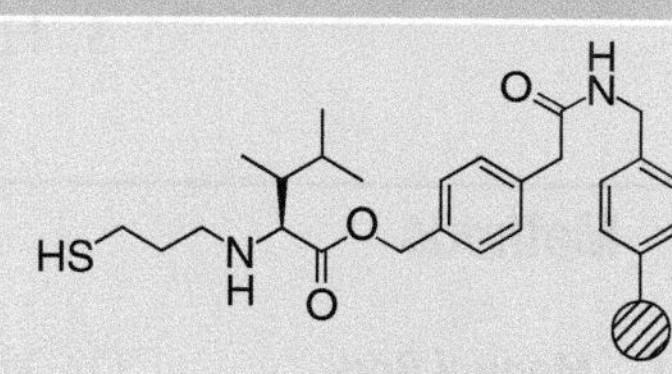

to the support or a preloaded resin can be used; if a preloaded resin is used, appropriate measures should be taken, such as Boc-deprotection, to prepare this for further synthesis. For the accompanying peptide synthesis, stepwise deprotection of the Nα-Boc protecting group is accomplished by treatment of the resin bound amino acid with TFA. The incoming Nα-Boc protected amino acid is coupled utilizing a pre-activated species or in situ activation. Upon complete chain assembly, the resin bound peptide is deprotected and detached from the solid support via HF cleavage. This is generally achieved using HF together with appropriate selection scavengers, as described later. TFMSA or TMSOTf can also be used, but they lack general applicability [4].

2 Materials

2.1 Boc—SPPS

1. Solid support—Merrifield or preloaded phenylacetamidomethyl (PAM) resin.
2. *N*α-Boc protected amino acids.

3. *N*-[(1H-benzotriazol-1-yloxy)(dimethylamino)methylene]-*N*-methylmethanaminium hexafluorophosphate (HBTU).
4. *N,N*-diisopropylethylamine (DIEA).
5. *N,N*-Dimethylformamide (DMF).
6. Dichloromethane (DCM).
7. Trifluoroacetic acid (TFA).
8. Glass reaction vessel complete with bottom filter, Teflon cap, and stopcock.
9. Shaker (wrist-action shaker, Fig. 3).
10. Draining/vacuum accessories.
11. Reagents for Kaiser test. Reagent A: 0.5 g ninhydrin in 10 mL EtOH (w/v); reagent B: 80 % Phenol in EtOH (w/v); reagent C: 0.4 mL of 0.001 M aq. KCN in 20 mL pyridine.

2.2 HF Cleavage

1. Hydrogen fluoride (anhydrous).
2. Scavengers—*see* flow chart (Fig. 4).
3. Hydrogen fluoride apparatus (Fig. 5) and accessories.

3 Methods

3.1 Manual Boc-Based SPPS

The single wrist-action shaker combined with a glass reaction vessel is a simple and convenient equipment to synthesize peptides on solid phase (Fig. 3). The wrist-action shaker ensures the reaction suspension to be in fluid motion without grinding the resin particles. The rotation can be adjusted to allow thorough mixing preventing the resin from caking. The glass reaction vessel can be closed in each of the ends by winding on the handle of the stopcock and sealing with a Teflon lid. The vessel also has a filter in one end allowing draining without wasting the resin. To allow excess reagents to be rapidly drained and facilitate efficient washing of the resin using a flow wash, light suction may be applied and the flow-through collected into a connected waste flask. Other manual systems include the MiniBlock system (Mettler Toledo, Columbia MD, USA), where multiple syntheses can be made in parallel and the reaction vessels are disposable and the shaking and draining can be performed directly on the block. Automated Boc peptide synthesis is possible using a variety of commercially available automated peptide synthesizers, such as Symphony (Protein technologies, Inc. Tucson, AZ, USA), Apex 396 (AAPPTec, Louisville, KY, USA) and Liberty (CEM, Matthews, NC, USA).

3.1.1 Preparing the Resin

Before synthesis can begin, the appropriate resin/linker has to be chosen (Table 3) depending on the nature of the desired peptide. The first AA can in some cases be troublesome to couple

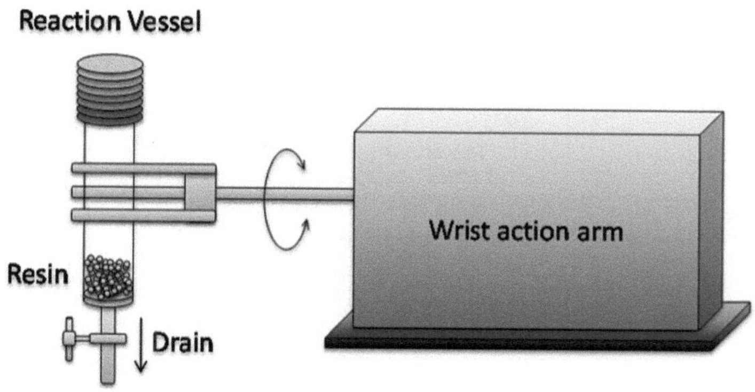

Fig. 3 Illustration of the wrist arm instrument

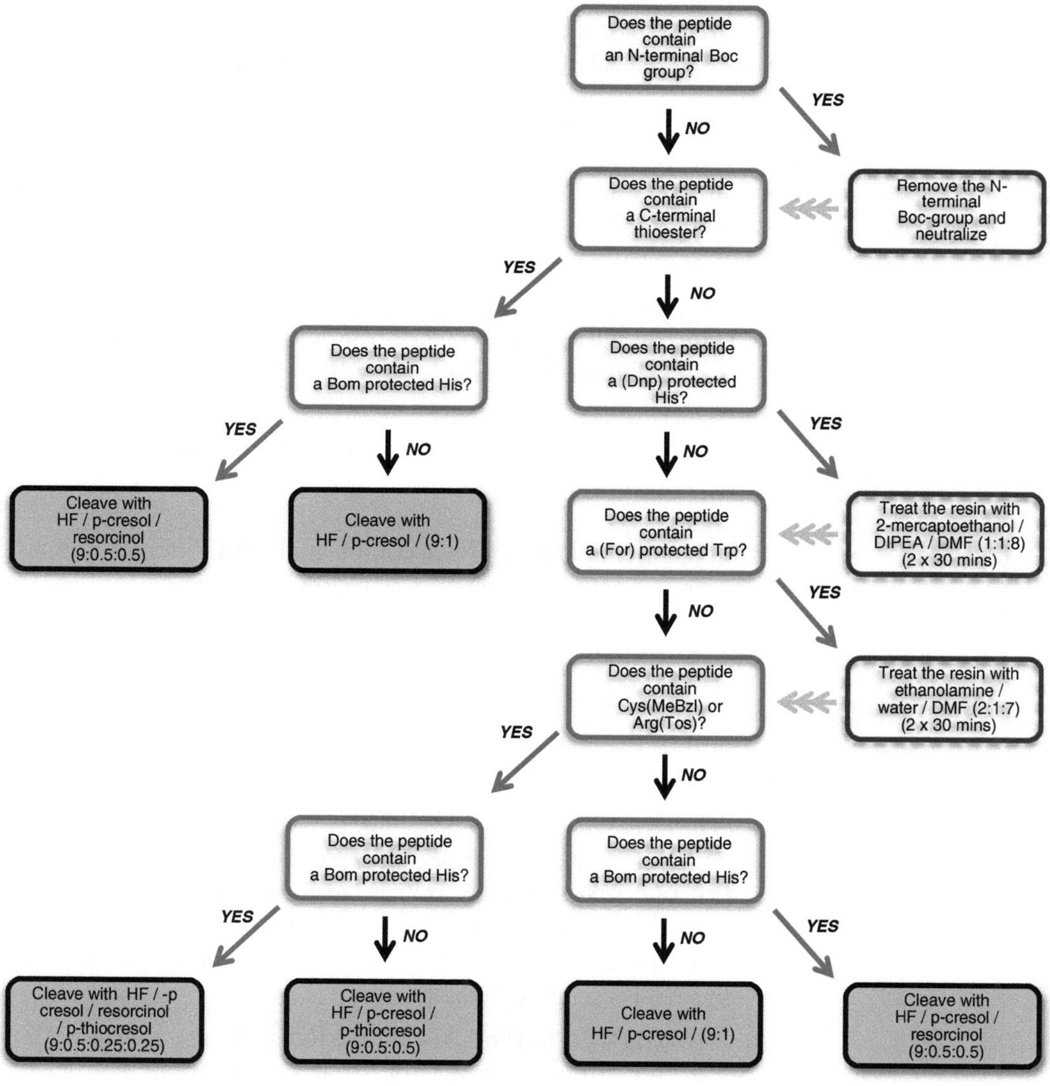

Fig. 4 Flowchart for selection of scavengers in the HF cleavage

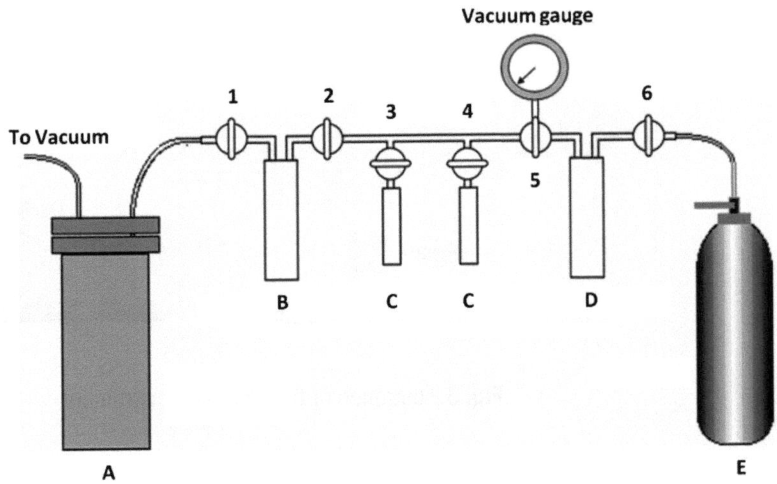

Fig. 5 Illustration of the HF apparatus. (A) Calcium oxide tower; (B) large reaction vessels (extra); (C) reaction vessel; (D) distillation vessel; (E) HF pressure cylinder

on the resin so having the resin preloaded with an AA can be advantageous:

1. Weigh out the resin in the desired scale and place it into a fritted reaction vessel containing a stopcock and a Teflon screw cap (Fig. 3). Close the stopcock and add DMF until the solvent is 2 cm above the resin level. Allow the resin to swell for at least 20 min.
2. If the resin is preloaded and protected, proceed directly to the deprotection step.

3.2 Coupling

Amino acid coupling can be accomplished with different coupling reagents and activating bases. The following scheme (Table 4) can be used for 0.25 mmol scale synthesis using HBTU and DIEA as reagents:

1. Use a 4:4:8 (AA:HBTU:DIEA) stoichiometric ratio in proportion to the resin throughout the couplings.
2. Prepare a 0.5 M HBTU stock in DMF by dissolving HBTU in peptide synthesis grade DMF (gentle heating may be required).
3. Weigh 4 equivalents of Boc-protected amino acid into a 25 mL sealable glass scintillation vial. Add the required a volume of the HBTU stock solution corresponding to the correct stoichiometric ratio. Close the vial and shake the suspension until the AA is completely dissolved.
4. Add the DIEA to the AA-solution and shake the mixture. Allow 2–3 min for activation (note that the solution may change color, indicating formation of the activated ester). Transfer the activated amino acid to the reaction vessel with the resin. The

Table 4
Scheme for 0.25 mmol scale synthesis using HBTU and DIEA as reagents

	Eqv.	mmol	Volume
Boc-AA-OH	4	1	–
HBTU (0.5 M)	4	1	2 mL
DIEA	8	2	0.348 mL

amount of solution should sufficiently swell the resin such that it flows freely upon agitation. If not, additional DMF may be added and the coupling time prolonged if necessary.

5. Close the reaction vessel with the Teflon cap and start the mixing by turning on the rotation of the wrist-arm. The speed should be adjusted so the resin suspension is in continues motion without being shaken vigorously (*see* **Note 1**).
6. Coupling is achieved within 10–20 min and the coupling efficiency can be checked by Kaiser test (*see* **Note 2**; *see* also Chapter 2). If the coupling is incomplete, the beads and solution become blue, and the coupling step is repeated. If the coupling is judged to be complete (no color change), the top cap is removed, the stopcock opened, and suction is applied to remove the reagent. The resin is flow washed with DMF for 1 min.

3.2.1 N^{α}-Deprotection

1. Flow wash the resin for 30 s with DCM to remove excess DMF. Close the stopcock and add sufficient TFA (neat) to the resin to allow fluid motion of the resin when agitated. Replace the cap and apply rotation. Deprotect for 1 min and open the stopcock remove the TFA. Avoid sucking excess air through the resin, particularly with peptides with oxygen sensitive moieties (e.g., methionine, cysteine, and thioesters).
2. Flow wash for 1 min with DCM.
3. Repeat the deprotection step to ensure complete removal of the Boc group.
4. Flow wash for 1 min with DMF to quench and remove remaining TFA.

3.2.2 Storage

If the synthesis is stopped before completion, wash the resin thoroughly with DCM before transferring the reaction vessel to a desiccator.

3.3 HF Cleavage

3.3.1 Scavenger Selection and Amount

The amino acid protection groups and peptide type will determine the scavenger reagents needed during HF cleavage. The scavenger flow chart (Fig. 4) can be used for this purpose. The amount of scavenger can be read from Table 5.

Table 5
Guide to how much scavenger should be used

Resin	Scavenger (mL)	Total volume[a] (mL)
<150 mg	0.5	5
150 mg < resin < 500 mg	1	10
>500 mg	1.5	15

[a]10 % scavenger/90 % HF mix

3.3.2 Cleavage Time

One hour reaction time is usually enough unless the peptide contains any of the following: Arg(Tos), Cys(Meb), MBHA resin, or more than 20 amino acid residues. In such cases, 2 h reaction time is recommended.

3.3.3 HF Cleavage Procedure

The HF apparatus is a completely closed system consisting of three main parts: A HF pressure cylinder, connected distillation tubings/vessels with valves, and a calcium tower connected to a vacuum pump (Fig. 5). All parts exposed to HF are either made of Teflon or has a Teflon surface. The HF always moves from the HF cylinder towards the calcium oxide tower and never in the opposite direction. The main driving force of the HF movement is a temperature difference applied to the vessels either by water bath or acetone/dry ice in Dewars to facilitate evaporation and condensation:

1. Switch on the vacuum pump and open valves 1, 2, and 6 on the HF manifold to evacuate the system. Place dry ice/acetone Dewar under the distillation vessel.
2. Remove the reaction vessels from the manifold and weigh resin into the reaction vessel. Add scavengers to the resin (Fig. 4; *see* **Note 3**). Freeze the resin-scavenger slurry using a dry ice/acetone bath. Add a stirring bar to the frozen resin/scavenger mixture, place an anti-bumping stopper cap onto the reaction vessel and attach onto the manifold. Open valves 3 and 4 to evacuate the reaction vessels. Place the smaller acetone/dry ice Dewars onto each of the reaction vessels.
3. After 15 min, close valve 3, 4, and 5. Open the HF cylinder and wait for a sufficient quantity of HF to transfer into the distillation vessel. The volume of HF can be checked by removing the Dewar and shining light from behind the vessel to observe the contents. If more HF is required, replace the Dewar and wait a further 2–3 min and check again. If HF transfer is very slow, it may be necessary to evacuate the distillation vessel again. To do this, close valve 6 and then briefly open valve 5 (2–3 s). Open valve 6 again and wait 1–3 min before checking again.

4. When sufficient HF has been transferred into the distillation vessel, close the HF cylinder valve, and then close valve 6. Remove the Dewar flask from the distillation vessel, replace with a 60 °C temperature controlled water bath and allow to stir for 10 min.
5. Close valve 2 and then open valve 5. Wait for 10–15 s and then open the reaction vessels (valves 3 and 4). After 1–2 min, check the level of HF in the reaction vessels. When sufficient HF has been transferred into the reaction vessels, close the reaction vessels (valves 3 and 4) and open valve 2 to allow any excess HF inside the manifold to be evacuated. Remove the Dewars and replace with an ice/salt bath on a magnetic stirrer. Ensure that the temperature is maintained during the course of the cleavage to prevent deleterious side reactions from occurring.
6. After sufficient cleavage time, close valve 5 and then *very slowly* turn each of the reaction vessel taps (valves 3 and 4) until slight bubbling occurs in each reaction vessel. Initially, the bubbles should be no more than 2–3 cm above the liquid level. As the HF evaporation progresses, the valves can be fully opened to allow complete evaporation of the HF.
7. Check if remaining HF is left in the reaction vessel by closing the valves 3 and 4, waiting 15 s to allow the vacuum to recover, then quickly opening the taps again. If bumping is observed, wait for another 1–2 min and repeat until no more bumping is observed (this is usually repeated two to three times). Then close valves 3 and 4.
8. Remove the magnetic stirrer and ice/salt bath and place at the base of the fume hood.
9. Carefully remove the reaction vessel and place in the ice/salt bath (*This stage is particularly hazardous since residual HF can splash out of the reaction vessel*). Carefully remove the anti-bumping stopper and drop into the ice bath to quench any residual HF. Add ¾ of the reaction vessel volume with cold diethyl ether to precipitate the peptide and allow stirring for a further 5–10 min.
10. Meanwhile, open valves 1, 2, 5, and 6 on the HF manifold and allow any residual HF to evacuate from the lines for a further 10–15 min. Then open the three-way valve on the vacuum line to allow air to purge through the pump for 5 min before switching off the pump. *Ensure all valves are closed and double check that the HF cylinder valve is closed.*
11. Filter the peptide/resin using a polyethylene filter frit from the diethyl ether and wash with an additional ¾ of the reaction vessel volume of cold ether.

12. Separate the peptide from the resin by dissolving the peptide/resin aggregate using ¾ of the reaction vessel volume of H_2O/Acetonitrile/TFA (1:1:0.002) buffer. Filter the slurry into a round bottom flask and lyophilize the filtrate.
13. Rinse the reaction vessels with DCM/methanol (1:1) and wash with soapy water (pay particular attention to the screw thread where resin can accumulate), rinse with acetone, and then thoroughly dry before placing back onto the manifold.

    Safety measures:

    - Make sure there is someone else nearby (preferably inside the lab) when performing all transfers.
    - Be sure you know where the calcium gluconate gel is before you start working with HF, in the rare event that there is a HF burn.
    - Wear proper rubber gloves, apron, and a full face mask when performing HF transfers.
    - Keep the fume hood tidy and organized at all times when working with HF to avoid accidents.

3.4 Example of the Synthesis of a Peptide with a C-Terminal Thioester

A peptide sequence containing a thioester was synthesis using standard Nα-Boc SPPS (Fig. 6). Preloaded trityl protected TAMPAM resin (0.25 mmol) was swelled for 20 min and deprotected by TFA treatment twice for 1 min. The first AA-coupling was performed using side-chain and BOC-protected Lys, HBTU, and DIEA for 15 min. The reaction mix was removed and the terminal BOC group removed with TFA. The following couplings were carried out using standard conditions and monitored using Kaiser test. After end synthesis, the resin was transferred into a HF reaction vessel and cleaved with 15 mL HF for 1 h. After evaporation the peptide was precipitated using ether and dissolved in water/acetonitrile (1:4). The peptide was purified using a C18 preparative HPLC column which generated a >99 % pure product judged by UV trace from an analytical HPLC (Yield 150 mg, 21.1 %).

4 Notes

1. Intermittent agitation can be achieved using a glass rod, if a wrist-action shaker is not available.
2. A small amount of beads is placed in a glass container and washed several times with EtOH. Two drops of reagent A, four drops of reagent B, and two drops of reagents C are added, and the suspension is warmed at 120 °C for 5 min.
3. It may be necessary to melt the scavengers with mild heat.

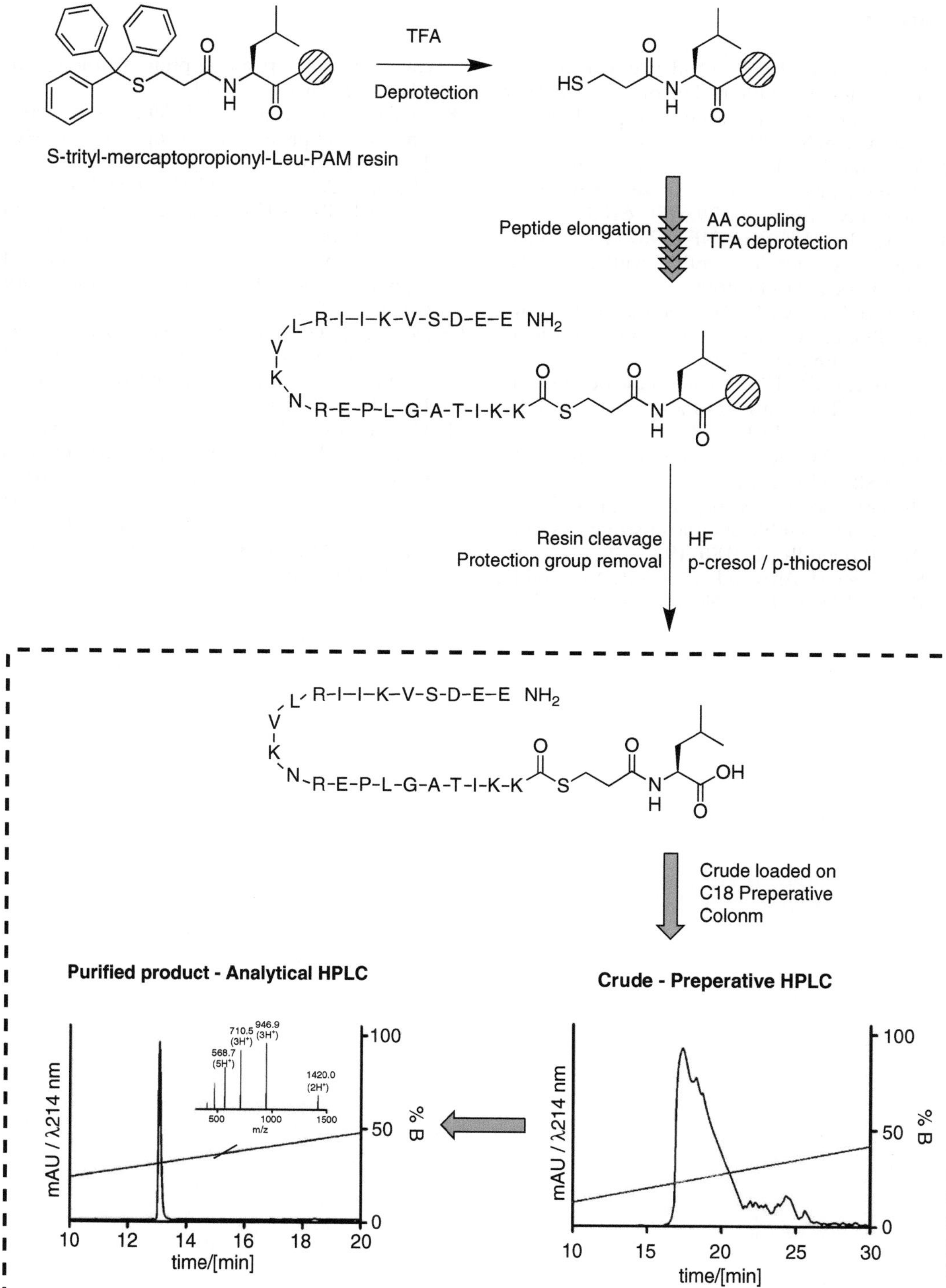

Fig. 6 Example of the synthesis, cleavage, and purification of a thioester containing peptide

References

1. Carpino LA (1957) Oxidative reactions of hydrazines. IV. Elimination of nitrogen from 1,1-disubstituted-2-arenesulfonhydrazides. J Am Chem Soc 79:4427–4431
2. Anderson GW, McGregor AC (1957) t-Butyloxycarbonylamino acids and their use in peptide synthesis. J Am Chem Soc 79:6180–6183
3. McKay FC, Albertson NF (1957) New amine-masking groups for peptide synthesis. J Am Chem Soc 79:4686–4690
4. Stewart JM (1997) Cleavage methods following Boc-based solid-phase peptide synthesis. Methods Enzymol 289:29–44
5. Merrifield RB (1963) Solid phase peptide synthesis. I. The synthesis of a tetrapeptide. J Am Chem Soc 85:2149–2154
6. Schnölzer M, Alewood P, Jones A, Alewood D, Kent SBH (1992) In situ neutralization in Boc-chemistry solid phase peptide synthesis. Rapid, high yield assembly of difficult sequences. Int J Pept Protein Res 40:180–193
7. Schnölzer M, Alewood P, Jones A, Alewood D, Kent SBH (2007) Neutralization in Boc-chemistry solid phase peptide synthesis. Int J Pept Res Ther 13:31–44
8. Isidro-Llobet A, Álvarez M, Albericio F (2009) Amino acid-protecting groups. Chem Rev 109:2455–2504
9. Mitchell AR, Kent SBH, Engelhard M, Merrifield RB (1978) A new synthetic route to tert-butyloxycarbonylaminoacyl-4-(oxymethyl) phenylacetamidomethyl-resin, an improved support for solid-phase peptide synthesis. J Org Chem 43:2845–2852
10. Matsueda GR, Stewart JM (1981) A *p-methylbenzhydrylamine* resin for improved solid-phase synthesis of peptide amides. Peptides 2:45–50
11. Hojo H, Kwon Y, Kakuta Y, Tsuda S, Tanaka I, Hikichi K, Aimoto S (1993) Development of a linker with an enhanced stability for the preparation of peptide thioesters and its application to the synthesis of a stable-isotope-labelled Hu-type DNA-binding protein. Bull Chem Soc Jpn 66:2700–2706

Chapter 5

Sequential Formation of Regioselective Disulfide Bonds in Synthetic Peptides with Multiple Disulfide Bonds

Fazel Shabanpoor, Mohammed Akhter Hossain, Feng Lin, and John D. Wade

Abstract

Numerous methods have been developed for the formation of disulfide bonds in recombinant DNA-derived and chemically synthesized peptides and proteins, but only a few have found widespread acceptance. The choice of method(s) for formation of disulfide in synthetic peptides and proteins needs to be tailored for each individual polypeptide in such a way so that the reaction conditions are selective, efficient, and safe and give the maximum yield. Here we describe the sequential formation of three disulfide bonds regioselectively which has been optimized for the synthesis of two-chained, heterodimeric polypeptide members of the insulin–relaxin superfamily.

Key words Disulfide bond, Regioselective, Orthogonal S-protecting groups, Relaxin, Insulin

1 Introduction

Disulfide bridges are formed between the thiol side chains of two cysteines and are present in many bioactive peptides and proteins. They are primarily formed in the endoplasmic reticulum of eukaryotic cells [1] and prior to secretion of the peptide or protein into the extracellular compartments. Many peptides and proteins rely innately upon disulfide bonds for their correct conformational folding, stability, and subsequent biological activities. Disulfide bonds have also been purposely introduced into peptides as artificial structural motifs in order to constrain and reduce their conformational freedom [2]. Over the past two decades, the development of synthetic methodologies for the formation of disulfide bonds has received prime attention, and significant advances have been made [3]. This is principally due to the identification via genomic sequences of a large number of peptide hormones and natural products that contain disulfide bridges. In order to study the biological function of these disulfide-containing peptides and proteins as well as their de novo

Knud J. Jensen et al. (eds.), *Peptide Synthesis and Applications: Second Edition*, Methods in Molecular Biology, vol. 1047, DOI 10.1007/978-1-62703-544-6_5,

designed analogues, they are often chemically synthesized. Their synthesis is often a challenging task, especially for those peptides possessing multiple disulfide bonds. One approach to overcome this problem exploits the use of various thiol-protecting groups which have been developed and allows the formation of multiple regioselective disulfide bonds in an orthogonal fashion [4–9].

We have a highly optimized protocol for the stepwise regioselective disulfide bond formation in two-chained peptide members of the insulin–relaxin superfamily (Fig. 1). This uses three different side-chain protecting groups for cysteines to form one intrachain and two interchain disulfide bonds regioselectively using Fmoc/*t*Bu strategy. We have routinely used this method for over a decade in our laboratory. It is very effective and the final product can be obtained in good overall yield. It can be readily applied in the preparation of any of the ten members of the insulin–relaxin superfamily of polypeptides hormones. Furthermore, two of the three steps can be readily and successfully used with single-chain peptides bearing one or two disulfide bonds.

2 Materials

All solutions need to be prepared using deionized water. Trifluoromethanesulfonic acid (TFMSA), thioanisole, and other reagents have to be of analytical and chromatography grades:

1. 100 mM NH_4HCO_3 buffer pH 8: Add 7.905 g of NH_4HCO_3 to a 900 mL of water in a graduated cylinder or 1 L beaker. Stir until completely dissolved and adjust the pH with 1 M NaOH. Make up to 1 L with water and store at room temperature.
2. 1 mM solution of 2,2′-dispyridyl disulfide (DPDS): Add 88.1 mg of DPDS to 100 mL of methanol in a beaker and stir until dissolved. Store at room temperature (*see* **Note 1**).
3. 6 M guanidine hydrochloride (GnHCl)/0.1 M Gly-NaOH, pH 8.5: Weigh 286.6 g of GnHCl and 3.75 g of glycine in a beaker, add 500 mL of water, stir to dissolution, and then adjust the pH with NaOH. Store at room temperature.
4. 6 M guanidine hydrochloride (GnHCl), pH less than 7: Weigh 286.6 g of GnHCl in a beaker, dissolve it in 500 mL of water. Store at room temperature.
5. 60 mM HCl: Add 0.5 mL of 12 M HCl to 9.5 mL of water. Store at room temperature.
6. 20 mM iodine (I_2) solution: Weigh 253.8 mg of I_2 and stir it in 100 mL of acetic acid until completely dissolved (*see* **Note 2**). Store at room temperature.
7. 20 mM ascorbic acid: Dissolve 176.1 mg of ascorbic acid in 50 mL of water (*see* **Note 3**).

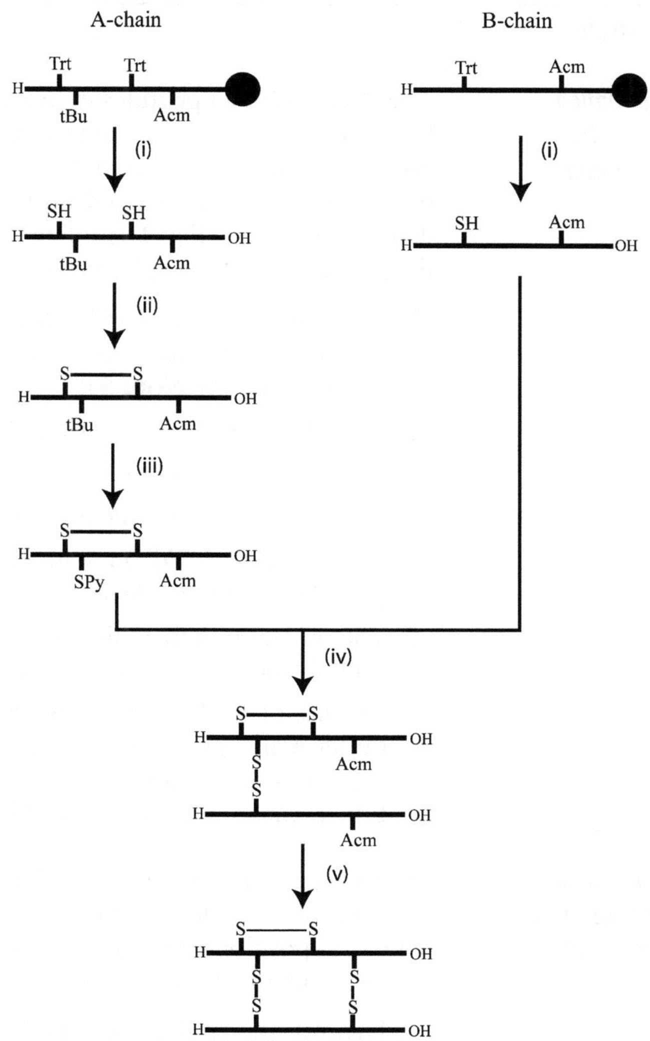

Fig. 1 Sequential formation of intra- and interchain disulfide bonds. (*i*) Cleavage of peptide from solid support using TFA. The S-*t*Bu and S-Acm-protecting groups are not cleaved by TFA. (*ii*) The first intra-A-chain disulfide bond is formed using DPDS. (*iii*) The Cys (*t*Bu) is deprotected and converted to Cys(Pys) using TFMSA in the presence of DPDS. (*iv*) The A-chain with Cys(Pys) is combined with Cys(SH) on the B-chain. (*v*) Simultaneous removal of S-Acm and formation of disulfide bond using iodine

8. RP-HPLC buffers 0.1 % trifluoroacetic acid (TFA) in water (A) and in acetonitrile (B): Add 1 mL of TFA to 1 L of water for buffer A and similarly add 1 mL of TFA to 1 L of acetonitrile for buffer B (*see* **Note 4**). Both solvents need to be degassed either by sonication or by helium sparging.

3 Methods

3.1 Formation of Intra-A-chain Disulfide Bond

The A-chains of all peptides in insulin–relaxin superfamily contain four cysteines. The two cysteines that are involved in intra-A-chain disulfide bond are protected with the S-trityl (S-Trt) group. This protecting group is removed during the peptide–resin cleavage/deprotection step (Fig. 1(i)). The cysteine numbering is based on the primary structure of human relaxin-3.

1. The crude cleaved [$Cys^{10,15}$(S-thiol), Cys^{11}(tBu), Cys^{24}(Acm)] A-chain can be dissolved in either of the following solutions at 1 mg/3 mL (*see* **Note 5**):
   (a) 0.1 M NH_4HCO_3 buffer, pH 8.
   (b) 6 M GnHCl, pH 6.
   (c) 20–50 % acetonitrile in water.
2. Add 0.2 mL of DPDS solution per 1 mg of peptide to the peptide solution while stirring (Fig. 1(ii)).
3. The oxidation of free thiols is usually very rapid in the present of DPDS. The reaction usually goes to completion within 5–10 min.
4. An aliquot of the mixture is acidified (*see* **Note 6**) by adding buffer A and analyzed by RP-HPLC. The oxidized form of the peptide is often eluted from the reverse phase column at earlier time compared to the reduced form.

3.2 Formation of First Interchain Disulfide Bond

In order to form the first interchain disulfide bond between the A- and B-chains, the S-*tert*-butyl (S-*t*Bu) protecting group of cysteine at position 11 of the A-chain is converted to the 2-pyridylsulfenyl derivative (S-Pyr) (Fig. 1(iii)):

1. Dissolve the A-chain and DPDS (*see* **Note 7**) (1:4 μmol) in TFA and thioanisole (9:1 v/v) (50 μg/μL) (*see* **Note 8**). Chill the reaction mixture on ice to about 4–0 °C.
2. Add equal volume of TFMSA in TFA (1:4 v/v) to the above mixture and stir the reaction for 30 min (*see* **Note 9**).
3. Add the whole reaction mixture into a 50 mL tube containing 20–30 mL of pre-chilled diethyl ether. Centrifuge the precipitated peptide and wash the pellet 4–5 times with cold diethyl ether to remove excess DPDS (*see* **Note 10**).
4. Dissolve the purified [$Cys(S\text{-}Pyr)^{11}$] A-chain in NH_4HCO_3, pH 8 [10, 11]. In the case of poorly soluble peptides, these can be dissolved in either 6 M GnHCl, pH 8.5 or 6 M GnHCl/0.1 M Gly-NaOH buffer, pH 8.5 [5] (*see* **Note 11**). Dissolve an equimolar amount of B-chain with $Cys(SH)^{10}$ in either water or 6 M GnHCl, pH 6 (*see* **Note 12**). Add the B-chain solution dropwise to the stirring A-chain solution (Fig. 1(iv)). The progress of the reaction can be monitored by RP-HPLC (*see* **Note 13**).

3.3 Formation of the Second Interchain Disulfide Bond

The second interchain disulfide bond is carried by simultaneous removal of S-acetamidomethyl (S-Acm) and oxidation of cysteines on A- and B-chain using iodine (Fig. 1(i)) [5, 10, 12, 13]:

1. Dissolve the combined A- and B-chain peptide each with a single S-Acm group in glacial acetic acid (1.5–2 mg/mL) (*see* **Note 14**).
2. Add 60 mM HCl (not more than 5 % of the total volume).
3. Immediately add iodine solution (20 mM in acetic acid, 20–25 eq/Acm) to the peptide solution dropwise while stirring. Allow the reaction to proceed for 45–60 min at room temperature. Due to the light sensitivity of the iodine solution, the reaction is carried out in a dark glass vial.
4. To monitor the reaction progress, take an aliquot of the mixture and add similar volume of the 20 mM ascorbic acid to quench the iodine. Dilute the mixture and analyze by RP-HPLC and mass spectrometry.
5. The disappearance of the starting S-Acm-containing peptide (*see* **Note 15**) indicates the completion of the disulfide bond formation. The whole reaction mixture is poured into a 50 mL tube containing pre-chilled diethyl ether and left on ice for further 3 min. The precipitated peptide is pelleted by centrifugation [14]. The ether should be carefully decanted and the residual ether evaporated. Dissolve the resulting solid in appropriate solvent and add 20 mM ascorbic acid to quench the residual iodine (*see* **Note 16**).

4 Notes

1. The solution can be kept on the shelf for several months.
2. The iodine solution can be kept at room temperature for several weeks in a dark container, but it is best to prepare it fresh.
3. Ascorbic acid solution needs to be prepared fresh each time as it can degrade over time.
4. This needs to be done in the fume hood wearing protective goggles and gloves using a glass pipette.
5. The peptide solution should be dilute in order to prevent A-chain dimerization. Some peptides are also not soluble at high pH's. In that case, the reaction can also be carried out at lower pHs, and it will still be very rapid in the presence of DPDS [15].
6. As the pH of the peptide solution is about 8 and most of the silica-based RP-HPLC columns are not resistant to high pH's, it is best to lower the pH of the solution before loading it onto HPLC. This is especially important if there is a large volume of peptide solution that needs to be purified.
7. Excessive use (e.g., more than 5 equiv.) of DPDS should be avoided as it can co-elute with the peptide during the RP-HPLC purification.

8. It is better to purify the A-chain after the formation of the intra-A-chain disulfide bond, especially if buffer with high concentration of salts has been used.
9. TFMSA is an extremely strong acid. The reaction must be carried out in the fume hood with appropriate protective clothing and eyewear.
10. With the initial washes, the intense yellow color of ether indicates the presence of DPDS. As the peptide is washed several times, the yellow color fades which is indicative of the reduced amount of DPDS. The remainder can be removed during HPLC purification.
11. Some peptides are not soluble in NH_4HCO_3 buffer at pH 8. In those cases, we have used 6 M GnHCl at pH 8.5 alone or a mixture of 6 M GnHCl/0.1 Gly-NaOH buffer, pH 8.5 to dissolve the peptides. The excess use of Gly should be avoided as its removal can become problematic during the RP-HPLC purification. It may co-elute with the desired product.
12. Do not dissolve the B-chain with free thiol in high pH buffers as it will lead to the formation of B-chain dimer. The use of a highly hydrophobic solvent such acetonitrile should be avoided during the chain combination. We have found that addition of excess acetonitrile leads to sluggish A- and B-chain combination reaction.
13. The A-chain usually elutes 3–4 min earlier than the B-chain, and the combined product elutes in between the two peaks. The chain combination reaction is very rapid, and it usually goes to completion in 15–20 min.
14. The use of methanol is not recommended for the peptides in insulin–relaxin family as they all contain either Trp, His, or Tyr. Use of MeOH can lead to iodination of these sensitive residues.
15. The reaction time needs to be optimized for each individual peptide. We have found that generally the reaction goes to completion in 30–60 min.
16. Once the peptide is solubilized, the solution has a light yellow color due to the presence of residual iodine. Add ascorbic acid dropwise until the solution becomes clear.

Acknowledgements

Part of the work described herein was funded by NHMRC (Australia) project grants 350284 and 508995 to JDW. Studies at the FNI are also supported by the Victorian Government's Operational Infrastructure Support Program.

References

1. Sevier CS, Kaiser CA (2002) Formation and transfer of disulphide bonds in living cells. Nat Rev Mol Cell Biol 3:836–847
2. Li P, Roller PP (2002) Cyclization strategies in peptide derived drug design. Curr Top Med Chem 2:325–341
3. Witt D (2008) Recent developments in disulfide bond formation. Synthesis 16:2491–2509
4. Boulègue C, Musiol H-J, Prasad V, Moroder L (2006) Synthesis of cystine-rich peptides. Chim Oggi 24:24–36
5. Bathgate RA, Lin F, Hanson NF, Otvos L Jr, Guidolin A, Giannakis C, Bastiras S, Layfield SL, Ferraro T, Ma S, Zhao C, Gundlach AL, Samuel CS, Tregear GW, Wade JD (2006) Relaxin-3: improved synthesis strategy and demonstration of its high-affinity interaction with the relaxin receptor LGR7 both in vitro and in vivo. Biochemistry 45:1043–1053
6. Bathgate RA, Samuel CS, Burazin TC, Layfield S, Claasz AA, Reytomas IG, Dawson NF, Zhao C, Bond C, Summers RJ, Parry LJ, Wade JD, Tregear GW (2002) Human relaxin gene 3 (H3) and the equivalent mouse relaxin (M3) gene. Novel members of the relaxin peptide family. J Biol Chem 277:1148–1157
7. Bullesbach EE, Schwabe C (1991) Total synthesis of human relaxin and human relaxin derivatives by solid-phase peptide synthesis and site-directed chain combination. J Biol Chem 266:10754–10761
8. Lin F, Otvos L Jr, Kumagai J, Tregear GW, Bathgate RA, Wade JD (2004) Synthetic human insulin 4 does not activate the G-protein-coupled receptors LGR7 or LGR8. J Pept Sci 10:257–264
9. Sieber P, Kamber B, Hartmann A, Johl A, Riniker B, Rittel W (1977) Total synthesis of human insulin. IV. Description of the final steps (author's transl). Helv Chim Acta 60:27–37
10. Shabanpoor F, Hughes RA, Bathgate RA, Zhang S, Scanlon DB, Lin F, Hossain MA, Separovic F, Wade JD (2008) Solid-phase synthesis of europium-labeled human INSL3 as a novel probe for the study of ligand–receptor interactions. Bioconjug Chem 19:1456–1463
11. Zhang S, Hughes RA, Bathgate RA, Shabanpoor F, Hossain MA, Lin F, van Lierop B, Robinson AJ, Wade JD (2010) Role of the intra-A-chain disulfide bond of insulin-like peptide 3 in binding and activation of its receptor, RXFP2. Peptides 31:1730–1736
12. Hossain MA, Zhang S, Lin F, Ferraro T, Bathgate RAD, Tregear GW, Wade JD (2006) Regioselective disulfide solid phase synthesis, chemical characterization and in vitro receptor binding activity of equine relaxin. Int J Pept Res Ther 12:211–215
13. Akhter Hossain M, Bathgate RA, Kong CK, Shabanpoor F, Zhang S, Haugaard-Jonsson LM, Rosengren KJ, Tregear GW, Wade JD (2008) Synthesis, conformation, and activity of human insulin-like peptide 5 (INSL5). Chembiochem 9:1816–1822
14. Zhang S, Lin F, Hossain MA, Shabanpoor F, Tregear GW, Wade JD (2008) Simultaneous post-cysteine(S-Acm) group removal quenching of iodine and isolation of peptide by one step ether precipitation. Int J Pept Res Ther 14:301–306
15. Maruyama K, Nagasawa H, Suzuki A (1999) 2,2′-Bispyridyl disulfide rapidly induces intramolecular disulfide bonds in peptides. Peptides 20:881–884

Chapter 6

Synthesis of Cyclic Disulfide-Rich Peptides

Muharrem Akcan and David J. Craik

Abstract

In this chapter we describe two SPPS approaches for producing cyclic disulfide-rich peptides in our laboratory, including cyclotides from plants, cyclic conotoxins from cone snail venoms, chlorotoxin from scorpion venom, and the sunflower trypsin inhibitor peptide, SFTI-1.

Key words Cyclic peptides, Cyclotides, Solid-phase peptide synthesis

1 Introduction

Our laboratory mainly focuses on cyclic disulfide-rich peptides called cyclotides [1], which are important in drug design applications and also in agriculture [2–5]. They occur naturally in plants as host defense peptides [6]. In addition to a head-to-tail cyclic backbone, they contain six cysteine residues forming three disulfide bonds with a Cys^{I}–Cys^{IV}, Cys^{II}–Cys^{V}, and Cys^{III}–Cys^{VI} connectivity referred to as a cyclic cystine knot (CCK) motif [1]. Their circular topology leads cyclotides to have six peptide loops in the backbone between successive cysteine residues. Cyclotides are divided into two main subfamilies: Möbius and bracelet [7]. The prototypic cyclotide kalata B1 is an example from the Möbius subfamily as it has a *cis*-proline peptide bond in loop 5 of its sequence, whereas bracelet subfamily cyclotides have a conventional all-*trans* peptide backbone. The trypsin inhibitory peptides MCoTI-I and MCoTI-II from the tropical plant *Momordica cochinchinensis* can be regarded as a third subfamily of cyclotides as they have a cyclic backbone and a cystine knot motif [8–10]. However, they have no sequence homology to the Möbius and bracelet subfamilies.

The CCK motif provides extraordinary stability to cyclotides [11], making them highly suitable scaffolds for bioengineering applications such as the grafting of peptide fragments onto the cyclic backbone. For example, grafting of the fragments of the extracellular matrix proteins laminin and osteopontin and a

Knud J. Jensen et al. (eds.), *Peptide Synthesis and Applications: Second Edition*, Methods in Molecular Biology, vol. 1047, DOI 10.1007/978-1-62703-544-6_6,

sequence from vascular endothelial growth factor (VEGF) onto MCoTI-II and another cyclic peptide framework, sunflower trypsin inhibitor-1 (SFTI-1), were reported recently [12]. The grafted peptides had higher stability than the parent linear epitopes and showed in vivo angiogenic activity at nanomolar concentrations.

In addition to these types of grafting studies, the discovery of naturally occurring cyclic peptides has stimulated the idea of reengineering conventional (acyclic) disulfide-rich linear peptides to make them backbone-cyclized to enhance their stabilities [13–15]. For example, backbone cyclization of a linear two-disulfide-bonded cone snail venom peptide, conotoxin Vc1.1, and a four-disulfide-bonded scorpion venom peptide, chlorotoxin, were reported recently and shown to enhance the stabilities of the linear peptides [15, 16]. Vc1.1 is a 16 amino acid peptide that targets nicotinic acetylcholine receptors (nAChR) and gamma-aminobutyric acid B ($GABA_B$) receptors. It was found that the backbone cyclized Vc1.1 is more stable than its linear form and also more potent at $GABA_B$ receptors [15]. Chlorotoxin is a 36 amino acid peptide with four disulfide bonds [17] that binds to glioma cells, a type of tumor cells in the brain, but not to normal cells [18]. This selective targeting feature has made chlorotoxin an attractive molecule to develop imaging agents to distinguish tumors from normal tissue [19, 20], multifunctional nanoparticles [21], and targeted gene therapy tools [22, 23].

Solid-phase peptide synthesis (SPPS) allows us to synthesize these various cyclic peptides. In SPPS the peptide sequence is assembled onto a solid polymer resin by coupling each amino acid one by one [24, 25]. To prevent couplings between monomeric amino acids during the peptide chain assembly, one of two amine-protecting groups, *tert*-butoxycarbonyl (Boc) or fluoren-9-ylmethoxycarbonyl (Fmoc), are used. Boc is an acid-labile group and has to be removed before the coupling of the next amino acid by trifluoroacetic acid (TFA). On the other hand, a piperidine/dimethylformamide (DMF) mixture is used to remove the base-labile Fmoc group.

In our laboratory we use two different SPPS-based approaches to synthesize cyclic disulfide-rich peptides. Most commonly we use the intramolecular native chemical ligation (NCL) reaction between a thioester group at the C-terminus and a cysteine residue at the N-terminus [16, 26–29]. For this approach the coupling of the thioester linker requires Boc-SPPS protection chemistry as the thioester is unstable in the basic conditions used for Fmoc chemistry. The first amino acid coupled after the thioester linker could be any residue; however, Gly, Cys, and His are most favorable, and Leu, Thr, Val, Ile, and Pro are the least favorable amino acids, as reflected in their ligation rates with the N-terminal cysteine [26]. The second SPPS approach used in our laboratory to synthesize the cyclic disulfide-rich peptides involves an Fmoc-based method, in

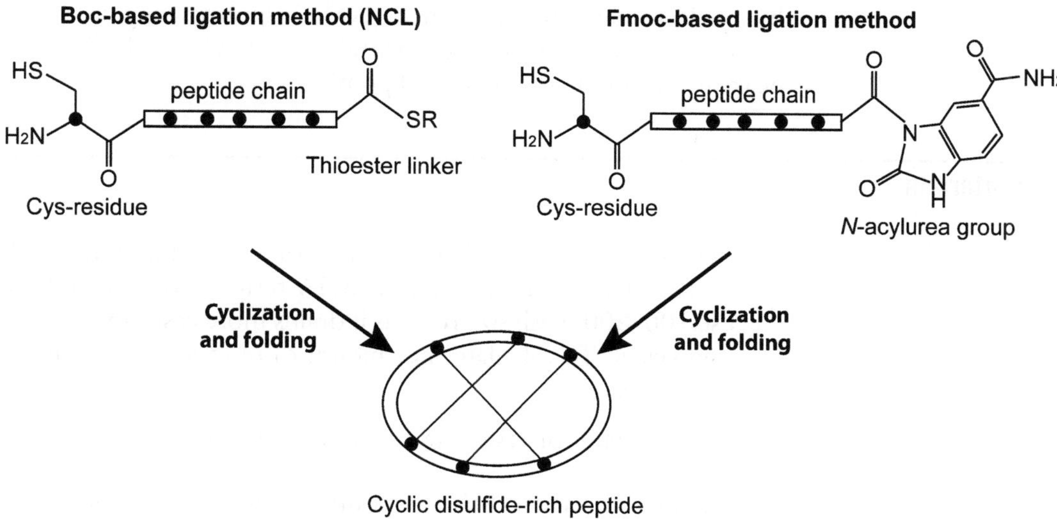

Fig. 1 Boc- and Fmoc-based ligation methods. In Boc-based method, the intramolecular ligation occurs between a C-terminal thioester and an N-terminal cysteine residue. In Fmoc-based method, the intramolecular ligation reaction occurs between a C-terminal N-acylurea group and an N-terminal cysteine residue. Six cysteine residues are shown as *black dots* and the three disulfide bonds are shown as *lines*

which the intramolecular ligation reaction occurs between a C-terminal N-acylurea group and an N-terminal cysteine residue [30]. In the Fmoc method it is necessary to use Boc-Cys(Trt) as the N-terminal amino acid instead of a Fmoc-protected Cys (Fig. 1).

Peptides synthesized with the Boc-SPPS and Fmoc-SPPS methods can be cleaved from the resin by hydrogen fluoride (HF) and TFA, respectively. Although it is necessary to use hazardous HF during the cleavage step after Boc-SPPS, it is the preferred method in our laboratory because of its higher deprotection and coupling yields compared to the Fmoc-SPPS [31]. In the next sections, the materials and detailed protocols for Boc-SPPS and Fmoc-SPPS approaches for cyclic disulfide-rich peptides are described. The main aim of the chapter is to describe robust and routinely used disulfide-rich peptide synthesis strategies that included peptide chain assembly and cyclic backbone formation. Oxidative folding to form disulfide bonds is also an essential step in the synthesis but tends to be a procedure that needs to be done on a trial and error basis for each peptide, and so we have provided indicative procedures only for this step. It is possible to use either orthogonal protection of pairs of Cys residues to force a particular disulfide connectivity or to do a general oxidation and then purify the desired isomer if a mixture is formed. The best strategy varies from peptide to peptide. In general, we prefer not to use orthogonal protection for peptides with the CCK motif because cyclization and folding at the same time often results in peptides with native disulfide connectivity in high yields. Peptide folding is also checked by nuclear magnetic resonance (NMR) spectroscopy in which correctly folded disulfide-rich peptides often give well-dispersed peaks in the amide region (7–10 ppm).

2 Materials

1. Resins: Boc-glycine-phenylacetamidomethyl (Boc-Gly-PAM) resin and Fmoc-Rink amide MBHA resin. It is preferable to use 100–200 mesh size resin as it drains more easily in the reaction vessel. To calculate the amount of resin for the synthesis, Eq. 1 is used:

$$\text{Mass of resin} = (\text{Scale in mmol}) / (\text{SV}). \quad (1)$$

   Equation 1, where SV (substitution value) is the number of millimole per gram of resin (this value is given on the side of the resin bottle).
2. Reaction vessel: the resin is placed in a reaction vessel as illustrated in Fig. 2. It has a sintered glass filter which allows reagents to be drained away while the resin is retained in the vessel.
3. 0.5 M HBTU (*N*-[(1*H*-benzotriazol-1-yl)(dimethylamino) methylene]-*N*-methylmethanaminium hexafluorophosphate *N*-oxide) (MW = 379.3 g/mol). Dissolve 94.82 g of HBTU in 500 mL of DMF (*see* **Note 1**). Store at 4 °C in the dark.

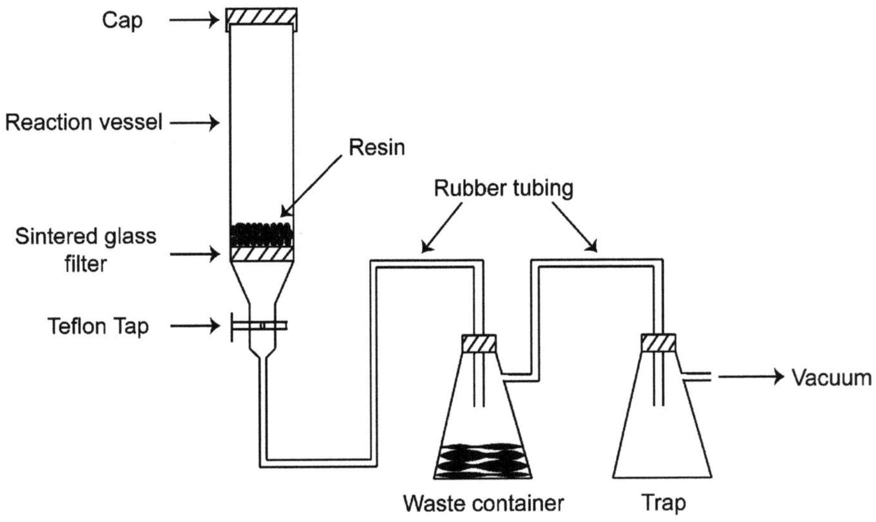

Fig. 2 Solid-phase peptide synthesis apparatus. The reaction vessel has a sintered glass filter which allows reagents to be drained away while the resin is retained in the vessel. The drained liquid mixture is collected in a waste container which is connected to the vessel by acid and base resistant rubber tubing. A second waste container is also used as a trap to prevent the leakage of any liquid to the vacuum source

4. *N,N*-Diisopropyl ethylamine (DIEA). Use concentrated and store at 4 °C in the dark.
5. 20 % Piperidine solution. Add 100 mL of piperidine in a graduated cylinder and add DMF to a volume of 500 mL. Store at 4 °C in the dark.
6. Cleavage agents: Liquid hydrogen fluoride (HF), *p*-cresol, TFA, and triisopropylsilane (TIS).
7. Buffer solutions A and B: Solution A—add 500 μL of TFA to 1 L of distilled water. Solution B—add 100 mL of distilled water and 450 μL of TFA to 900 mL of acetonitrile. Make 150 mL of Solution A/Solution B mixture (50:50, v/v) and store at room temperature.
8. Quantitative Kaiser ninhydrin test solutions: 76 % (w/w) phenol in ethanol. 0.2 mM KCN in pyridine: dissolve 1.3 mg of KCN in 100 mL of pyridine. 0.28 M ninhydrin in ethanol: dissolve 4.98 g of ninhydrin in 100 mL of ethanol. Store all solutions separately at 4 °C in dark. 60 % ethanol in water (*see* **Note 3**).
9. 0.1 M of NH_4HCO_3 (pH 8.2): dissolve 0.796 g of NH_4HCO_3 in 100 mL of H_2O.
10. 0.1 M Tris–HCl, 0.2 M NaCl, and 5 mM reduced glutathione/0.5 mM oxidized glutathione (pH 7.8): dissolve 1.570 g of Tris–HCl, 0.580 g of NaCl, 0.015 g of reduced glutathione, and 0.030 g of oxidized glutathione in 100 mL of water.
11. 100 mM ammonium acetate (CH_3COONH_4) and 20 mM dithiothreitol (DTT): dissolve 0.770 g of CH_3COONH_4 and 0.308 g of DTT in 100 mL of H_2O.

3 Methods

Figure 3 summarizes the steps involved in cyclic peptide synthesis and these are described in more detail in Subheadings 3.1 and 3.2 for Boc and Fmoc methods, respectively. Carry out all procedures at room temperature and in a well-ventilated fume hood unless otherwise specified. The amounts of reagents are given based on a 0.5 mmol scale synthesis, which typically yields >10 mg of final product after purification. These procedures have been used in our laboratory to synthesize several classes of disulfide-rich cyclic peptides, including SFTI-1 (14 amino acids, one disulfide bond), cyclic Vc1.1 (22 amino acid, two disulfide bonds), kalata B1 (29 amino acids, three disulfide bonds), MCoTI-II (34 amino acids, three disulfide bonds), and cyclic chlorotoxin (43 amino acids, four disulfide bonds) (Fig. 4). The procedures are described for manual synthesis but are also applicable to automated SPPS.

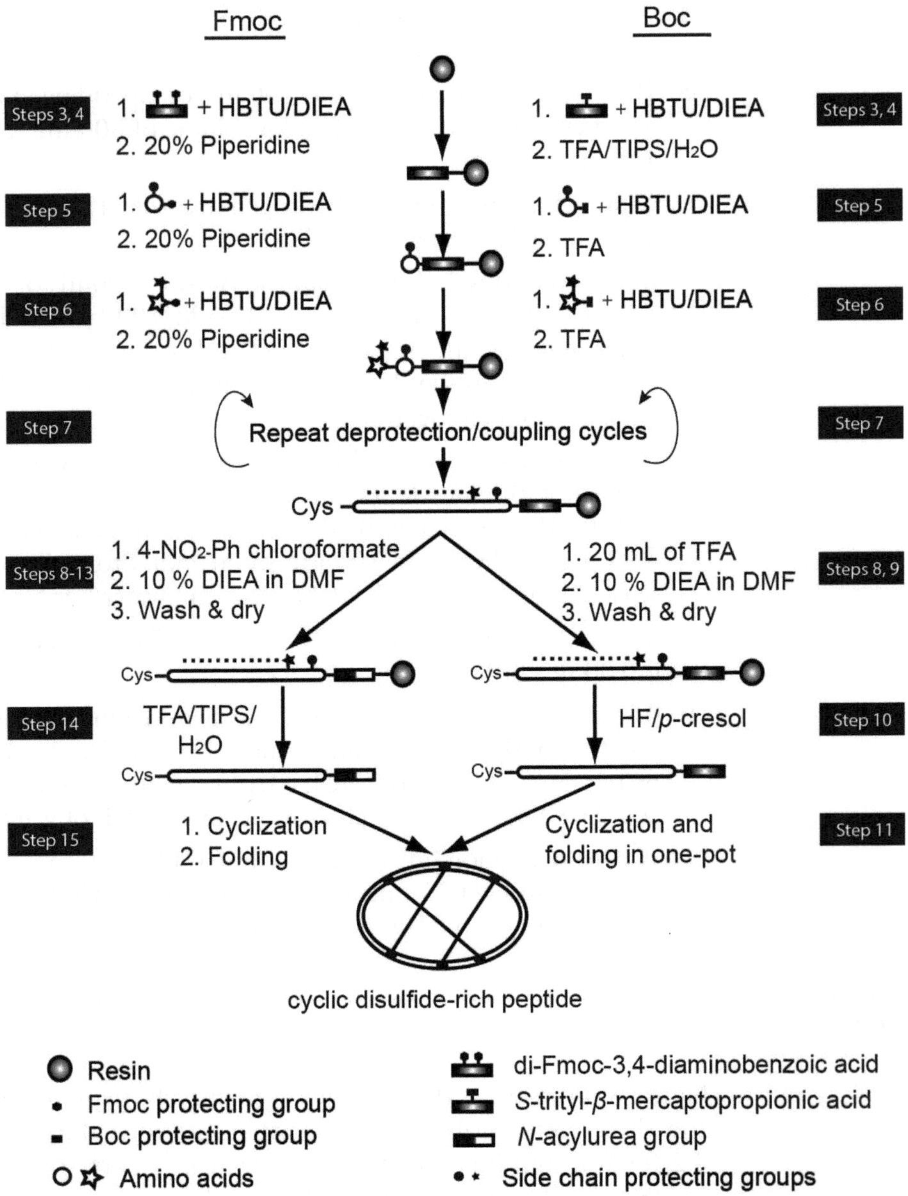

Fig. 3 Synthesis of cyclic disulfide-rich peptides by Fmoc-SPPS and Boc-SPPS approaches. Both methods include repetitive deprotection and amino acid coupling steps. C-terminal di-Fmoc-3,4-diaminobenzoic acid and S-trityl-β-mercaptopropionic acid are used as linkers for the Fmoc-SPPS and Boc-SPPS approaches, respectively. Deprotection of Fmoc and Boc groups can be achieved by treating the reaction mixture with piperidine and TFA as shown on the *left* and *right* sides of the reaction scheme, respectively. Each deprotection step is followed by amino acid couplings activated by HBTU/DIEA. The coupling yields can be calculated by a quantitative ninhydrin test. After completion of the couplings, the peptide is cleaved from the resin by using TFA for Fmoc or HF for the Boc-SPPS method. Cyclization and folding of peptides synthesized with the Boc-SPPS approach can be achieved in one pot, whereas peptides synthesized with the Fmoc-SPPS approach need to be cyclized first and then folded. Detailed descriptions of the various reactions are described in the text, with the steps indicated on the figure. Preparatory steps 1 and 2 are not shown in the figure

Peptide	Sequence	# of aa	# of SS
SFTI-1	GRCTKSIPPICFPD	14	1
cyclic Vc1.1	GCCSDPRCNYDHPEICGGAAGG	22	2
kalata B1	GLPVCGETCVGGTCNTPGCTCSWPVCTRN	29	3
MCoTI-II	GGVCPKILKKCRRDSDCPGACICRGNGYCGSGSD	34	3
cyclic chlorotoxin	MCMPCFTTDHQMARKCDDCCGGKGRGKCYGPQCLCRGAGAAGG	43	4

Fig. 4 Sequences of the cyclic disulfide-rich peptides synthesized by Boc-SPPS and Fmoc-SPPS: SFTI-1, cVc1.1, kalata B1, MCoTI-II, and chlorotoxin with disulfide bonds. The numbers of amino acids (aa) and disulfide bonds (SS) are given on the right

3.1 Synthesis of Cyclic Disulfide-Rich Peptides with Boc-SPPS

1. Swell 0.83 g of PAM-Gly resin (0.60 mmol/g) in 10 mL of DMF at least 1 h prior to the first coupling (*see* **Note 2**).
2. Drain DMF and add 10 mL of TFA to the reaction vessel to remove the Boc protecting group from the resin. Wait 1 min and wash with 10 mL of DMF. Repeat this step once and wash the resin with 200 mL of DMF.
3. Coupling of the thioester linker: after removing the Boc group, wash the resin with DMF. Dissolve 261 mg of *S-trityl-β-*mercaptopropionic acid in 1.4 mL of 0.5 M HBTU in DMF and 145 μL of DIEA. Add this solution to the resin and wait 30 min and wash the resin with 20 mL of DMF. Repeat the coupling of the thioester linker by following the coupling procedure just described. Check coupling yield with a ninhydrin test (*see* **Note 3**).
4. To remove the trityl group on the linker, wash the resin (10× 1 min) with 10 mL of TFA/TIS/H_2O (96:2:2) mixture (*see* **Note 4**).
5. Coupling of the first amino acid: wash resin thoroughly with 200 mL of DMF. Activate the first amino acid by dissolving in 4 mL of 0.5 M HBTU and 460 μL DIEA. Add the solution to the reaction vessel and allow it to react for 1 h. For the five peptides (Fig. 3), suggested starting residues are SFTI-1, R3; cyclic Vc1.1, G1; kalata B1, G18; MCoTI-II, A20; cyclic chlorotoxin, M1 (e.g., the assembly order for SFTI-1 is R, G, D, P, F, C, I, P, P, I, S, K, T, C, remembering the chain is assembled from C- to N-terminus).

6. Wash resin with 20 mL of DMF to remove excess reagents and by-products. Treat the resin with 20 mL of TFA (2 × 1 min) to remove the Boc protecting group. Wash resin thoroughly with 200 mL of DMF to remove TFA. Activate the second amino acid by dissolving in 4 mL of 0.5 M HBTU and 460 μL DIEA. Add the solution to the reaction vessel and allow it to react for 10 min. Check the coupling yield with a ninhydrin test. If the yield is greater than 99.6 % then couple the next amino acid (*go to* **step** 7). If it is lower than 99.6 %, wash the resin with 20 mL of DMF and repeat **step 6** for a second coupling of the same residue (*see* **Note 5**)
7. Couple all the amino acids of the peptide sequence one-by-one as in **step 6**. Do not forget to remove the Boc group between each coupling step.
8. After finishing the assembly of the peptidyl resin, treat the resin with 20 mL of TFA for 1 min. Drain and treat with a fresh 20 mL of TFA for another 5 min. Drain and wash with 200 mL of DMF and then treat with 10 mL of 10 % DIEA in DMF for 1 min. Wash with 20 mL of DMF and add another 10 mL of 10 % DIEA in DMF for 1 min. Wash thoroughly with 100 mL of DMF.
9. Dry resin by washing with DCM (~100 mL) and then dry with vacuum suction under nitrogen gas. The peptide is now ready for HF cleavage.
10. Peptide cleavage from the resin is achieved using hydrogen fluoride with *p*-cresol as scavenger [9.5/1.0 (v/v) HF/*p*-cresol mixture] at –4 to 0 °C for 1.5 h and lyophilized. 1 mL of *p*-cresol is added to 0.5 g of resin and the slurry is frozen. 10 mL of HF is added and incubated with stirring at –4 to 0 °C (salt water ice bath) for 1.5 h. HF is evaporated off and the peptide is precipitated in ice cold ether. This solution is filtered and the filtrate is resuspended in Solution A/B (50:50, v/v) before being filtered to remove the resin. Residual diethyl ether is removed by a rotary evaporator before freeze drying (*see* **Note 6**). Purify the crude linear peptide with reversed phase-high performance chromatography (RP-HPLC).
11. Both cyclization and folding of the synthesized peptide is carried out in a buffer solution with a pH range of 7.5–8.5 (*see* **Note** 7). The buffer solutions for folding of each peptide are given in Table 1 with reaction temperature and time. For example, for the cyclization and folding of kalata B1, dissolve 50 mg of dry peptide in a 100 mL of buffer solution containing 0.1 M of NH_4HCO_3 (pH: 8.2) and 50 % isopropanol. Stir the reaction mixture at room temperature over 24 h. Filter the reaction mixture, dilute and purify with RP-HPLC.

Table 1
Buffer solutions for the cyclization and folding of disulfide-rich cyclic peptides

Peptide	Buffer solution	Temperature (°C)	Time (h)
SFTI-1	0.1 M NH_4HCO_3 (pH 8.2)	25	24
Cyclic Vc1.1	0.1 M NH_4HCO_3 (pH 8.2)	25	24
Kalata B1	0.1 M NH_4HCO_3 (pH 8.2) and 50 % isopropanol	25	24
MCoTI-II	0.1 M NH_4HCO_3 (pH 8.2) and 50 % isopropanol	25	24
Cyclic chlorotoxin	0.1 M Tris–HCl, 0.2 M NaCl, 5 mM reduced glutathione/0.5 mM oxidized glutathione (pH 7.8)	25	24

3.2 Synthesis of Cyclic Disulfide-Rich Peptides with Fmoc-SPPS

1. Swell 0.85 g of Fmoc-Rink amide MBHA resin (0.59 mmol/g) in 10 mL of DMF at least 1 h prior to the coupling of the linker di-Fmoc-3,4-diaminobenzoic acid (*see* **Note 2**).
2. Add 10 mL of 20 % piperidine in DMF to the reaction vessel containing MBHA resin and wait 10 min to remove the Fmoc group from the resin. Wash with 10 mL of DMF and add another 10 mL of 20 % piperidine in DMF for another 10 min.
3. Coupling of linker: wash the resin with 50 mL of DMF and add di-Fmoc-3,4-diaminobenzoic acid (1.19 mg) in 4 mL of 0.5 M HBTU and 460 μL DIEA. Wait 30 min and do a second coupling of the linker for another 30 min. Check the coupling yield with a ninhydrin test (*see* **Note 3**).
4. To remove the Fmoc groups on the linker, repeat **step 2**.
5. Coupling of the first amino acid: wash resin thoroughly with 200 mL of DMF. Activate the amino acid by dissolving in 4 mL of 0.5 M HBTU and 460 μL DIEA. Add the solution to the reaction vessel and allow the solution to react for 10 min. Check the coupling yield with a ninhydrin test. If the yield is greater than 99.6 %, then couple the next amino acid (*go to* **step 6**). If it is lower than 99.6 %, wash the resin with 20 mL of DMF and repeat **step 5** (*see* **Note 5**). For the five peptides shown in Fig. 3, suggested starting residues are SFTI-1, R3; cyclic Vc1.1, G1; kalata B1, G18; MCoTI-II, A20; and cyclic chlorotoxin, M1 (e.g., the assembly order for SFTI-1 is R, G, D, P, F, C, I, P, P, I, S, K, T, and C, remembering the chain is assembled from C-terminus to N-terminus).
6. Wash resin thoroughly with 200 mL of DMF. Activate the second amino acid by dissolving in 4 mL of 0.5 M HBTU and 460 μL DIEA. Add the solution to the reaction vessel and allow the solution to react for 10 min. Check the coupling yield with a ninhydrin test.
7. Couple all the amino acids of the peptide sequence one-by-one as in **step 6** (*see* **Note 8**).

8. After finishing the assembly of the peptide chain, add 10 mL of DCM, stir for 30 s, and then remove the solvent by vacuum filtration. Repeat this step four times.
9. Dissolve 200 mg of 4-nitrophenyl chloroformate in 10 mL of DCM, add this solution to the resin, stir and stand for 1 h, and then remove by vacuum filtration.
10. Wash the resin with DCM (10 mL×4) for 30 s.
11. Wash the resin with DMF (10 mL×6), stir for 30 s, and then remove the solvent by vacuum filtration.
12. Add 10 mL of 10 % DIEA/DMF (v/v), stir, and stand for 15 min before removing the excess solution by vacuum filtration.
13. Wash the resin thoroughly with DMF and then with DCM. Dry with vacuum suction under nitrogen.
14. Cleavage of the peptide from the resin: weigh about 500 mg of dried resin into an Erlenmeyer flask and add 50 mL of TFA/TIS/water (96:2:2) mixture (*see* **Note 9**). Stir the reaction mixture for at least 2 h at room temperature. Filter the reaction mixture and evaporate almost all the TFA by rotary evaporator. Add 50 mL of cold diethyl ether and extract three times with 50 mL of solution A/solution B mixture (50:50, v/v) (solution A: H_2O/0.05 % TFA; solution B: 90 % acetonitrile/10 % H_2O/0.045 % TFA). Collect the water layer and lyophilize for purification. Purify the crude linear peptide with RP-HPLC before cyclization and folding.
15. Cyclization and folding: prepare a buffer solution containing 100 mM ammonium acetate (CH_3COONH_4) and 20 mM DTT. The peptide concentration should be ~1 mg/mL. Stir the solution and monitor the reaction by mass spectrometry. When all the starting material is consumed, purify the cyclic-reduced peptide by RP-HPLC. Folding of the cyclized peptide is carried out in the buffer solutions given in Table 1 for each peptide.

4 Notes

1. Weigh HBTU carefully in a well-ventilated fume hood as it is hazardous if inhaled.
2. Generally, it is much better to weigh the required amount of resin and leave it soaking in DMF overnight.
3. Quantitative ninhydrin test: weigh 3–5 mg of resin into a test tube and add two drops of 76 % (w/w) phenol in ethanol, four drops of 0.2 mM KCN in pyridine, and two drops of 0.28 M ninhydrin in ethanol. A blank, containing no resin, is also

prepared. The samples are incubated at 100 °C for 5 min and then 2.8 mL of 60 % ethanol in water added. The samples are briefly spun to settle the resin and the absorbance of the solution read against the reagent blank at 570 nm. The percent coupling yield is then calculated using Eq. 2:

$$\% \text{ yield} = 100 \times \left[1 - \frac{A_{570} \times 200}{\text{SV} \times \text{mass of resin}}\right] \quad (2)$$

Equation 2, where A_{570} is the absorbance at 570 nm and SV is the substitution value.

The ninhydrin test, which detects the amount of primary amine remaining, cannot be used to test the coupling yield after the coupling of Pro. Therefore, routinely do a second coupling after a Pro residue as a precaution against incomplete coupling. A qualitative ninhydrin test is described in Chapter 2.

4. Minimize the amount of air sucked through the resin at this stage as the free thiol may oxidize.
5. If the ninhydrin test remains positive after second coupling of the residue, repeat **step 5** again and leave the reaction for longer (i.e. from 1h or up to overnight). Do another ninhydrin test. If it is still positive, block the remaining free amine groups at the N-terminus of the peptide chain by acetylation.

   Acetylation protocol: Prepare a mixture of 870 μL of acetic anhydride, 470 μL of DIEA, and 15 mL of DMF. Treat the resin with this mixture 2 × 5 min. Then go to **step 6**. *See* Chapter 2 for additional acylations procedures.
6. As HF is highly hazardous and corrosive, only certified persons should carry out HF cleavage reactions.
7. To determine the specific folding conditions for each peptide, firstly do the folding reaction on a small scale (100 μM of peptide in 200 μL of buffer solution) in an Eppendorf tube at room temperature or 4 °C. Analyze the reactions with analytical RP-HPLC after 24, 48, 72 h, etc. to determine the best folding condition. Then carry out the folding on a larger scale. Purify the folded peptides with RP-HPLC.
8. Use Boc-Cys(Trt) as the N-terminal amino acid instead of a Fmoc-protected Cys.
9. Wait approximately 10 min. If the reaction mixture is still yellow, add more TIPS until the yellow color has disappeared.

Acknowledgements

D.J.C. is a NHMRC Professorial Fellow. Research in our group on cyclic peptides is supported by the Australian Research Council (ARC) and the National Health and Medical Research Council (NHMRC).

References

1. Craik DJ, Daly NL, Bond T, Waine C (1999) Plant cyclotides: a unique family of cyclic and knotted proteins that defines the cyclic cystine knot structural motif. J Mol Biol 294:1327–1336
2. Ireland DC, Wang CKL, Wilson JA, Gustafson KR, Craik DJ (2008) Cyclotides as natural anti-HIV agents. Biopolymers 90:51–60
3. Henriques ST, Craik DJ (2010) Cyclotides as templates in drug design. Drug Discov Today 15:57–64
4. Garcia AE, Camarero JA (2010) Biological activities of natural and engineered cyclotides, a novel molecular scaffold for peptide-based therapeutics. Curr Mol Pharmacol 3:153–163
5. Burman R, Herrmann A, Tran R, Kivela JE, Lomize A, Gullbo J, Goransson U (2011) Cytotoxic potency of small macrocyclic knot proteins: structure-activity and mechanistic studies of native and chemically modified cyclotides. Org Biomol Chem 9:4306–4314
6. Craik DJ (2009) Circling the enemy: cyclic proteins in plant defence. Trends Plant Sci 14:328–335
7. Rosengren KJ, Daly NL, Plan MR, Waine C, Craik DJ (2003) Twists, knots, and rings in proteins – structural definition of the cyclotide framework. J Biol Chem 278:8606–8616
8. Heitz A, Hernandez JF, Gagnon J, Hong TT, Pham TT, Nguyen TM, Le-Nguyen D, Chiche L (2001) Solution structure of the squash trypsin inhibitor MCoTI-II. A new family for cyclic knottins. Biochemistry 40:7973–7983
9. Felizmenio-Quimio ME, Daly NL, Craik DJ (2001) Circular proteins in plants: solution structure of a novel macrocyclic trypsin inhibitor from *Momordica cochinchinensis.* J Mol Biol 276:22875–22882
10. Hernandez JF, Gagnon J, Chiche L, Nguyen TM, Andrieu JP, Heitz A, Trinh Hong T, Pham TT, Le Nguyen D (2000) Squash trypsin inhibitors from *Momordica cochinchinensis* exhibit an atypical macrocyclic structure. Biochemistry 39:5722–5730
11. Craik DJ (2006) Chemistry - Seamless proteins tie up their loose ends. Science 311:1563–1564
12. Chan LY, Gunasekera S, Henriques ST, Worth NF, Le SJ, Clark RJ, Campbell JH, Craik DJ, Daly NL (2011) Engineering pro-angiogenic peptides using stable, disulfide-rich cyclic scaffolds. Blood 118:6709–6717
13. Clark RJ, Fischer H, Dempster L, Daly NL, Rosengren KJ, Nevin ST, Meunier FA, Adams DJ, Craik DJ (2005) Engineering stable peptide toxins by means of backbone cyclization: stabilization of the alpha-conotoxin MII. Proc Natl Acad Sci U S A 102:13767–13772
14. Craik DJ, Daly NL, Waine C (2001) The cystine knot motif in toxins and implications for drug design. Toxicon 39:43–60
15. Clark RJ, Jensen J, Nevin ST, Callaghan BP, Adams DJ, Craik DJ (2010) The engineering of an orally active conotoxin for the treatment of neuropathic pain. Angew Chem Int Ed 49:6545–6548
16. Akcan M, Stroud MR, Hansen SJ, Clark RJ, Daly NL, Craik DJ, Olson JM (2011) Chemical re-engineering of chlorotoxin improves bioconjugation properties for tumor imaging and targeted therapy. J Med Chem 54:782–787
17. DeBin JA, Maggio JE, Strichartz GR (1993) Purification and characterization of chlorotoxin, a chloride channel ligand from the venom of the scorpion. Am J Physiol 264:C361–C369
18. Lyons SA, O'Neal J, Sontheimer H (2002) Chlorotoxin, a scorpion-derived peptide, specifically binds to gliomas and tumors of neuroectodermal origin. Glia 39:162–173
19. Ferdani R, Zheleznyak A, Jacoby DB, Sentissi A, Sherman C, Anderson CJ (2009) 64Cu labeled chlorotoxin as a potential imaging agent for glioma. J Labelled Comp Radiopharm 52:S35
20. Veiseh M, Gabikian P, Bahrami SB, Veiseh O, Zhang M, Hackman RC, Ravanpay AC, Stroud MR, Kusuma Y, Hansen SJ, Kwok D, Munoz NM, Sze RW, Grady WM, Greenberg NM, Ellenbogen RG, Olson JM (2007) Tumor paint: a chlorotoxin: Cy5.5 bioconjugate for intraoperative visualization of cancer foci. Cancer Res 67:6882–6888
21. Sun CR, Du K, Fang C, Bhattarai N, Veiseh O, Kievit F, Stephen Z, Lee DH, Ellenbogen RG, Ratner B, Zhang MQ (2010) PEG-mediated synthesis of highly dispersive multifunctional superparamagnetic nanoparticles: their physicochemical properties and function in vivo. ACS Nano 4:2402–2410
22. Kievit FM, Veiseh O, Fang C, Bhattarai N, Lee D, Ellenbogen RG, Zhang MQ (2010) Chlorotoxin labeled magnetic nanovectors for targeted gene delivery to glioma. ACS Nano 4:4587–4594
23. Veiseh O, Kievit FM, Gunn JW, Ratner BD, Zhang MQ (2009) A ligand-mediated nanovector for targeted gene delivery and transfection in cancer cells. Biomaterials 30:649–657
24. Merrifield RB (1963) Solid phase peptide synthesis. 1. Synthesis of a tetrapeptide. J Am Chem Soc 85:2149–2154
25. Nishiuchi Y, Kumagaye K, Noda Y, Watanabe TX, Sakakibara S (1986) Synthesis and secondary

structure determination of omega-conotoxin GVIA - A 27-peptide with 3 intramolecular disulfide bonds. Biopolymers 25:S61–S68
26. Hackeng TM, Griffin JH, Dawson PE (1999) Protein synthesis by native chemical ligation: expanded scope by using straightforward methodology. Proc Natl Acad Sci U S A 96:10068–10073
27. Dawson PE, Muir TW, Clark-Lewis I, Kent SB (1994) Synthesis of proteins by native chemical ligation. Science 266:776–779
28. Clark RJ, Craik DJ (2010) Native chemical ligation applied to the synthesis and bioengineering of circular peptides and proteins. Biopolymers 94:414–422
29. Gunasekera S, Foley FM, Clark RJ, Sando L, Fabri LJ, Craik DJ, Daly NL (2008) Engineering stabilized vascular endothelial growth factor-A antagonists: synthesis, structural characterization, and bioactivity of grafted analogues of cyclotides. J Med Chem 51:7697–7704
30. Blanco-Canosa JB, Dawson PE (2008) An efficient Fmoc-SPPS approach for the generation of thioester peptide precursors for use in native chemical ligation. Angew Chem Int Ed 47:6851–6855
31. Schnölzer M, Alewood P, Jones A, Alewood D, SBH K (2007) In situ neutralization in Boc-chemistry solid phase peptide synthesis: rapid, high yield assembly of difficult sequences. Int J Pept Res Ther 13:31–44

Chapter 7

Preparation of C-terminally Modified Chemokines by Expressed Protein Ligation

Lars Baumann, Max Steinhagen, and Annette G. Beck-Sickinger

Abstract

In order to link structural features on a molecular level to the function of chemokines, site-specific modification strategies are strongly required. These can be used to incorporate fluorescent dyes and/or physical probes to allow investigations in a wide range of biological and physical techniques, e.g., nuclear magnetic resonance (NMR) spectroscopy, fluorescence microscopy, fluorescence resonance energy transfer (FRET), or fluorescence correlation spectroscopy (FCS). Only a limited number of functional groups within the 20 canonical amino acids allow ligation strategies that can be helpful to introduce novel functionalities, which in turn expand the scope of chemoselective and orthogonal reactivity of (semi)synthetic chemokines. In the present chapter we mainly focus on the fabulous history of native chemical ligation (NCL) and provide a general protocol for the preparation of C-terminally modified SDF-1α including tips and tricks for practical work. We believe that this protocol can be easily adapted to other chemokines and many proteins in general.

Key words Expressed protein ligation, Native chemical ligation, Protocol, Protein ligation, Chemokines, SDF-1α, Stromal cell-derived factor-1, CXCL12

1 Introduction

Solid-phase peptide synthesis (SPPS) revolutionized peptide synthesis with respect to the handling of the growing peptide chain, the possibility of automatization as well as efficient purification of intermediates and the final product. By introducing orthogonally protected amino acids, selective demasking and modification of functional groups within the peptide chain can be easily realized. However, SPPS and all its recent improvements give access to peptides up to a length of about 50 amino acids in reasonable yield and purity, both depends on the primary sequence, hydrophobicity, and aggregation tendency on solid support. Preparation of larger, modified peptides or even entire proteins requires condensation steps of pre-synthesized segments, at best without protection.

Knud J. Jensen et al. (eds.), *Peptide Synthesis and Applications: Second Edition*, Methods in Molecular Biology, vol. 1047, DOI 10.1007/978-1-62703-544-6_7, © Springer Science+Business Media New York 2013

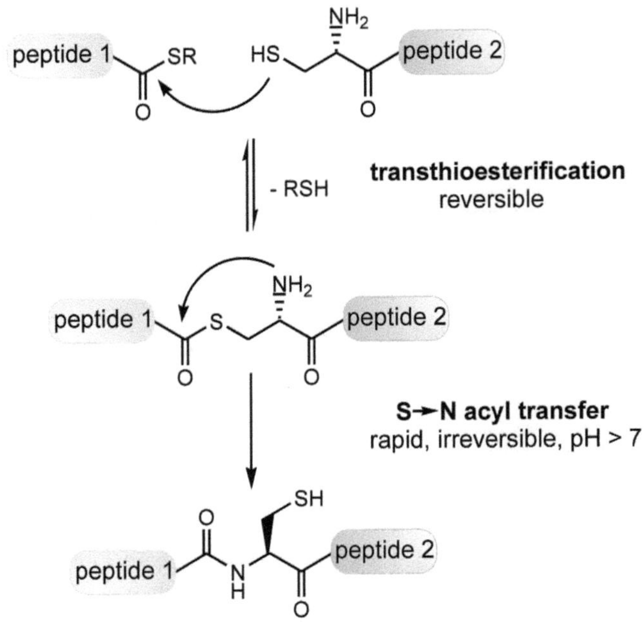

Fig. 1 *Native chemical ligation (NCL)*—a reaction scheme

Chemical ligation (CL) describes the coupling of unprotected peptide fragments by a chemoselective reaction of functional groups without affecting other functionalities resulting in backbone-modified analogs. In 1992, Schnölzer et al. presented the first fully synthetic HIV protease through the reaction between an α-peptide thioacid (1–50,Gly-αCOSH) and an N-terminal bromoacetyl-peptide [$BrCH_2CO$(53–99)] fragment by replacing the native Gly^{51}-Gly^{52} peptide bond through a peptide thioester bond [1]. Two years later, in 1994, Dawson et al. published the first preparation of interleukine-8 by chemoselective condensation of two unprotected peptide segments leading to the native protein [2]. According to the full-amide backbone, the authors named this technique native chemical ligation (NCL). Because of simple prerequisites, an N-terminal fragment as peptide α-thioester and a C-terminal fragment bearing an N-terminal cysteine residue, the method turned out to be versatile and applicable for a wide range of midsized proteins.

Mechanistically, NCL is a two-step process, which is initiated by a reversible transthioesterification step followed by a spontaneous, irreversible *S*- to *N*-acyl transfer (Fig. 1). The reactivity is affected by several factors such as concentration of the reactants, temperature, pH, and choice of an appropriate thiol. Also the C-terminally thioesterified amino acid at the ligation site has to be considered carefully [3]. In fact, NCL relies basically on an observation of Wieland et al., who first demonstrated the transfer of a valyl residue from a reactive thioester onto a cysteine under slightly basic conditions followed by a rearrangement and yielding a Val-Cys dipeptide [4].

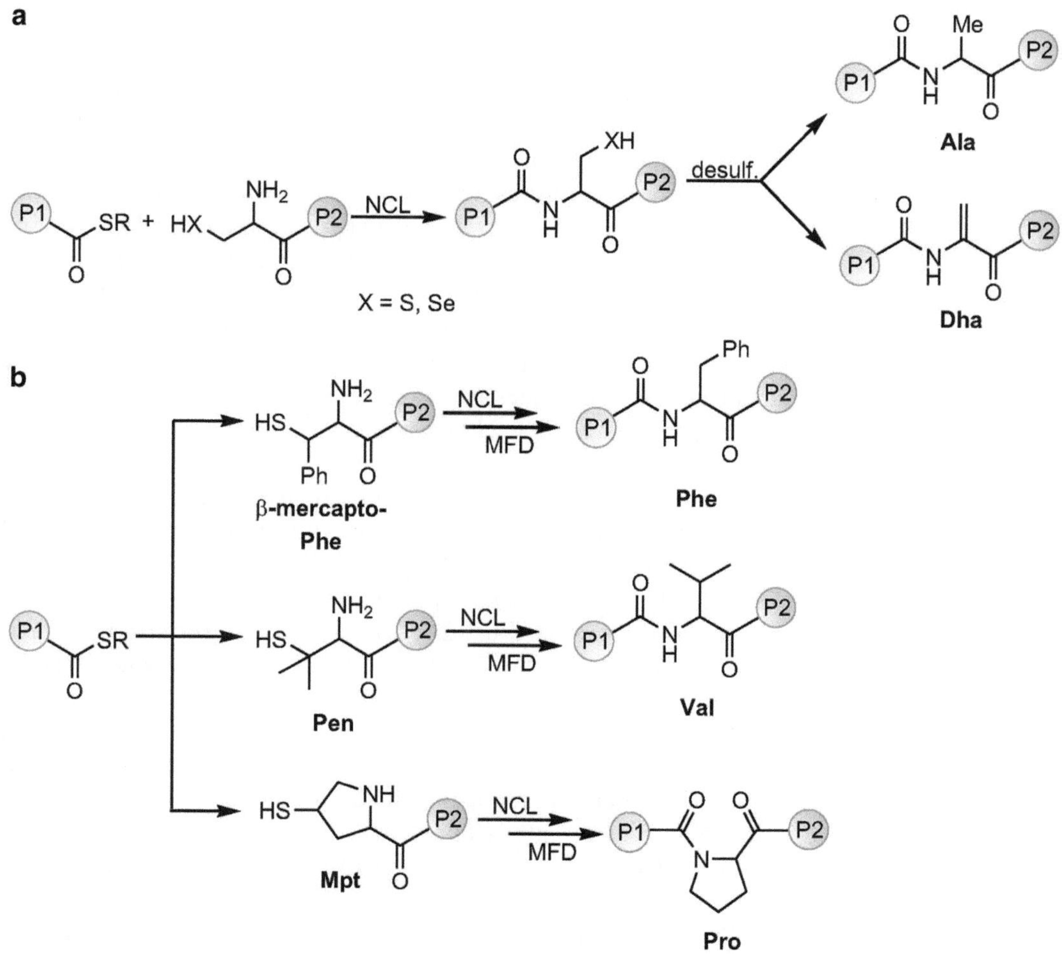

Fig. 2 Advanced ligation strategies. (**a**) NCL and subsequent desulfurization of the unprotected cysteine/selenocysteine. DHA (dehydroalanine) is thiol reactive and can be further modified with SH-bearing compounds. (**b**) Examples for ligation-desulfurization strategies leaving amino acids different from cysteine at the junction site. *MFD* metal-free desulfurization, *Pen* penicillamine, *Mpt* mercapto-pyrrolidine-2-carboxylic acid

Since the beginning, NCL has undergone a rapid evolution. Up to now, several strategies have been developed to apply NCL to proteins, which do not contain any cysteine residues or cysteines at unfavorable sites (Fig. 2). The first attempts toward cysteine-free ligations were made by Tam et al. who chose an N-terminal histidine as acyl acceptor during the ligation, however, with moderate success [5]. Selenocysteine, the 21st natural amino acid, basically reacts the same way as cysteine but exhibits increased nucleophilicity that leads to accelerated reaction rates [6].

Substantial progress has been made with the concept to convert the required selenocysteine/cysteine residue upon ligation into an alanine [7–9] or dehydroalanine [10] by a process called desulfurization. In turn, the generated dehydroalanine moiety can then further react with other thiol-functionalized compounds leading to modifications addressed to the former ligation site. Further development of a mild metal-free desulfurization procedure [11] had extensively expanded the scope of the NCL reaction to junction sites different from cysteine. This is due to efficient removal of thiol groups even in the presence of additional protected cysteine residues without time-consuming optimization. Recently, several NCL strategies with phenylalanine [12], valine [13, 14], lysine [15], leucine [16], and threonine [17] at the junction site have been developed. Latest progress in this vibrating field is a ligation technique involving thioprolines followed by desulfurization [18]. For complex proteins, especially glycoproteins, more advanced NCL strategies are available, e.g. sugar-assisted glycopeptide ligation (SAL) [19–21] or side-chain assisted ligation (SCAL).

The NCL methodology was further extended by Chong et al. who made use of an engineered intein, which allows the single-column purification of proteins and thiol-induced cleavage and affinity tag removal leaving the protein as reactive α-thioester [22, 23]. Inteins are internal self-splicing protein fragments, discovered in yeast in the early 1990s [24, 25]. Kane et al. reported the protein biosynthesis of the *TFP1* gene from *Saccharomyces cerevisiae* results in a single translational product that cleaves itself posttranslationally, independent of other factors. In analogy to introns and exons on RNA level, the self-splicing element is termed intein (internal protein sequence) which is flanked by an N- and C-terminal extein (external protein sequence), respectively. During intein-mediated, autocatalytic splicing the N- and C-extein are being ligated while the intein is released (Fig. 3).

Protein splicing is initiated by spontaneous formation of a (thio)ester through a *N*- to *S*- or *N*- to *O*-acyl shift. X-ray spectroscopic studies on the GyrA intein from *Mycobacterium xenopi* (*Mxe*) revealed the close proximity of the intein's ends and *cis-conformation* of the peptide bond which connects the *N*-extein to the intein. The unfavorable *cis*-conformation destabilizes the amide bond and facilitates the initial acyl shift by an nucleophilic attack of the cysteine's sulfhydryl group [26]. This reversible thioester formation is followed by an acyl transfer of the N-extein onto the N-terminal amino acid (Cys, Ser or Thr) of the C-extein. Then, in the most critical step, the C-terminal Asn of the intein forms a succinimide that leads to the release of the intein, while the N-extein/C-extein are being ligated and rearrange in a *S*- to *N*-acyl or *O*- to *N*-acyl shift forming the native protein structure [27]. The engineered intein introduced by Chong et al. bears an alanine instead of the essential Asn and is therefore not able to perform the splicing reaction; however, the initial *N*- to *S*-acyl transfer is not affected. Hence, the main advantage arises from recombinant preparation of reactive protein α-thioesters by thiol-induced cleavage and their subsequent use in NCL reactions in order to site-specifically label proteins of nearly any size. Owing to the recombinant origin of at

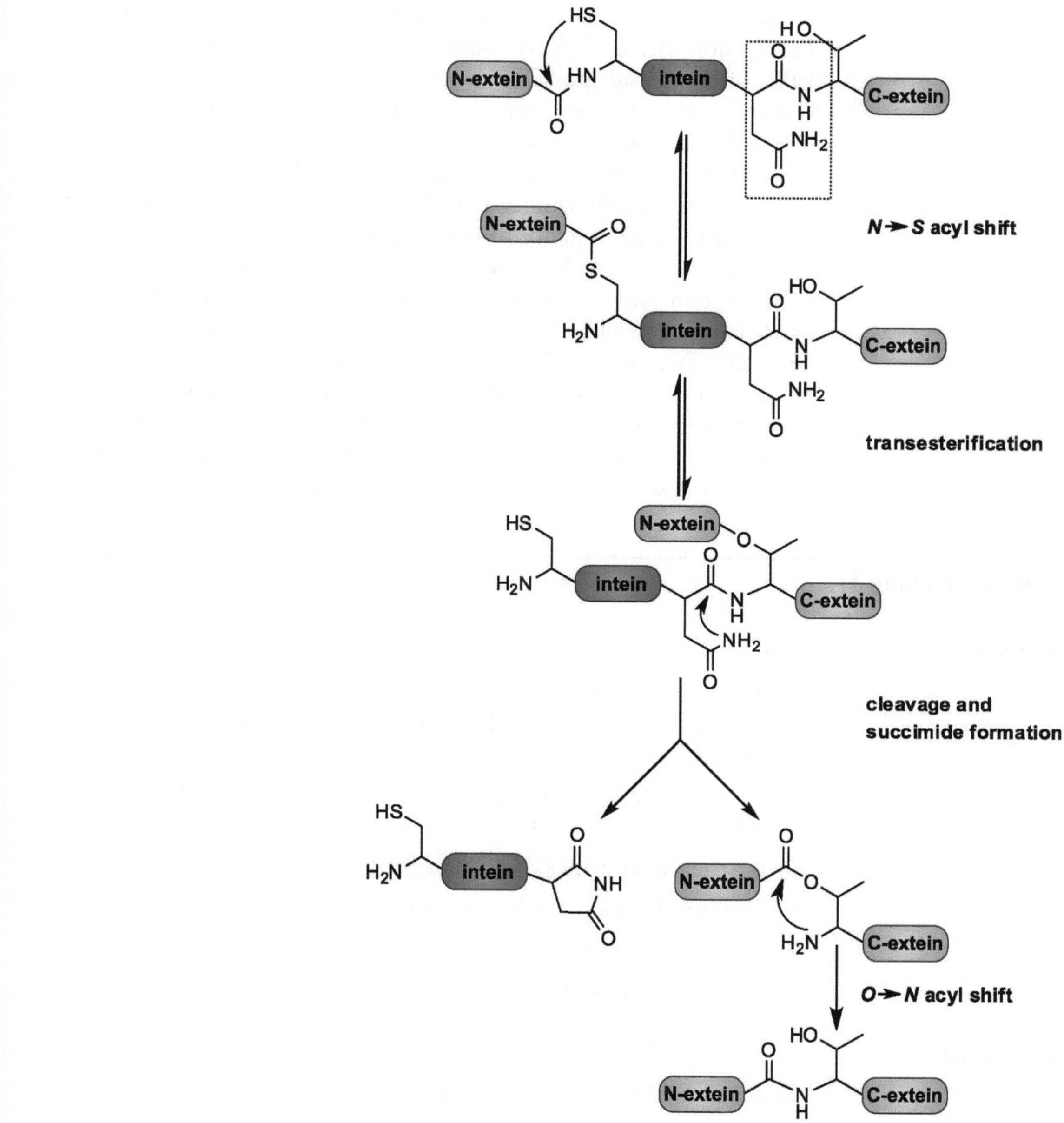

Fig. 3 Protein splicing mechanism of the *Mxe* GyrA intein. The asparagine (*dashed box*) is essential for cleavage, thus, replaced by an alanine residue in engineered intein suitable for protein α-thioester preparation

least one of the segments, Muir et al. referred this method expressed protein ligation (EPL) [28]. By exploiting inteins, it is also possible to prepare proteins bearing an N-terminal Cys. Therefore, the C-terminus of an engineered intein can be used and the protein of interest is released by a shift in temperature and pH [29]. The Intein-based technology is commercially available and *New England Biolabs* offers various kit solutions kits for individual requirements under the name IMPACT™ (Intein-Mediated Purification with an Affinity Chitin-Binding Tag) system. Although EPL offers broad possibilities to label proteins either N- or C-terminally, the splicing reaction requires slightly basic (pH 7.5–8.5) and non-denaturating conditions, thus proteins with an isoelectric point (pI) in this range are hardly accessible by this methodology. Furthermore, misfolded and unfolded target proteins might be aggregation prone and this makes optimization of proper cleavage and elution conditions necessary. The robustness of the intein and chitin-binding domain structure allow readily harsh conditions like high salt concentrations and tolerate detergents like Tween 20 and triton X-100 up to at least 0.2 %.

In conclusion, not all of the herein mentioned strategies will fit to your target protein, but each of them represents a solution for the individual problem. Therefore, it is important to plan the work carefully and select the best strategy in advance.

2 Material (Note 1)

2.1 Cloning

1. pTXB1 vector (NEB).
2. SDF-1(1–49) cDNA.
3. Forward primer: 5′-GGT GGT CAT ATG AAG CCC GTC AGC CTG AG-3′.
4. Reverse primer: 5′-GGT GGT TGC TCT TCC GCA CGC TTG TCT GTT GTT GTT C-3′.
5. Restriction enzymes: NdeI and SapI.
6. Shrimp alkaline phosphatase (SAP).
7. DNA preparation kit.
8. PCR/gel clean-up kit.

2.2 Bacteria

1. *E. coli* DH5α: competent cells.
2. *E. coli* ER2566: competent cells.

2.3 Protein Chemistry

1. Luria broth (LB) media.
2. Isopropyl *β*-D-1-thiogalactopyranoside (IPTG): prepare stock solution with final concentration of 1 M and store aliquots at −20 °C.
3. Ampicillin: prepare stock solution from sodium salt in water with final concentration of 100 mg/mL. Store aliquots at −20 °C. Make sure having the ampicillin sodium salt!
4. Basis buffer (BB): 20 mM HEPES, 0.5 M NaCl, 1 mM EDTA, pH 8.
5. Wash buffer (WB): BB containing 2 M urea, 0.2 % Tween 20, pH 8.

6. Solubilization buffer (SB): BB containing 8 M urea, pH 8.
7. Column buffer: BB containing 3 M urea, pH 8.
8. Cleavage buffer (CB): BB containing 2 M urea, 0.2 % Tween 20, 0.2 M MESNA, pH 8.
9. Ligation buffer: 100 mM sodium phosphate, 6 M guanidinium chloride, pH 6.
10. Refolding buffer: 100 mM sodium phosphate buffer, 0.3 M sucrose, 0.3 mM GSH/0.3 mM GSSG (1:1), pH 7.4.
11. Chitin beads (NEB).
12. HCl (37 %).
13. 0.3 M NaOH.
14. 1 M $MgCl_2$.
15. DNase I.
16. Tris(2-carboxyethyl)phosphine (TCEP).

2.4 Peptide Synthesis

1. Fmoc-N^{α}-amino acids and resins.
2. Dimethylformamide (DMF).
3. Dichloromethane (DCM).
4. Piperidine.
5. Hydrazine monohydrate.
6. Diethyl ether (Et_2O).
7. Diisopropylcarbodiimide (DIC).
8. Hydroxybenzotriazole monohydrate (HOBt).
9. Trifluoroacetic acid (TFA): HPLC grade
10. Thioanisole (TA).
11. Ethanedithiol (EDT).
12. *Kaiser* test solution I: 1 g ninhydrin in 20 mL EtOH.
13. *Kaiser* test solution II: 80 g phenol in 20 mL EtOH.
14. *Kaiser* test solution III: 0.4 mL aqueous KCN solution (1 mM) in 20 mL pyridine.

2.5 Analytical and Preparative Work

1. Agarose.
2. Rotiphorese® Gel 30: ready-to-use solution (Roth).
3. *N,N,N′,N′*-tetramethyl-ethane-1,2-diamine (TEMED).
4. Ammonium persulfate (APS): Prepare 10 % stock solution (w/v). Store aliquots at −20 °C.
5. Sodium dodecyl sulfate (SDS): Prepare 10 % stock solution (w/v) and store at room temperature.
6. Reducing SDS-PAGE sample buffer: 62.5 mM Tris, 20 % glycerine (v/v), 5 % β-mercaptoethanol, 1 % SDS (w/v), 0.0025 % bromophenol blue (w/v), pH 7.8 adjusted with HCl.

7. Staining solution: 1.5 mM amido black 10B, 50 % methanol (v/v), and 10 % acetic acid (v/v).
8. Destaining solution: Aqueous solution containing 50 % methanol (v/v) and 10 % acetic acid (v/v).
9. Acetic acid: Prepare a 10 % stock solution (v/v) and store at room temperature.
10. Acetonitrile (ACN): HPLC grade.
11. Water: ultrapure.

2.6 Consumables and Equipment

1. Amicon Ultra-15 Centrifugal Filter Units (MW cut-off: 3kDa).
2. Chromatography column: glass or plastic, diameter 2–3 cm and length around 20 cm.
3. 2 mL Syringes (with Luer connection) equipped with teflon frit and stamp.
4. Luer stopper.
5. Electrophoresis device.
6. Analytical and preparative high pressure liquid chromatography (HPLC) system equipped with C18 columns (*see* **Note 2**).
7. Mass spectrometer (*see* **Note 3**).
8. Cuvette photospectrometer.
9. *French press.*
10. 0.45 μm filter.

3 Methods

In this section we provide you a general protocol to prepare functional human SDF-1α modified on position 56. In a previous study, we identified this particular position by introducing the photocleavable 6-nitroveratryloxycarbonyl (Nvoc) group, and we found no differences in signal transduction experiments compared to recombinant M-SDF-1a (a menas alpha, greek letter), neither prior nor after UV irradiation [30]. This strategy allows direct comparison of the modified and native protein within the same molecule. Motivated by this study, the position was further evaluated by attaching a carboxyfluorescein moiety which was used for microscopic CXC receptor 4 (CXCR4) internalization studies [31].

SDF-1α is a member of the CXC family of chemokines consisting of 68 amino acids. As nearly all chemokines, SDF-1α possesses four highly conserved cysteine residues (Cys 9, 11, 34, and 50) and adopts the typical tertiary structure characterized by a flexible N-terminus and central antiparallel three-stranded *β*-sheet which is overlaid by an α-helical segment. For practical reasons we decided to use the peptide bond Val^{49}–Cys^{50} as junction site for ligation, in

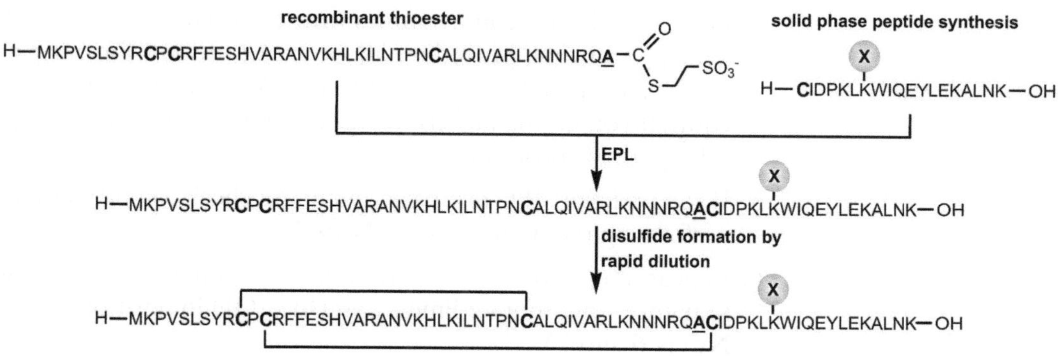

Fig. 4 Schematic workflow to prepare SDF-1α analogs. The N-terminal methionine originates from bacterial expression and is not processed. The beneficial mutation Val[49]Ala is shown underlined. X = any modification you want to incorporate into the protein, e.g., a fluorescent dye or a biophysical probe

order to prepare the longer N-terminal fragment SDF-1(1–49) recombinantly and keep the chemically synthesized and modified C-terminal fragment SDF-1(50–68) as short as possible (Fig. 4). Further, we mutated the valine on position 49 to an Ala because the reactivity of the protein thioester depends on the side chain of the last amino acid. Previous studies have shown that valine, leucine, and proline are less favorable at the ligation site [3]. Since protein splicing is initiated by the reverse acyl shift reaction (*N* to *S*), basically the same holds true for intein-based purification techniques. Furthermore, aspartic acid at this particular position frequently leads to substantial *in vivo* cleavage of the fusion protein during protein thioester preparation. For both reasons you should avoid these amino acids at the C-terminus of the N-terminal fragment. The best compromise seems to be alanine at this position but has to be tested for each individual protein in terms of function and bioactivity.

3.1 Synthesis and Purification of the Modified C-terminal Fragment by Fmoc-Based SPPS

1. Peptide is assembled stepwise by automated Fmoc-/*tBu*-based solid-phase peptide synthesis (SPPS); however, it can also be easily done manually (*see* **Note 4**).
2. Weigh in 25 μmol Fmoc-Lys(Boc)-OH preloaded Wang resin in a syringe. Pre-swell the resin in 1 mL DMF for 15 min.
3. Fmoc deprotection. Treat the resin with 0.5 mL 20 % piperidine/DMF solution (v/v) for 20 min. Aspirate solution and wash resin once with 1 mL DMF. Repeat deprotection step for quantitative removal of the Fmoc group and wash resin five times with 1 mL DMF.
4. Amino acid double coupling. Weight 7 equiv. Fmoc-N$^{\alpha}$-amino acid and 7 equiv. HOBt in a tube and dissolve in DMF to a final concentration of 0.5 M. Vortex solution, add 7 equiv. DIC, and vortex again. Pour mixture onto the pre-swollen resin and allow the suspension to react 45 min. Aspirate solution, wash once with 1 mL DMF, and repeat coupling the same way.

Finally, wash the resin five times with 1 mL DMF. The equivalents given in the procedure above are optimized for our particular automated peptide synthesizer. You can also use just 5 equiv. if you do manual couplings or adopt your standard peptide synthesis procedures.

5. If you perform SPPS manually, you migth check completeness by performing the *Kaiser* test (*see* **Note 5**).
6. Repeat **steps 3–5** for each individual amino acid of the sequence. Incorporate Fmoc-Lys(Dde)-OH in position 56 (*see* **Notes 6** and 7).
7. Remove Dde protecting group through treatment with 0.5 mL of 2 % hydrazine/DMF (v/v) solution for 10 min, 10 times. Check completeness of deprotection by spectroscopic measurement at 301 nm (*see* **Note 8**).
8. Coupling of amino-reactive probe/dye at the liberated amino group. Activate building block or use preactivated building blocks. Coupling conditions must be optimized individually.
9. Full cleavage of the peptide was performed by treatment with 1 mL trifluoroacetic acid (TFA), thioanisole (TA), and ethanedithiol (EDT) (TFA/TA/EDT - 90/7/3) for 3 h at room temperature.
10. Precipitate peptide from ice-cold Et_2O. Peptide appears as white, cloudy solid after a few minutes. If not, allow the solution to stand overnight at −20 °C. Centrifuge and wash pellet at least five times with 1 mL ice-cold Et_2O until no thiol odor is detectable.
11. Evaporate residual Et_2O and analyze the peptide by RP-HPLC and mass spectrometry.
12. Lyophilize the peptide from 0.1 % aqueous TFA and store powder at 4 °C.

3.2 Cloning

Preliminary to your protein work, you have to construct your expression vector carrying the sequence of [Ala^{49}]-SDF-1(1–49) in the same open reading frame as the intein and the chitin-binding domain. We recommend the pTXB1 vector from *New England Biolabs* (NEB). Procedures are not described in detail, because of various suppliers and devices.

1. Perform a standard PCR to amplify [Ala^{49}]-SDF-1(1–49) cDNA with NdeI and SapI restriction sites flanking the encoding sequence. Consider: Primers introduce a codon which encodes for an Ala at position 49.
2. Digest your plasmid as well as your PCR product with NdeI and SapI, respectively. Afterwards, dephosphorylate your digested plasmid DNA and run a 1 % agarose gel to purify

linearized plasmid and insert DNA. Cut out gel slices of interest and extract DNA (e.g., PCR/gel clean-up kit).

3. Ligate pure plasmid and insert DNA and transform *E. coli* DH5α cells directly. Plate cells on agar containing ampicillin100 μg/mL and incubate overnight at 37 °C.
4. Pick individual clones, extract plasmid DNA (e.g., DNA preparation kit) and verify correct sequence, and open reading frame by DNA sequencing.
5. Select a positive clone and transform *E. coli* ER2566 cells. Plate cells on agar containing ampicillin100 μg/mL and incubate overnight at 37 °C. Prepare a glycerol stock and store at −80 °C.

3.3 Protein Thioester Preparation

1. Precultivate pTXB1-[Ala^{49}]-SDF-1(1–49) transformed *E. coli* ER2566 cells in 200 mL LB media containing 100 μg/mL ampicillin overnight at 37 °C.
2. Spin down pre-cultured cells 5 min at 5,000 × *g* and resuspend them in LB media containing ampicillin (100 μg/mL). Adjust 4 L LB media (containing 100 μg/mL ampicillin) in flasks with the prepared suspension to an optical density at 600 nm (OD_{600}) of 0.1. Incubate culture at 37 °C until OD_{600} reaches 0.8–1 and induce protein expression with 1 mM IPTG.
3. Allow bacteria cells to express protein for 4–6 h. Take out samples (100 μL) hourly and store them at −20 °C.
4. Centrifuge cell suspension (5,000 × *g*, 4 °C) and discard supernatant. Resuspend pellet in 80 mL BB.

   Perform following steps on ice or at 4 °C unless stated otherwise.
5. Break down cells by applying a *French press* device until homogeneity (*see* **Note 9**).
6. DNA digestion. In order to reduce viscosity, treat suspension with DNase I in the presence of 5 mM $MgCl_2$. Incubate suspension for 30 min at room temperature under shaking.
7. Centrifuge suspension (12,000 × *g*, 45 min), take out a sample (100 μL), and freeze supernatant. Wash pellet twice with 10 mL WB.
8. Solubilize inclusion bodies with 20 mL SB and stir suspension moderately overnight.
9. Centrifuge suspension (208,000 × *g*, 35 min), take out sample (50 μL), and freeze supernatant.
10. Repeat solubilization with 10–20 mL SB and store separately.
11. Run a 12 % SDS-PAGE to check samples for expression progress and location of the fusion protein. Expect the protein in inclusion bodies; however, it might be found in the soluble fraction as well. Don't throw away supernatant until you are sure where your protein is expressed!

12. Prepare two chitin column for affinity-chromatographic purification of expressed fusion protein. Therefore, fill slurry of chitin beads in a column and allow them to decant (final bed volume 15 mL). Wash chitin beads with water (20 bed volumes) and equilibrate with column buffer containing 3 M urea (10 bed volumes).
13. Slowly dilute solubilized inclusion bodies to a final concentration of 3 M urea with BB and load the solution onto the prepacked chitin column. Flow rate should not exceed 1 mL/min. Further dilute flow-through with BB down to 2 M urea and perform a second loading step. Take sample (100 μL).
14. Wash loaded column with 20 bed volumes WB and drain remaining buffer.
15. Thiol-induced cleavage of the protein. Add two volumes of CB and allow them to flow through rapidly. Add one more bed volume of CB and incubate overnight at 4 °C.
16. On the next day, elute protein thioester in 5 mL fractions and add 3 more bed volumes WB to wash the column (collect in 5 mL fractions, too). Decrease pH of the fractions to 7 and store them separately. Take out samples (50 μL). Repeat **step 15** if induced cleavage was not satisfying.
17. Analyze samples by 15 % SDS-PAGE to check loading and cleavage efficiency. Combine protein thioester containing fractions and verify identity with mass spectrometry. Include always a sample of beads, because protein thioester might precipitate that will in turn reduce yield, if not realized.
18. Concentrate protein thioester by application of Amicon Ultra-15 filter device (MW cut-off 3 kDa). Exchange buffer to ligation buffer by three repeating cycles of adding ligation buffer and concentrate to the minimum volume (*see* **Note 10**).
19. Store aliquots of around 500–1,000 μL at −80 °C until usage. The average yield of purified [Ala^{49}]-SDF-1(1–49)-MESNA is around 4 mg/L expression culture.

3.4 Ligation

1. Thaw one aliquot of dissolved thioester [Ala^{49}]-SDF-1(1–49)-MESNA (Subheading 3.3) on ice and determine protein concentration by spectrometric measurement (dilution in BB, ε_{280} = 1,490/M^{-1}/cm^{-1}). Therefore dilute a sample immediately before measurement in order to prevent aggregation. Alternatively, determine protein amount by performing a *Bradford* assay. Concentration ranges between 5 and 10 mg/mL.
2. Add 2–3 equiv. C-terminal peptide (Subheading 3.1) to the chilled thioester solution and stir gently. Add TCEP to a final concentration of 5 mM and incubate 30 min at room temperature.
3. Allow solution to warm up to 30 °C and start reaction by raising pH to 7.8. Incubate mixture under gently stirring at 30 °C.

4. To monitor reaction progress, take out samples (5 μL), dilute 1:10 with 10 % acetic acid, and analyze by analytical RP-HPLC (*see* **Note 11**). Verify product formation by mass spectrometry.
5. When reaction completed, dilute reaction mixture to 5 mL with 10 % acetic acid, filtrate with a 0.45 μm filter, and purify mixture by using preparative RP-HPLC.
6. Lyophilize product containing fractions.

3.5 Refolding

1. Modified [Ala^{49}]-SDF-1α is refolded by rapid dilution. Therefore, dissolve the protein to a final concentration of 5–10 mg/mL in 100 mM phosphate buffer, 6 M guanidinium chloride, pH 6.
2. Prepare refolding buffer (100 volumes) *without GSH/GSSG redox system* and allow the solution to reach 4 °C under moderate stirring. Afterwards, add redox system (0.3 mM each), control pH, and readjust if necessary.
3. Add protein solution slowly, dropwise to the refolding buffer. After 100 μL wait 30 min and add next 100 μL. Make sure that drops disrupt properly and distribute homogenously before adding the next.
4. Incubate refolding mixture under the same conditions at least for 16 h. Monitor refolding progress by RP-HPLC. Typically, a shift in retention time is observed (*see* **Note 12**).
5. Acidify solution (pH 3) with HCl and concentrate by using Amicon Ultra-15 filter device to around 5 mL.
6. Filtrate solution (0.45 μm filter) and separate refolding products with RP-HPLC.
7. Pool fractions and analyze by high resolution mass spectrometric techniques as FT-ICR-ESI-MS.
8. Lyophilize correct folded protein and store at 4 °C.
9. Protein is reconstituted with favored buffer or simply water.

The method was used, as described or adapted, to prepare human IL-8 [32, 33] and SDF-1 analogs [30, 31]. For ligation of enzymes, which require native conditions, check reference [34].

4 Notes

1. *Precaution*: Solvents and reagents might be harmful! Please read material safety data sheets (MSDS) carefully and follow recommended safety instructions. Consider national regulations for biological work and handling of genetically modified organisms.
2. In our lab, an analytical HPCL system *LaChrome* (Merck Hitachi) composed of a diode array detector L7455, autosampler L7200, pump 7100, column oven L2350, and a analytical

C18 column (250 mm×4.6 mm, 5 μm, 300 Å, Grace Vydac) is operated for routine sample measurement. Our preparative HPLC system (Shimadzu) is equipped with a preparative (250 mm×21 mm, 10 μm, 90 Å) and a semi-preparative (250 mm×10 mm, 5 μm, 300 Å) C18 column (Phenomenex).

3. Mass spectrometric analysis is routinely performed on an *Ultraflex III* MALDI-TOF/TOF mass spectrometer (Bruker Daltonics).
4. In the general procedure manual peptide synthesis is described. *Kaiser* test (*see* **Note 5**) should be performed after each amino acid coupling ensuring completeness of the reaction. In our lab, we use a *Syro II* robot (MultiSynTech) for automated SPPS. If you are equipped with a robot, use your established standard protocols for peptide preparation, because the sequence is not known to be difficult.
5. *Kaiser* test is a rapid, qualitative ninhydrin-based test for detection of free primary amines. Put a few beads of your resin in a 1.5 mL tube and add 1–2 drops of *Kaiser* test solutions I-III. Incubate 5 min at 95 °C. Blue solution indicates the presence of free amines. Notice: The test does not work in case of proline, because of its nature as secondary amine!
6. You can modify any position different from Lys^{56}. As an alternative we suggest lysine on position 68 [35]. Consider for structural reasons that the modification should face the solution. Nonetheless, incorporate Fmoc-Lys(Dde)-OH moiety at the position you later want to introduce the modification!
7. The Dde protection group can be cleaved selectively and orthogonal to other protecting groups using 2 % hydrazine. The resulting amino group at the lysine side chain can be used for further modifications, e.g., introduction of dyes, further peptidic structures, or posttranslational modifications.
8. To check for complete Dde deprotection, perform absorbance measurements at 301 nm. Collect cleavage mixture after first deprotection step, wash the resin with 1 mL DMF, and combine both solutions (final volume 1.5 mL). Do the same after the last deprotection step. High absorbance values (>10) are expected for the first cleavage step, whereas absorption should not exceed 0.1 for the last. Important: Unsaturated hydrocarbons, like the alkene functionality of the Aloc protecting group, are reported to be effected by this cleavage reaction [36]. Therefore, the addition of 200 equiv. allyl alcohol prevents these moieties from reduction.
9. If you do not own a *French press,* apply sonication or break down cells by repeating freeze-and-thaw cycles. Avoid lysozyme for cell lysis, because it will damage and drastically reduce the life-span of your chitin beads!
10. Other protein concentrator devices might be useful as well, especially for larger volumes. For that we use the Minimate

tangential flow filtration (TFF) system (Pall Corporation) to concentrate volumes up to 500 mL.

11. Sometimes ligation proceeds slow and therefore give low yields. To accelerate reactivity, add catalytical amounts (0.1 equiv.) of sodium thiophenolate which forms more reactive thioesters *in situ*. To avoid rapid hydrolysis perform reaction at pH 7–8. Increase temperature while decreasing pH and vice versa.
12. If the given rapid dilution protocol does not work out, reduce the final protein concentration (e.g., double refolding buffer to 200 volumes). Make sure your unfolded protein is pure (≥90 %). Alternatively, refolding by dialysis can be tried [30, 31], however, we obtained our best result with the described rapid dilution procedure. If the protocol is transferred to other proteins, optimization of refolding work-up might be necessary. We recommend screening various concentrations and ratios of salts, supplements, pH values, and redox systems in analytical scale, similar as described in [37].

References

1. Schnolzer M, Kent SB (1992) Constructing proteins by dovetailing unprotected synthetic peptides: backbone-engineered HIV protease. Science 256:221–225
2. Dawson PE et al (1994) Synthesis of proteins by native chemical ligation. Science 266:776–779
3. Hackeng TM, Griffin JH, Dawson PE (1999) Protein synthesis by native chemical ligation: expanded scope by using straightforward methodology. Proc Natl Acad Sci U S A 96:10068–10073
4. Wieland T et al (1953) Ueber Peptidsynthesen. 8. Bildung von S-haltigen Peptiden durch intramolekulare Wanderung von Aminoacylresten. Liebigs Ann Chem 583:129–149
5. Zhang LS, Tam JP (1997) Orthogonal coupling of unprotected peptide segments through histidyl amino terminus. Tetrahedron Lett 38:3–6
6. Raines RT, Hondal RJ, Nilsson BL (2001) Selenocysteine in native chemical ligation and expressed protein ligation. J Am Chem Soc 123:5140–5141
7. Bang D et al (2006) Dissecting the energetics of protein alpha-helix C-cap termination through chemical protein synthesis. Nat Chem Biol 2:139–143
8. Bang D et al (2005) Total chemical synthesis and X-ray crystal structure of a protein diastereomer: [D-Gln 35]ubiquitin. Angew Chem Int Ed Engl 44:3852–3856
9. Yan LZ, Dawson PE (2001) Synthesis of peptides and proteins without cysteine residues by native chemical ligation combined with desulfurization. J Am Chem Soc 123:526–533
10. Bernardes GJ et al (2008) Facile conversion of cysteine and alkyl cysteines to dehydroalanine on protein surfaces: versatile and switchable access to functionalized proteins. J Am Chem Soc 130:5052–5053
11. Wan Q, Danishefsky SJ (2007) Free-radical-based, specific desulfurization of cysteine: a powerful advance in the synthesis of polypeptides and glycopolypeptides. Angew Chem Int Ed Engl 46:9248–9252
12. Crich D, Banerjee A (2007) Native chemical ligation at phenylalanine. J Am Chem Soc 129:10064–10065
13. Chen J et al (2008) Native chemical ligation at valine: a contribution to peptide and glycopeptide synthesis. Angew Chem Int Ed Engl 47:8521–8524
14. Haase C, Rohde H, Seitz O (2008) Native chemical ligation at valine. Angew Chem Int Ed Engl 47:6807–6810
15. Liu XW et al (2009) Dual native chemical ligation at lysine. J Am Chem Soc 131:13592–13593
16. Harpaz Z et al (2010) Protein synthesis assisted by native chemical ligation at leucine. ChemBioChem 11:1232–1235
17. Chen J et al (2010) A program for ligation at threonine sites: application to the controlled total synthesis of glycopeptides. Tetrahedron 66:2277–2283
18. Danishefsky SJ et al (2011) An advance in proline ligation. J Am Chem Soc 133:10784–10786

19. Brik A et al (2006) Sugar-assisted glycopeptide ligation. J Am Chem Soc 128:5626–5627
20. Ficht S et al (2007) Second-generation sugar-assisted ligation: a method for the synthesis of cysteine-containing glycopeptides. Angew Chem Int Ed Engl 46:5975–5979
21. Yang YY et al (2007) Sugar-assisted ligation in glycoprotein synthesis. J Am Chem Soc 129:7690–7701
22. Chong S et al (1997) Single-column purification of free recombinant proteins using a self-cleavable affinity tag derived from a protein splicing element. Gene 192:271–281
23. Chong S et al (1998) Utilizing the C-terminal cleavage activity of a protein splicing element to purify recombinant proteins in a single chromatographic step. Nucleic Acids Res 26:5109–5115
24. Hirata R et al (1990) Molecular structure of a gene, VMA1, encoding the catalytic subunit of H(+)-translocating adenosine triphosphatase from vacuolar membranes of Saccharomyces cerevisiae. J Biol Chem 265:6726–6733
25. Kane PM et al (1990) Protein splicing converts the yeast TFP1 gene product to the 69-kD subunit of the vacuolar H(+)-adenosine triphosphatase. Science 250:651–657
26. Klabunde T et al (1998) Crystal structure of GyrA intein from *Mycobacterium xenopi* reveals structural basis of protein splicing. Nat Struct Biol 5:31–36
27. David R, Richter MP, Beck-Sickinger AG (2004) Expressed protein ligation. Method and applications. Eur J Biochem 271:663–677
28. Muir TW, Sondhi D, Cole PA (1998) Expressed protein ligation: A general method for protein engineering. Proc Natl Acad Sci U S A 95:6705–6710
29. Southworth MW et al (1999) Purification of proteins fused to either the amino or carboxy terminus of the *Mycobacterium xenopi* gyrase A intein. Biotechniques 27:110–114, 116, 118–120
30. Baumann L, Beck-Sickinger AG (2010) Identification of a potential modification site in human stromal cell-derived factor-1. Biopolymers 94:771–778
31. Bellmann-Sickert K, Baumann L, Beck-Sickinger AG (2010) Selective labelling of stromal cell-derived factor 1alpha with carboxyfluorescein to study receptor internalisation. J Pept Sci 16:568–574
32. David R, Beck-Sickinger AG (2007) Identification of the dimerisation interface of human interleukin-8 by IL-8-variants containing the photoactivatable amino acid benzoylphenylalanine. Eur Biophys J 36:385–391
33. David R, Machova Z, Beck-Sickinger AG (2003) Semisynthesis and application of carboxyfluorescein-labelled biologically active human interleukin-8. Biol Chem 384:1619–1630
34. Steinhagen M et al (2011) Simultaneous “one pot” expressed protein ligation and CuI-catalyzed azide/alkyne cycloaddition for protein immobilization. ChemBioChem 12:2426–2430
35. Bellmann-Sickert K, Beck-Sickinger AG (2011) Palmitoylated SDF1alpha shows increased resistance against proteolytic degradation in liver homogenates. ChemMedChem 6:193–200
36. Dumy P et al (1998) Hydrazinolysis of Dde: complete orthogonality with Aloc protecting groups. Tetrahedron Lett 39:1175–1178
37. Dechavanne V et al (2011) A high-throughput protein refolding screen in 96-well format combined with design of experiments to optimize the refolding conditions. Protein Expr Purif 75:192–203

Chapter 8

Synthesis of C-Terminal Peptide Thioesters Using Fmoc-Based Solid-Phase Peptide Chemistry

Pernille Tofteng Shelton and Knud J. Jensen

Abstract

This chapter provides two protocols for the solid-phase synthesis of peptide thioesters using N^{α}-*Fmoc-protected* amino acids. The first protocol is based on a so-called safety-catch linker, while the second relies on a backbone amide linker.

Key words Thioesters, Safety-catch linker, Backbone amide linker, BAL

1 Introduction

C-terminal peptide thioesters are used extensively in synthetic protein chemistry, especially for native chemical ligation (NCL) and other chemoselective reactions, which has inspired a search for robust synthetic strategies. Initially, peptide thioesters were mainly prepared using solid-phase peptide synthesis with amino acids N^{α}-*protected* with Boc (Boc-SPPS) [1–3], see Chapter 4. However, this technique requires specialized equipment for handling of hydrofluoric acid (HF) for release of the peptide from the resin, and it is therefore currently not used in many laboratories. Furthermore, the HF treatment is incompatible with many post-translational modifications such as glycosylations or phosphorylations [4]. Boc-SPPS is described thoroughly in Chapter 4.

The sp^2-carbon in the resin-bound thioesters is susceptible to nucleophilic attack when treated repeatedly with piperidine; thus, the thioester can normally not be maintained during Fmoc-SPPS. However, modified Fmoc-SPPS have been reported where the thioester moiety either is masked during SPPS or is being installed in the final step. Initially, non-nucleophilic bases for the cleavage of the Fmoc groups were used [5–7]. By applying these alternative deprotection agents, the thioesters were not decomposed by aminolysis during the N^{α} deprotection, and the thioester was retained.

Knud J. Jensen et al. (eds.), *Peptide Synthesis and Applications: Second Edition*, Methods in Molecular Biology, vol. 1047, DOI 10.1007/978-1-62703-544-6_8, © Springer Science+Business Media New York 2013

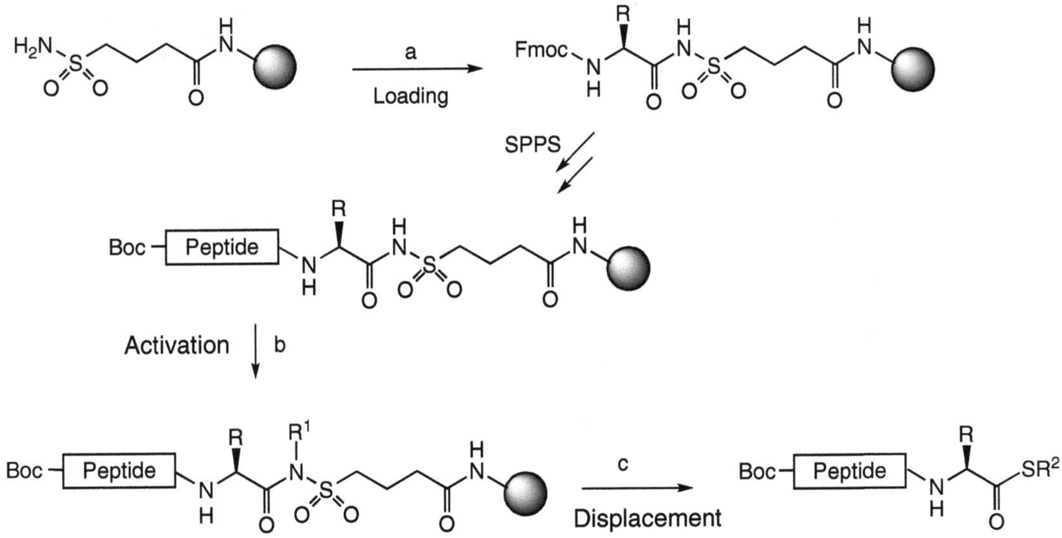

Fig. 1 The SCL strategy for the synthesis of peptide thioesters, (*a*) Fmoc-Aaa-OH, PyBOP, DIEA; (*b*) TMS-CHN_2 or ICH_2CN/DIEA; (*c*) HSR^2, NaSPh

However, utilizing these modified deprotection conditions did not completely eliminate thioester degradation, and the method has not been widely used for the assembly of longer peptide thioesters [8]. The synthesis of fully protected peptides and their subsequent conversion to thioesters in solution have been performed using the acid-labile 2-chlorotrityl resin in good yields [9, 10]. Unfortunately, this approach can lead to epimerization of the C-terminal residue, when it is activated and converted to an active ester prior to formation of the thioester.

A broad selection of strategies for the synthesis of peptide thioesters has been described in the literature, which includes the use of the aryl hydrazine linker [11]; the *N*-acylurea approach [12, 13]; side-chain anchoring of some trifunctional C-terminal residues, i.e., Lys, Glu, Gln, Asp or Asn, and Ser or Thr [14, 15]; and methodologies to obtain thioester intermediates through in situ *O*–*S* acyl shift [16–18] and *N*–*S* acyl shift [19–21]. Each of these methods overcomes some of the difficulties involved in the synthesis of peptide thioesters by the Fmoc strategy. Unfortunately, many of them are not generally applicable and new, more general methods are of interest. In 2009 our group published an approach for thioester synthesis by Fmoc chemistry utilizing the formation of a pyroglutamyl imide structure on resin followed by thiolysis to yield the thioester [22].

Strategies for the synthesis of peptide thioesters utilizing special linkers have been reported, of which the most widely used is the safety-catch linker (SCL). Upon activation and nucleophilic displacement, it provides fully protected peptide thioesters in solution [23] (Fig. 1). It is commonly used in Fmoc-SPPS and is the

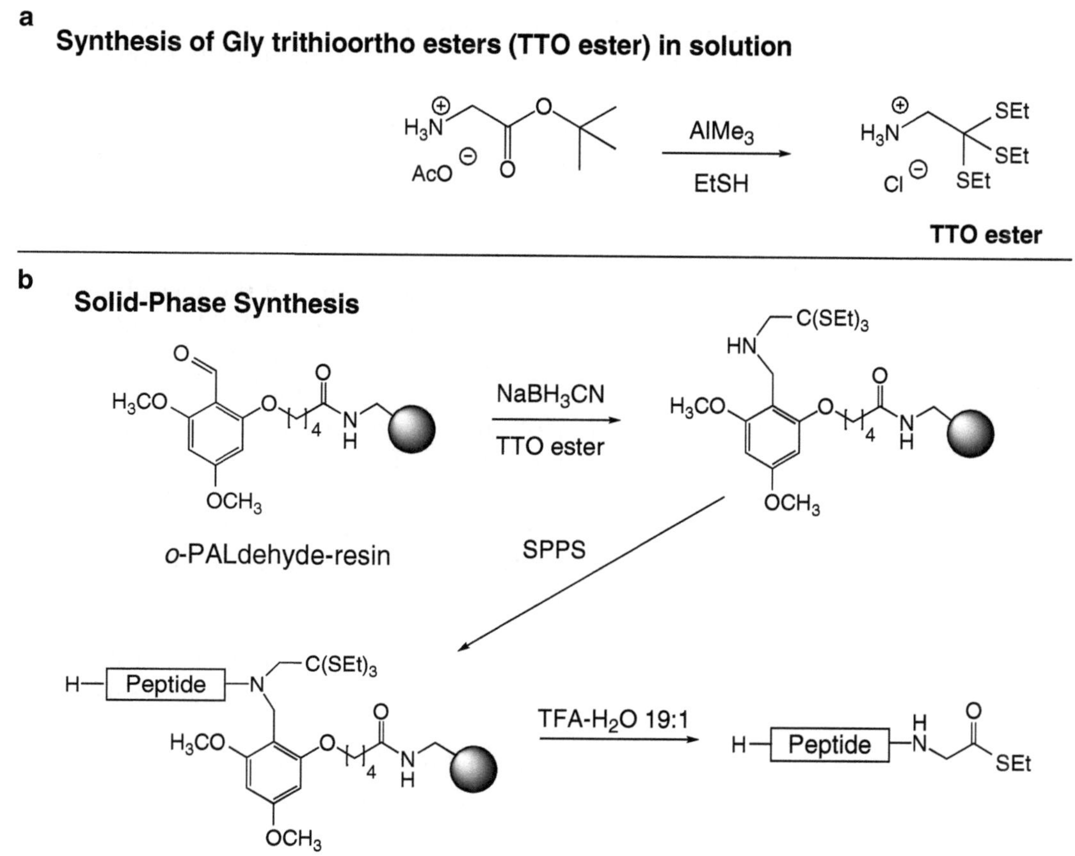

Fig. 2 Synthesis of peptide thioesters with a BAL TTO strategy (**a**) Synthesis of TTO reagent in solution; (**b**) anchoring of TTO reagent through BAL handle

main focus of this chapter. Furthermore, the chapter includes a protocol for the formation of C-terminal thioesters using the backbone amide linker (BAL) strategy (Chapter 9) which anchors the peptide through the backbone nitrogen and thereby leaves the C-terminal free for modification [24] (Fig. 2). The thioester is masked as a trithioortho ester during synthesis. This BAL strategy is limited to the synthesis of peptide thioesters with C-terminal Gly residues; however, a C-terminal Gly is often desirable.

1.1 The Safety-Catch Linker Strategy

The safety-catch linker (SCL) strategy utilizes a modified version of Kenner's acylsulfonamide linker [23, 25–27]. An aliphatic sulfonamide has proven superior to the aryl sulfonamide, due to the enhanced nucleophilicity which correlates with the increased basicity of the aliphatic acylsulfonamide anion [28]. The 3-carboxypropane-sulfonamide linker is stable under basic and strongly nucleophilic conditions but can be activated by the reaction with trimethylsilane-diazomethane (TMS-diazomethane) or iodoacetonitrile to form the *N*-alkyl acylsulfonamide. Nucleophilic attack by a thiol creates with the thioester with concomitant release as fully protected peptide thioester, which subsequently can be side chain deprotected in solution (Fig. 1). By changing the nucleophile in the displacement step, the strategy can also be applied for the synthesis of peptides with other C-terminal modifications such as amides or esters [29]. Numerous peptide thioesters have been prepared by the SCL strategy including a 24-residue glycopeptide thioester [4, 30] and phosphopeptide α-thioesters [31, 32], among others [29].

There are two reagents for activation of the acylsulfonamide linker by alkylation, iodoacetonitrile/DIEA and TMS-diazomethane, with the latter providing superior yields [23, 33]. The subsequent nucleophilic displacement has been performed using a large variety of thiols, with benzyl mercaptan and ethyl-3-mercaptopropionate being the most common. Sodium thiophenolate, which was originally recommended by Pessi et al., has only been used in a few cases, but in our laboratories, it provided increased yields compared to the most common [23, 33]. A variation of the sulfamylbutyryl resin has been suggested where the linker is combined with the acid-labile Rink Amide AM resin. This enables sequential cleavage with TFA and facilitates more control of the synthesis in particular the alkylation reaction [30, 34].

Loading of the first building block unto the safety-catch resin was initially reported to give excellent yields (64–93 %, except for Fmoc-Thr(O*t*Bu)-OH and Fmoc-Pro-OH), with a single PyBOP treatment [27]. A PyBOP coupling at −20 °C with a reaction time of <8 h was used in order to minimize epimerization. Nevertheless, insufficient loading level has been a problem, and alternative coupling conditions such as use of Fmoc amino acid fluorides have been suggested [35]. Fortunately, preloaded safety-catch linker resins are today commercially available.

1.2 The BAL TTO Strategy

Peptides with a C-terminal Gly thioester can be prepared by a BAL strategy, where the sp^2 carbonyl carbon of the Gly thioester is masked as the sp^3 carbon of a trithioortho (TTO) ester during peptide chain assembly. The key TTO derivative is obtained by treatment of the Gly ester with "$Al(SEt)_3$" in solution. The resulting glycine TTO ester, $HCl{\cdot}H_2NCH_2C(SEt)_3$, is subsequently anchored to resin-bound trialkoxybenzyl BAL by reductive amination. After peptide chain assembly, the final treatment with TFA-H2O will convert the TTO ester to the thioester with simultaneous release from the BAL handle (Fig. 2). This is a general approach to the synthesis of peptide thioesters with a C-terminal Gly residue; peptide thioesters used in native chemical ligation often incorporate a sterically undemanding C-terminal Gly residue.

2 Materials

2.1 The Safety-Catch Linker Strategy

1. 4-Sulfamylbutyryl AM NovaGel (Loading 0.67 mmol/g, Novabiochem by Merck Millipore).
2. Fmoc- and Boc-protected amino acids.
3. Solvents:
   (a) Dichloromethane (DCM).
   (b) *N,N*-Dimethylformamide (DMF).
   (c) *N*-Methyl-2-pyrrolidinone (NMP).
   (d) Methanol (MeOH).
   (e) Tetrahydrofuran (THF).
   (f) Trifluoroacetic acid (TFA).
   (g) Piperidine.
4. Coupling reagents:
   (a) Diisopropylcarbodiimide (DIC).
   (b) *N*-[(1*H*-Benzotriazol-1-yl)(dimethylamino)methylene]-*N*-methylmethanaminium hexafluorophosphate *N*-oxide (HBTU).
   (c) 1-[(1-(Cyano-2-ethoxy-2-oxoethylideneaminooxy)-dimethylamino-morpholinomethylene)]methanaminium hexafluorophosphate (COMU).
   (d) *N*-[(Dimethylamino)-1*H*-1,2,3-triazolo[4,5-*b*]pyridin-1-ylmethylene]-*N-methylmethanaminium* hexafluorophosphate *N*-oxide (HATU).
   (e) 1-Hydroxybenzotriazole (HOBt).
   (f) Ethyl(2-cyano-2-(hydroxyimino)acetate) (Oxyma).
   (g) Benzotriazol-yloxytris(pyrrolidino)phosphonium hexafluorophosphate (PyBOP).
5. Other reagents:
   (a) *N,N*-Diisopropylethylamine (DIEA).
   (b) Triethylsilane (TES).
   (c) 1 M TMS-CHN_2 in hexane-THF (1:1) (Sigma-Aldrich).

2.2 The Trithioortho Ester Strategy

1. Aminomethylated Tentagel resin, loading 0.29 mmol/g (Rapp Polymere).
2. *o*-PALdehyde: 5-(2-formyl-3,5-dimethoxyphenoxy)pentanoic acid (Peptides International).
3. Fmoc- and Boc-protected amino acids.
4. Glycine trithioortho esters: not commercially available but can be synthesized according to a protocol in Subheading 3.2.1 [24].
5. Sodium cyanoborohydride ($NaBH_3CN$).

6. Coupling reagents: DIC, HBTU, HOBt, Oxyma (*see* Subheading 2.1).
7. Solvents: DMF, NMP, DCM, TFA, DIEA, piperidine (*see* Subheading 2.1).
8. Triethylsilane (TES).
9. Microwave instrument: Biotage Initiator using Biotage glass vials or their new filtered glass vial which connects to the initiator peptide workstation.

3 Methods

In the following section, two strategies for peptide thioester synthesis will be described: (1) the safety-catch linker (SCL) strategy and (2) the backbone amide linker (BAL) strategy using trithioortho ester building blocks.

3.1 Synthesis of C-Terminal Peptide Thioesters by Kenner's Safety-Catch Linker Strategy

3.1.1 Loading of the First Amino Acid to the Safety-Catch Linker

This initial step can be omitted by purchasing preloaded resin from different manufactures such as Merck/Novabiochem.

1. Place the 4-sulfamylbutyryl AM NovaGel resin (loading 0.67 mmol/g; 0.1 mmol scale) in a 10 mL polypropylene syringe fitted with a polyethylene filter.
2. Add Fmoc-Aaa-OH (5 equiv.), DIEA (10 equiv.), and 2.5 mL of DCM and shake the resin for 20 min.
3. Cool the mixture to −18 °C using an ice bath.
4. When the resin mixture is sufficiently cooled, add PyBOP (5 equiv.) while mixing thoroughly.
5. Then the resin mixture is placed on the ice bath to keep the temperature down and left overnight (approx. 15–20 h).
6. Remove the excess reagents by filtration.
7. Wash the resin with DCM (3 × 2 mL), NMP (3 × 2 mL), and DCM (3 × 2 mL).
8. Dry a small amount of resin for Fmoc quantification (*see* **Notes 1** and **2**).
9. If the resin loading is below 30 %, repeat **steps 2–5**. The determined resin loading is used in the following chemical calculations.

3.1.2 Chain Elongation Using Standard SPPS Fmoc Chemistry

After the initial loading of the first amino acid onto the SCL resin, the remaining of the peptide assembly is well suited for automated peptide synthesis. The standard protocols for peptide chain assembly are described in Chapter 2.

1. Standard Fmoc chemistry, manual or automated (Chapter 2).
2. The N-terminal amino acid is incorporated as a *N*α-Boc-protected derivative.

3.1.3 Alkylation and Thioesterification

1. Place the resin (0.05 mmol) in a 10 mL syringe and wash with THF (2×2 mL).
2. Add THF (5 mL) to the resin and swell for 1 h.
3. Drain the resin by filtration and add 1 M TMS-CHN_2 in hexane-THF (1:1) (2.5 mL). The reaction mixture is shaken for 3 h before excess reagent is removed by filtration.
4. Wash the resin with THF (2×2 mL), DCM (2×2 mL), and DMF (2×2 mL).
5. Add ethyl 3-mercaptopropionate (50 equiv.), sodium thiophenolate (1 equiv.), and DMF (final concentration of 1 M) to the resin and mix well.
6. React for 18–24 h under shaking (*see* **Note 3**).
7. Collect the DMF-peptide filtrate in a round-bottom flask. Wash the resin with DMF (1×2 mL) and DCM (2×2 mL), collect the wash solution, and combine with the filtrate (*see* **Note 4**).
8. Remove the solvents (DMF, DCM, and thiol) by rotary evaporation. To remove traces of DMF, the round-bottom flask is placed under vacuum overnight (*see* **Note 5**).

3.1.4 Deprotection and Analysis

1. Add TFA-H_2O-TES (95:2.5:2.5) (3 mL) to the round-bottom flask and leave for 3 h.
2. Transfer the combined peptide/TFA solution to a 15 mL centrifuge tube and remove TFA by N_2.
3. When only a small amount of TFA is left, the peptide is precipitated with diethyl ether (8 mL), centrifuged before the diethyl ether is removed.
4. The peptide is air-dried for approx. 30 min before being dissolved in H_2O-CH_3CN and analyzed by LC-MS and lyophilized (*see* **Note 6**).

3.2 The BAL Strategy Using the Trithioortho Ester Building Block

The synthesis of C-terminal glycine peptide thioesters by the BAL strategy requires the trithioortho (TTO) ester building block. This building block is not commercially available but the synthesis is therefore described below. The chemistry below is almost identical to the chemistry described in the dedicated BAL chapter of this book (Chapter 9).

3.2.1 Synthesis of Trithioortho Esters in Solution [24]

1. Cool down dry DCM (50 mL) in a round-bottom flask to 0 °C under argon.
2. Add dropwise and carefully $AlMe_3$ in heptane (26 mmol) to the flask and leave for 15 min.
3. Dissolve H-Gly-OtBu, AcOH (2.6 mmol) in DCM (10 mL) and add to the flask.
4. Remove the cooling and stir for 18 h at room temperature.

5. Transfer to a large conical flask and quench excess reagent with ice (50 mL) while stirring.
6. Remove aqueous foam/layer in a separation funnel.
7. Extract the organic phase with 1 % HCl (4 × 50 mL).
8. Combine the aqueous phase and add solid NaOH until pH ~11.
9. Extract the basic solution with DCM (3 × 50 mL).
10. Combine the organic phases and dry with Na_2SO_3.
11. Remove the DCM to afford a yellow oil.
12. Dissolve the yellow oil in 1 % aqueous HCl and lyophilize to yield a white solid powder.
13. Yields of approx. 40–50 %. For analysis results by NMR and MS, *see* Brask et al. [24].

3.2.2 Anchoring of the o-PALdehyde Linker

1. Place the TG resin in a microwave vial with a filter (or a normal polystyrene syringe with a filter) (*see* **Note 7**) and swell with DCM for 20–30 min. Then wash the resin with DMF.
2. In a separate container (e.g., a 15 mL Falcon tube), measure out HBTU (1.8 equiv.), HOBt (2 equiv.) (*see* **Note 8**), o-PAldehyde (2 equiv.), and DIEA (4 equiv.) and add DMF to an overall concentration of 0.2 M. This means for a 0.1 mmol scale approx. 2 mL DMF.
3. Add the DMF mixture to the resin and close the vial with a septum before heating in the microwave for 10 min at 75 °C. The resin is filtrated to remove excess coupling reagent.
4. Repeat **steps 2–3**.
5. The resin is then washed with DMF (×3) and MeOH (×3).

3.2.3 Anchoring of the First Residue: Reductive Amination

1. In a microwave vial, suspend the o-PALdehyde TG resin in MeOH:AcOH (99:1) (approx. 7 mL for a 0.1 mmol scale) (*see* **Note 9**).
2. Add $NH_2CH_2C(SEt)_3{\cdot}HCl$ (5 equiv.) and $NaBH_3CN$ (5 equiv.) and seal the vial with a septum.
3. Heat in the microwave for 10 min at 60 °C before washing the resin thoroughly with MeOH (×3), DMF (×3) MeOH (×1).
4. Repeat **steps 2** and **3**, omitting the final MeOH wash.

3.2.4 Anchoring of the Second Residue: Acylation of Secondary Amine

1. Wash the resin with DCM.
2. Weigh out the Fmoc amino acid (10 equiv.) in a Falcon tube and add DMF (approx. 700 μL for 0.1 mmol scale). Then add DCM (6.3 mL for 0.1 mmol scale) and DIC (5 equiv.). Pre-activate for 10 min (*see* **Notes 10** and **11**).
3. Add the amino acid solution to the resin (in the microwave vial) and close with a septum.

4. React for 10 min at 60 °C in the microwave. Remove the excess coupling solution by filtration.
5. Repeat **steps 2–4**.
6. Wash the resin with DMF (×3) and DCM (×3).

3.2.5 Fmoc Quantification to Determine Loading of Dipeptide for Further Synthesis

In order to determine the quality of the dipeptidyl resin, a Fmoc quantification is performed (*see* Chapter 2).

3.2.6 Chain Elongation Using Standard SPPS: Suggestion for Automated Chemistry

After the on-resin synthesis of the first dipeptide, the remainder of the peptide assembly is well suited for automated synthesis. There are many different methodologies for Fmoc-SPPS, and for a more detailed description, *see* Chapter 2.

3.2.7 Release of Peptide in Solution

1. Wash resin with DCM and dry for 1–2 h under vacuum.
2. Add TFA-H_2O (19:1) (approx. 10 mL for a 0.1 mmol scale). React for 3 h (*see* **Note 12**).
3. Filtrate the resin and collect the TFA-peptide mixture. Wash the resin with TFA and DCM and collect the filtrates.
4. TFA and DCM can be removed under a stream of N_2 and the peptide precipitated with diethyl ether, centrifuged before the diethyl ether is removed. The ether treatment is repeated.
5. The peptide is air-dried for approx. 30 min before being dissolved in H_2O-CH_3CN and analyzed by LC-MS and lyophilized.

4 Notes

1. Fmoc quantification is a technique used to determine the amount of Fmoc present on the resin using UV absorptions (*see* Chapter 2 for more details). This will give a good estimate of the resin loading.
2. Different amino acids can be loaded with different efficiency on a 4-sulfamylbutyryl AM NovaGel resin. As an example, Fmoc-Ala-OH is loaded to approx. 50 % of original resin loading after one PyBOP treatment while Fmoc-Val-OH or Fmoc-Ile-OH is only loaded to 36 and 26 % of original resin loading.
3. The addition of the thiol gives a malodorous solution.
4. Remember that for this particular type of synthesis, the peptide is released by the addition of the thiol, and the peptide is therefore present in the DMF mixture at this stage.
5. The peptide can be analyzed at this stage, but solubility of fully protected peptides can be problematic. Furthermore, MS analysis of fully protected peptides is not always possible using ESI-MS.

6. Peptide thioesters in aqueous solution are prone to hydrolysis. Store peptide thioesters if possible as lyophilized powder.
7. The protocols are designed to use a microwave glass vial with a filter (Biotage Initiator peptide workstation). If this equipment is not available, then a normal polystyrene syringe with a filter can be used instead. The resin is then transferred to a microwave vial before each microwave-assisted reaction.
8. Access to HOBt has been limited due to restrictions on air transportation. As an alternative, Oxyma can be used as an additive. Alternatively, the use of additives can be eliminated completely and HBTU can be used alone. Furthermore, other coupling reagent such as HATU or COMU can also be used.
9. If the reaction is performed in an open microwave vial system, use DMF:AcOH (99:1) instead.
10. If the reaction is performed in an open microwave vial system, use DCE instead of DCM.
11. It can be problematic to dissolve some amino acids in DCM. Therefore 10 % DMF is added. Do not use more than 10 % DMF or it can negatively affect the reaction outcome.
12. The resin with the TFA cleavage mixture is stirred manually 3–5 times during the reaction.

References

1. Hojo H, Aimoto S (1991) Polypeptide synthesis using the S-Alkyl thioester of a partially protected peptide segment. Synthesis of the DNA-binding domain of *c*-Myb protein (142-193)-NH_2. Bull Chem Soc Jpn 64:111
2. Camarero JA, Cotton GJ, Adeva A, Muir TW (1998) Chemical ligation of unprotected peptides directly from a solid support. J Pept Res 51:303–316
3. Schnölzer M, Alewood P, Jones A, Alewood D, Kent SBH (1992) In situ neutralization in Boc-chemistry solid phase peptide synthesis. Int J Pept Protein Res 40:180–193
4. Shin Y, Winans KA, Backes BJ, Kent SBH, Ellman JA, Bertozzi CR (1999) Fmoc-based synthesis of peptide-αthioesters: application to the total chemical synthesis of a glycoprotein by native chemical ligation. J Am Chem Soc 121:11684–11689
5. Li X, Kawakami T, Aimoto S (1998) Direct preparation of peptide thioesters using an Fmoc solid-phase method. Tetrahedron Lett 39:8669–8672
6. Bu X, Xie G, Law CW, Guo Z (2002) An improved deblocking agent for direct Fmoc solid-phase synthesis of peptide thioesters. Tetrahedron Lett 43:2419–2422
7. Clippingdale AB, Barrow CJ, Wade JD (2000) Peptide thioester preparation by Fmoc solid phase peptide synthesis for use in native ligation. J Pept Sci 6:225–234
8. Alsina J, Albericio F (2003) Solid-phase synthesis of C-terminal modified peptides. Biopolymers 71:454–477
9. Futaki S, Sogawa K, Maruyama J, Asahara T, Niwa M (1997) Preparation of peptide thioesters using Fmoc-solid-phase peptide synthesis and its application to the construction of a template-assembled synthetic protein (TASP). Tetrahedron Lett 38:6237–6240
10. von Eggelkraut-Gottanka R, Klose A, Beck-Sickinger AG, Beyermann M (2003) Peptide αthioester formation using standard Fmoc-chemistry. Tetrahedron Lett 44:3551–3554
11. Camarero J, Hackel BJ, de Yoreo JJ, Mitchell AR (2004) Fmoc-based synthesis of peptide α-thioesters using an aryl hydrazine support. J Org Chem 69:4145–4151
12. Blanco-Canosa JB, Dawson PE (2008) An efficient Fmoc-SPPS approach for the generation of thioester peptide precursors for use in native chemical ligation. Angew Chem Int Ed 47:6851–6855

13. Mahto SK, Howard CJ, Shimko JC, Ottesen JJ (2011) A reversible protection strategy to improve Fmoc-SPPS of peptide thioesters by the N-acylurea approach. Chembiochem 12:2488–2494
14. Wang P, Miranda LP (2005) Fmoc-protein synthesis: preparation of peptide thioesters using a side-chain anchoring strategy. Int J Pept Res Ther 11:117–123
15. Li L, Wang P (2007) Transfer allyl esters to thioesters in solid phase condition: synthesis of peptide thioesters by Fmoc chemistry. Tetrahedron Lett 48:29–32
16. Botti P, Villain M, Manganiello S, Gaertner H (2004) Native chemical ligation through in situ O to S acyl shift. Org Lett 6: 4861–4864
17. Warren JD, Miller JS, Keding SJ, Danishefsky SJ (2004) Toward fully synthetic glycoproteins by ultimately convergent routes: a solution to a long-standing problem. J Am Chem Soc 126:6576–6578
18. Tofteng AP, Jensen KJ, Hoeg-Jensen T (2007) Peptide dithiodiethanol esters for in situ generation of thioesters for use in native ligation. Tetrahedron Lett 48:2105–2107
19. Ollivier N, Behr J-B, El-Mahdi O, Blanpain A, Melnyk O (2005) Fmoc solid-phase synthesis of peptide thioesters using an intramolecular N, S-acyl shift. Org Lett 7:2647–2650
20. Kawakami T, Sumida M, Nakamura K, Vorherr T, Aimoto S (2005) Peptide thioester preparation based on an N-S acyl shift reaction mediated by a thiol ligation auxiliary. Tetrahedron Lett 46:8805–8807
21. Nakamura K, Mori H, Kawakami T, Hojo H, Nakahara Y, Aimoto S (2007) Peptide thioester synthesis via an auxiliary-mediated N-S Acyl shift reaction in solution. Int J Pept Res Ther 13:191–202
22. Tofteng AP, Sørensen KK, Conde-Frieboes KW, Hoeg-Jensen T, Jensen KJ (2009) Fmoc solid-phase synthesis of C-terminal peptide thioesters by formation of a backbone pyroglutamyl imide moiety. Angew Chem Int Ed 48:7411–7414
23. Ingenito R, Bianchi E, Fattori D, Pessi A (1999) Solid phase synthesis of peptide C-terminal thioesters by Fmoc/*t*-Bu chemistry. J Am Chem Soc 121:11369
24. Brask J, Albericio F, Jensen KJ (2003) Fmoc solid-phase synthesis of peptide thioesters by masking as trithioortho esters. Org Lett 5:2951–2953
25. Kenner GW, McDermott JR, Sheppard RC (1971) J Chem Soc Chem Commun:636–637
26. Backes BJ, Virgilio AA, Ellman JA (1996) Activation method to prepare a highly reactive acylsulfonamide "safety catch" linker for solid phase synthesis. J Am Chem Soc 116:3055–3056
27. Backes BJ, Ellman JA (1999) An alkanesulfonamide "Safety-Catch" linker for solid-phase synthesis. J Org Chem 64:2322–2330
28. Bordwell FG (1988) Equilibrium acidities in dimethyl sulfoxide solutions. Acc Chem Res 21:456–463
29. Heidler P, Link A (2005) *N*-Acyl-*N*-alkyl-sulfonamide anchors derived from Kenner's safety-catch linker: powerful tools in bioorganic and medicinal chemistry. Bioorg Med Chem 13:585–599
30. Mezzato S, Schaffrath M, Unverzagt C (2005) An orthogonal double-linker resin facilitates the efficient solid-phase synthesis of complex type N-glycopeptide thioesters suitable for native chemical ligation. Angew Chem Int Ed 44:1650–1654
31. Flavell RR, Huse M, Goger M, Trester-Zedlitz M, Kuriyan J, Muir TW (2002) Efficient semisynthesis of a tetraphosphorylated analogue of the type I TGF β receptor. Org Lett 4:165–168
32. Huse M, Holford MN, Kuriyan J, Muir TW (2000) Semisynthesis of hyperphosphorylated type I TGF receptor: addressing the mechanism of kinase activation. J Am Chem Soc 122:8337
33. Biancalana S, Hudson D, Songster MA, Thompson SA (2001) Fmoc chemistry compatible thio-ligation assembly of proteins. Lett Pept Sci 7:291–297
34. Burlina F, Morris C, Behrendt R, White P, Offer J (2012) Simplifying native chemical ligation with an N-acylsulfonamide linker. Chem Commun 48:2579–2581
35. Ingenito R, Dreznjak D, Guffler S, Wenschuh H (2002) Efficient loading of sulfonamide safety-catch linkers by Fmoc amino acids Fluorides. Org Lett 4:1187–1188

Chapter 9

Backbone Amide Linker Strategy: Protocols for the Synthesis of C-Terminal Peptide Aldehydes

Pernille Tofteng Shelton and Knud J. Jensen

Abstract

In the backbone amide linker (BAL) strategy, the peptide is anchored not at the C-terminus but through a backbone amide, which leaves the C-terminal available for various modifications. This is thus a very general strategy for the introduction of C-terminal modifications. The BAL strategy was originally developed using a trisalkoxybenzyl linker, but since then range linkers (handles) with different properties have also been developed. The BAL anchoring is established by anchoring an aromatic aldehyde, typically a trisalkoxybenzaldehyde, to the solid support, followed by attachment of the first amino acid residue by reductive amination. This can be used as a general approach for the introduction of other C-terminal modifications as well as functionalities, such as fluorophors. The second step is an acylation of a secondary amine, followed by standard Fmoc-based solid-phase synthesis to assemble the final peptide. One useful application of this strategy is in the synthesis of C-terminal peptide aldehydes. The C-terminal aldehyde is masked as an acetal during synthesis and then conveniently demasked in the final cleavage step to generate the free aldehyde. Another application is in the synthesis of peptide thioesters with a C-terminal glycine.

Key words Backbone amide linker, Handle, Peptide, Aldehyde, Thioester, *N*-Alkyl amide, Cyclic peptide, Reductive amination

1 Introduction

In standard solid-phase peptide synthesis (SPPS), the growing peptide is anchored through the C-terminal carboxyl and then elongated in the N-terminal direction. However, this limits the kinds of C-termini that can be incorporated into peptides. Many biologically or medicinally interesting peptides require C-termini other than acids and simple amides, for example, aldehydes. The backbone amide linker (BAL) concept developed in the mid-1990s is a method to overcome this challenge by anchoring the peptide through the backbone nitrogen and leaving the C-terminal free for modifications [1–3] (Fig. 1). It has been comprehensively reviewed in [4].

Knud J. Jensen et al. (eds.), *Peptide Synthesis and Applications: Second Edition*, Methods in Molecular Biology, vol. 1047, DOI 10.1007/978-1-62703-544-6_9, © Springer Science+Business Media New York 2013

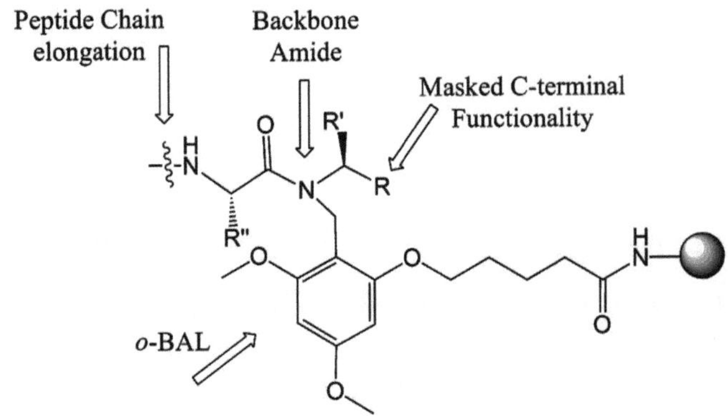

Fig. 1 Schematic outline of the concept of the backbone amide linker strategy shown for the *o*-BAL handle

A variety of linkers, starting with the tris(alkoxy)benzylamide system, have been developed for BAL-type chemistry. The most widely used linkers are 5-(4-formyl-3,5-dimethoxyphenoxy)-pentanoic acid (*p*-PALdehyde) or the butanoic acid derivative and the linker favored by our group, 5-(2-formyl-3,5-dimethoxyphenoxy) pentanoic acid (*o*-PALdehyde) [5–7]. An overview of the acid lability is presented in Table 1. In the BAL strategy, the first amino acid or amino acid derivative is anchored through a reductive amination (RA). This is followed by an acylation of the newly formed secondary amine before the peptide can be elongated through standard peptide amide couplings. The assembled peptide is often released by treatment with concentrated TFA which also removes acid-labile side-chain protecting groups; however, the *o-BAL* amide anchoring is very acid labile and can be cleaved with as little as 1% TFA in DCM, leaving the acid-labile protection groups intact on the peptide. Concomitant with the acidolytic release of the peptide, the C-terminal functionalities can be demasked to provide the desired functional groups. A variety of functional groups can be installed at the C-terminus, including alcohols, acids, ethers, esters, *N*-alkylamides, *N,N*-dialkylamides, hydrazides, trifluoromethyl ketones, and aldehydes [1, 8–10]. Examples of different C-terminal amino acid derivatives, which can be incorporated during synthesis, include aliphatic esters (for peptide esters), *t*-Bu ether (for peptide alcohols), *t*-Bu esters (for racemization-free peptide acids), dimethyl acetal (for peptide aldehydes), and trithioortho esters (for peptide thioesters, *see* Chapter 8). Furthermore, cyclic peptides can be synthesized by the use of an allyl ester [1]. A comprehensive review of the different BAL illustrated the widespread application of this technique which extends beyond the application in peptide chemistry [4].

One of the well-established applications of the BAL strategy is the synthesis of C-terminal peptide aldehydes [10]. Peptide aldehydes are the electrophilic partner in chemoselective reactions such

Table 1
The conditions for peptide release from the different BAL strategies

Linker\LG	Sulfonamide	Urea	Carbamate	Amide	Aniline	Amine
Monoalkoxy-benzyl BAL	–	–	–	HF-*p*-Cresol 10:1 or TFMSA-TFA 1:9, 2 h	–	–
Monoalkoxy-alkyl-benzyl BAL	–	–	–	TFA, reflux, 3 h	–	–
Dialkoxybenzyl BAL	TFA-DCM 5:95, 5 min	TFA-TES 97.5:2.5, 15 min	TFA-DCM 1:19	TFA-DCM 3:7, 10 min	–	–
Trialkoxy-benzyl BAL	TFA-$CHCl_3$-H_2O 50:50:1, 1 h		TFA-$CHCl_3$-H_2O 50:50:1, 1 h	TFA-DCM 5:95–1:99, 1–2 h	TFA-H_2O 19:1, 2 h	TFA-DCM 95:5, 16 h
Thiophene BAL	–	–	–	TFA-DCM 1:99, 2 h	–	TFA-DCM 1:1, 2 h, 60 °C
Indole BAL	TFA-DCM 1:99, 3 min	TFA-DCM 1:99, <2 min	TFA-DCM 1:99, 1 min	TFA-DCM 1:99, 8 h		

as oxime and hydrazone formation [11, 12]. They can also show inhibitory properties as transition state analogs for some proteolytic enzymes [13]. However, synthesis of these C-terminal peptide aldehydes is not a trivial synthetic challenge, and other approaches have been reported [14–17]. The BAL strategy for synthesis of C-terminal peptide aldehydes utilizes an acetal functionality which is demasked during cleavage. This strategy has in our group been used to synthesize an ensemble of different so-called carboproteins where a carbohydrate template, functionalized with four aminooxy groups, is reacted with peptide aldehydes in solution [18, 19]. A protocol for the synthesis of C-terminal peptide aldehydes using the BAL strategy is outlined in Subheading 3.

Another application of which the BAL linker strategy has proven useful is for the synthesis of C-terminal glycinal peptide thioesters [20]. This method is described in Chapter 8.

1.1 Anchoring of the First Residue: Reductive Amination

A main feature for all BAL-type linkers is the anchoring of the first residue by reductive amination through a resin-bound aldehyde functionality, which acts as the electrophile. The first building block, which has a primary amine, acts as the nucleophile. This is normally achieved by anchoring the first residue by RA to the resin-bound linker precursor (i.e., BAL handle).

The first building block to be incorporated is typically an amino acid derivative with either a free amine or as the corresponding HCl salt. The RA reaction is normally performed as a one-pot reaction imine formation and reduction under slightly acidic conditions. Often 1% (v/v) AcOH is added; however, HCL salts of amines can sometimes be used without added acid [1, 7]. The preferred solvent is in most cases MeOH, albeit when using polystyrene resins, DMF has proven more efficient. The RA is performed in a one-pot reaction using preferably $NaBH_3CN$ (tenfold excess) as reducing agent. Microwave heating to 60 °C during the RA step can increase the efficiency and reduce reaction time from overnight to approx. 10 min [18].

1.2 Anchoring of the Second Residue: Acylation of Secondary Amine

Acylation of a secondary amine requires special consideration in regards to the acylating agent. HATU activation or preformed symmetrical anhydrides are the most established methods. In our research group, symmetrical anhydride-mediated acylations has been the method of choice using DCM as the preferred solvent. For increased solubility of the amino acids, a DCM:DMF (9:1) mixture is often used, where the DMF is added prior to the DCM. The reaction can be performed at room temperature 2×2 h, but microwave heating 60 °C can reduce the reaction time down to minutes [21]. In this case DCM is substituted for the higher-boiling DCE.

1.3 Release of Peptide in Solution

All practical BAL handles release the assembled peptide from a resin by acidolysis. Upon treatment with, e.g., TFA, the BAL handle releases a secondary amide in the peptide backbone. As little as 1% TFA in DCM has been reported to release the peptide from the *o*-BAL handle, keeping the permanent side-chain protection group intact on the peptide in solution. Both BAL and C-terminal PAL anchoring generate the same trialkoxybenzyl carbenium ion upon acidolytic release of the peptide. However, there is a difference in the actual leaving group. For the PAL handle, this is the peptides with C-terminal primary amide (imidate), while the BAL handle releases a secondary amide (imidate) in the peptide backbone. Albericio and Barany have reported that the PAL handle requires concentrated TFA for its release; they summarize the requirement as 70–90% (v/v) TFA in DCM for efficient release of peptides, whereas lower concentrations of TFA in DCM (5–50%) only give partial release [5, 6]. In contrast hereto, the structurally analogous trialkoxybenzyl BAL handles can release peptides at significantly lower TFA concentrations (as little as 1% TFA in DCM). The higher acid lability of peptidyl-BAL linkages, compared to peptidyl-PAL, must be due to differences in the leaving group, i.e., a better leaving group capability of *N*-substituted amide vs. a non-substituted amide. This appears to be due to ground-state destabilization due to steric factors of peptides anchored to a BAL handle [22].

2 Materials

1. Aminomethylated Tentagel resin, loading 0.29 mmol/g (Rapp Polymere).
2. *o*-PALdehyde = 5-(2-formyl-3,5-dimethoxyphenoxy)pentanoic acid (Peptides International).
3. Fmoc-protected amino acids.
4. Aminoacetaldehyde dimethylacetal ($NH_2CH_2CH(OCH_3)_2$).
5. Amino acid *t*-Bu ester chloride salts.
6. $NaBH_3CN$: Sodium cyanoborohydride.
7. Coupling reagents: DIC, HBTU, HOBt, Oxyma.
8. Solvents: DMF, NMP, DCM, TFA, DIEA, piperidine.
9. Triethylsilane (TES).
10. Microwave instrument: Biotage Initiator using their glass vials or their new filtered glass vial which connect to the initiator peptide workstation.

3 Methods

3.1 Synthesis of C-Terminal Peptide Aldehydes

3.1.1 Anchoring of the o-PALdehyde Linker

The resin is anchored to the linker using standard chemistry for amide bond formation. Originally only two equivalents of *o-PALdehyde* were used; however, the coupling of the linker obviously proceeds faster with higher concentrations and larger equivalents. The protocol below was developed for use with glass vials with a filter polypropylene for microwave synthesis.

1. Place the resin in a microwave vial with a filter (or a normal polystyrene syringe with a filter *see* **Note 1**) and swell with DCM for 20–30 min. Then wash the resin with DMF.
2. In a separate container (e.g., a 15 mL Falcon tube), measure out HBTU (1.8 equiv.), HOBt (2 equiv., *see* **Note 2**), *o-PAldehyde* (2 equiv.), and DIEA (4 equiv.) and add DMF to an overall concentration of 0.2 M. This means, for a 0.1 mmol scale, approx. 2 mL DMF.
3. Add the DMF mixture to the resin and heat in the microwave for 10 min at 75 °C. Then the resin is filtrated to remove excess coupling reagent.
4. **Steps 2** and **3** can be repeated to achieve a higher incorporation (loading) by double coupling.
5. The resin is then washed with DMF (×3) and MeOH (×3).

3.1.2 Anchoring of the First Residue: Reductive Amination

The reductive amination is a highly optimized procedure and should be followed quite closely. For a discussion of the different developments, *see* Subheading 1.2. This is the step which determines the C-terminal functionality of the final peptide. The following protocol is for C-terminal peptide aldehydes only:

1. $NaBH_3CN$ (10 equiv.) and $NH_2CH_2CH(OCH_3)_2$ (10 equiv.) are added to the resin (in the microwave vial). MeOH:AcOH (99:1) is added (approx. 7 ml for a 0.1 mmol scale) and the vial is capped (*see* **Note 3**).
2. Heat in the microwave for 10 min at 60 °C before washing the resin thoroughly with MeOH (×3), DMF (×3) MeOH (×1).
3. **Steps 1** and **2** are repeated omitting the final MeOH wash.

3.1.3 Anchoring of the Second Residue: Acylation of Secondary Amine

Symmetric anhydride-mediated acylation is in our group used to incorporate the second residue on trialkoxybenzyl amines systems.

1. Wash the resin with DCM.
2. Measure out the Fmoc-amino acid (10 equiv.); add DMF (approx. 700 μL for 0.1 mmol scale). Then add DCM (6.3 mL for 0.1 mmol scale) and DIC (5 equiv.) (*see* **Note 4**). Preactivate for 10 min.

3. Add the solution to the resin (in the microwave vial) and reaction for 10 min at 60 °C. Then remove the excess coupling solution by filtration.
4. Repeat **steps 1** and **2**.
5. Wash the resin with DMF (×3) and DCM (×3).

3.1.4 Fmoc Quantification to Determine Loading of Dipeptide for Further Synthesis

In order to determine the quality of the dipeptidyl resin, an Fmoc quantification is carried out to determine the loading of the resin (*see* Chapter 2).

3.1.5 Chain Elongation Using Standard SPPS: Suggestion to Automated Chemistry

There are many different approaches of performing standard peptide couplings. After the on-resin synthesis of the first dipeptide, the remainder of the peptide assembly is well suitable for automated synthesis. A detailed description is outlined in Chapter 1.

3.1.6 Release of Peptide in Solution

The release of the peptides from the solid support is performed using TFA with water as scavenger. The use of TES/TIS is omitted in this type of cleavages due to the reducing effect these scavengers have on the C-terminal aldehyde functionality.

1. Wash resin with DCM and dry it down for 1–2 h under vacuum.
2. Add TFA-H_2O (19:1) (approx. 10 mL for a 0.1 mmol scale) and leave reacting for 3 h (*see* **Note 5**).
3. Filtrate the resin and collect the TFA-peptide mixture. Wash the resin with TFA and DCM and collect the filtrates.
4. TFA and DCM were removed by N_2 and the peptide precipitated with diethyl ether; centrifuged before the diethyl ether was removed. The ether treatment was repeated.
5. The peptide was air-dried for approx. 30 min before being dissolved in H_2O-CH_3CN and analyzed by LC-MS and then lyophilized.

4 Notes

1. The protocols are fitted to use a microwave glass vial with a filter (Biotage Initiator peptide workstation). If this equipment is not available, then a normal polystyrene syringe with a filter can be used instead. The resin is then transferred to a microwave vial before each microwave-assisted reaction.
2. Access to HOBt has been limited due to restrictions on air transportation. As an alternative Oxyma can be used as an additive. Alternatively, the use of additives can be eliminated completely and HBTU can be used alone. Furthermore, other coupling reagent such as HATU can also be used.

3. If the reaction is run in an open microwave vial system, use DMF:AcOH (99:1) instead.
4. If the reaction is run in an open microwave vial system, use DCE instead of DCM.
5. The resin with the TFA cleavage mixture is stirred manually 3–5 times during the reaction.

References

1. Jensen KJ, Alsina J, Songster M, Vagner J, Albericio F, Barany G (1998) Backbone amide linker (BAL) strategy for solid-phase synthesis of C-terminal-modified cyclic peptides. J Am Chem Soc 120:5441–5452
2. Jensen KJ, Songster MF, Vagner J, Alsina J, Albericio F, Barany G (1996) Innovation and Perspectives in Solid Phase Synthesis & Combinatorial Chemical Libraries. In: Epton R (ed) Collected Papers, Fourth International Symposium. Mayflower Scientific Ltd., Kingswinford, pp 187–190
3. Jensen KJ, Songster MF, Vágner J, Alsina J, Albericio F, Barany G (1996) In: PTP Kaumaya, RS Hodges (eds) Peptides—chemistry, structure and biology: proceedings of the fourteenth American peptide symposium (1995), Mayflower Scientific, Kingswinford, UK
4. Boas U, Brask J, Jensen KJ (2009) Backbone amide linker in solid-phase synthesis. Chem Rev 109:2092–2118
5. Albericio F, Barany G (1987) An acid-labile anchoring linkage for solid-phase synthesis of C-terminal peptide amides under mild conditions*. Int J Pept Protein Res 30:206–216
6. Albericio F, Kneib-Cordonier N, Biancalana S, Gera L, Masada RI, Hudson D, Barany G (1990) Preparation and application of the 5-(4-(9-fluorenylmethyloxycarbonyl) aminomethyl-3,5-dimethoxyphenoxy)-valeric Acid (PAL) handle for the solid-phase synthesis of C-terminal peptide amides under mild conditions. J Org Chem 55:3730–3743
7. Boas U, Brask J, Christensen J, Jensen KJ (2002) The ortho backbone amide linker (o-BAL) is an easily prepared and highly acid-labile handle for solid-phase synthesis. J Comb Chem 4:223–228
8. Alsina J, Jensen K, Albericio F, Barany G (1999) Solid-phase synthesis with Tris(alkoxy) benzyl backbone amide linkage (BAL). Chem Eur J 5:2787–2795
9. Alsina J, Yokum TS, Albericio F, Barany G (1999) Backbone amide linker (BAL) strategy for *N*α-9-fluorenylmethoxycarbonyl (Fmoc) solid-phase synthesis of unprotected peptide *p*-nitroanilides and thioesters. J Org Chem 64:8761–8769
10. Guillaumie F, Kappel JC, Nicholas MK, Barany G, Jensen KJ (2000) Solid-phase synthesis of C-terminal peptide aldehydes from amino acetals anchored to a backbone amide linker (BAL) handle. Tetrahedron Lett 41:6131–6135
11. Rose K (1994) Facile synthesis of homogeneous artificial proteins. J Am Chem Soc 116:30–33
12. Shao J, Tam JP (1995) Unprotected peptides as building blocks for the synthesis of peptide dendrimers with oxime, hydrazone, and thiazolidine linkages. J Am Chem Soc 117:3893–3899
13. Fehrentz JA, Paris M, Heitz A, Velek J, Liu C-F, Winternitz F, Martinez J (1995) Improved solid phase synthesis of C-terminal peptide aldehydes. Tetrahedron Lett 36:7871–7874
14. Leliévre D, Chabane H, Delmas A (1998) Simple and efficient solid-phase synthesis of unprotected peptide aldehyde for peptide segment ligation. Tetrahedron Lett 39:9675–9678
15. Ede NJ, Bray AM (1997) A simple linker for the attachment of aldehydes to the solid phase application to solid phase synthesis by Miltipin™ method. Tetrahedron Lett 38:7119–7122
16. Fehrentz JA, Paris M, Heitz A, Velek J, Winternitz F, Martinez J (1997) Solid phase synthesis of C-terminal peptide aldehydes. J Org Chem 62
17. Dinh TQ, Armstrong RW (1996) Synthesis of ketones and aldehydes via reactions of Weinreb-type amides on solid support. Tetrahedron Lett 37:1161–1164
18. Tofteng AP, Hansen TH, Brask J, Nielsen J, Thulstrup P, Jensen KJ (2007) Synthesis of functionalized de novo designed 8-16 kDa model proteins towards metal ion-binding and esterase activity. Org Biomol Chem 5:2225–2233
19. Brask J, Jensen KJ (2001) Carboproteins: a 4-α-helix bundle protein model assembled on

a D-galactopyranoside template. Bioorg Med Chem Lett 11:697–700

20. Brask J, Albericio F, Jensen KJ (2003) Fmoc solid-phase synthesis of peptide thioesters by masking as trithioortho esters. Org Lett 5:2951–2953
21. Brandt M, Gammeltoft S, Jensen K (2006) Microwave heating for solid-phase peptide synthesis: general evaluation and application to 15-mer phosphopeptides. Int J Pept Res Ther 12:349–357
22. Norrby P-O, Jensen K (2006) Steric effects in release of amides from linkers in solid-phase synthesis. Molecular mechanics modeling of key step in peptide and combinatorial chemistry. Int J Pept Res Ther 12:335–339

Chapter 10

Synthesis of N-Methylated Peptides: On-Resin Methylation and Microwave-Assisted Couplings

Renée Roodbeen and Knud J. Jensen

Abstract

N-Methylation may positively influence the pharmacokinetic properties of peptides by improving oral availability and in vivo half-life. Additionally, target affinity and specificity may be improved. Here, we describe the solid-phase N-methylation of peptides using direct alkylation. This method allows a rapid *N*-methyl scan of synthetic, bioactive peptides. Additionally, a microwave-enhanced method for the difficult coupling onto the methylated N terminus is provided.

Key words Solid-phase peptide synthesis, N-methylation, N-methylated peptides, Pharmacokinetics, Bioactive peptide, o-Nitrobenzenesulfonyl protecting group, Difficult couplings, Microwave heating, COMU

1 Introduction

Bioactive N-methylated peptides are often isolated from natural sources such as marine sponges [1, 2]. It is believed that methylation of the backbone nitrogens influences the conformation of peptides and can increase receptor selectivity and specificity [3]. N-Methylation was effective in the design of highly active somatostatin analogues which had an altered selectivity towards the somatostatin receptor subtype 2 and thereby reduced neurogenic inflammation [4]. Another example of the influence of N-methylation on activity and specificity is Cilengitide, a small cyclic methylated peptide discovered by Kessler and coworkers and developed by Merck-Serono which has been evaluated for treating glioblastoma (Fig. 1) [5, 6]. The IC_{50} value for the $\alpha v\beta 3$ integrin was lowered from 2.5 to 0.58 nM by methylation of a single residue.

To test if N-methylation is advantageous in the synthesis of a specific peptide, one could incorporate a Fmoc-*N*(methyl)-amino acid. However, most of these building blocks are expensive or need to be prepared from standard amino acids using solution-phase

Knud J. Jensen et al. (eds.), *Peptide Synthesis and Applications: Second Edition*, Methods in Molecular Biology, vol. 1047, DOI 10.1007/978-1-62703-544-6_10, © Springer Science+Business Media New York 2013

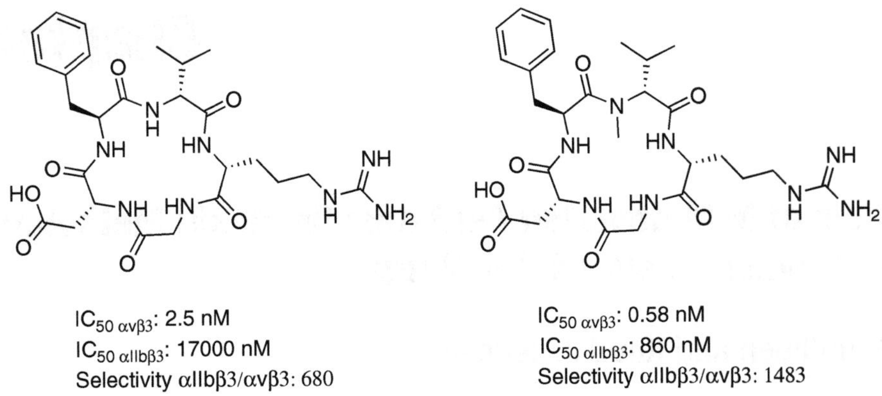

Fig. 1 Effect of N-methylation on IC_{50} values and receptor specificity for the cyclic peptide *c*(RGDfV). *Left*: Non-methylated peptide. *Right*: N-methylation of the Val residue, *c*(RGDf-N(Me)V-), with improved activity and selectivity of the Cilengitide peptide

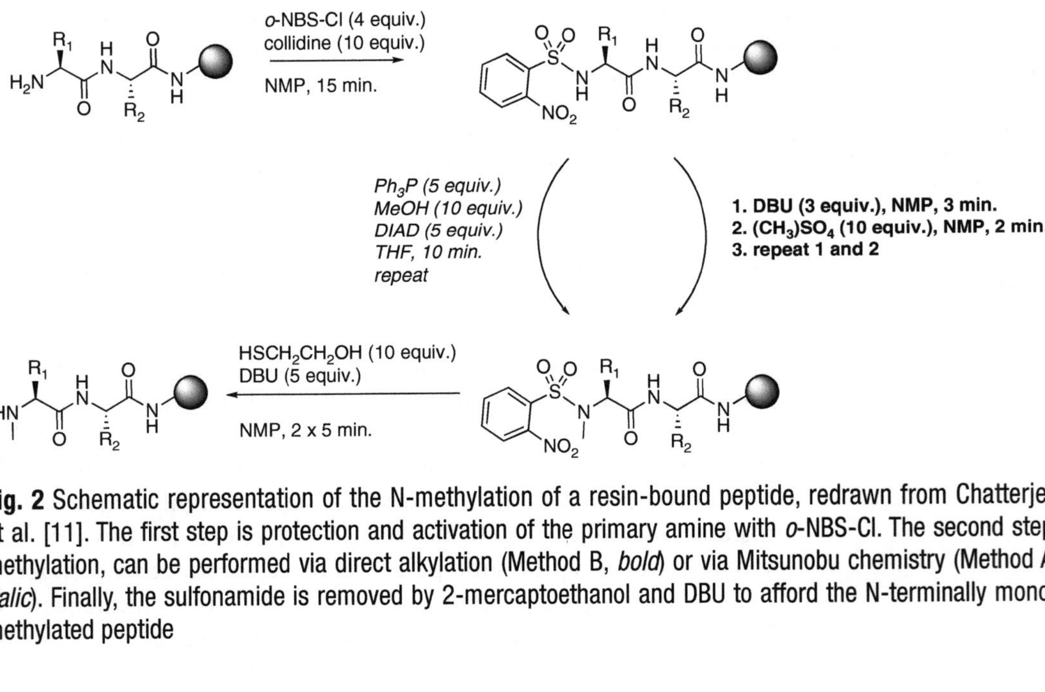

Fig. 2 Schematic representation of the N-methylation of a resin-bound peptide, redrawn from Chatterjee et al. [11]. The first step is protection and activation of the primary amine with *o*-NBS-Cl. The second step, methylation, can be performed via direct alkylation (Method B, *bold*) or via Mitsunobu chemistry (Method A, *italic*). Finally, the sulfonamide is removed by 2-mercaptoethanol and DBU to afford the N-terminally mono-methylated peptide

synthesis under harsh conditions [7–9]. Therefore, on-resin N-methylation is often preferred in the synthesis of selectively N-methylated peptides. The chemistry used to achieve this must be compatible with Fmoc-SPPS to avoid premature deprotection of the side-chain protecting groups and release of the peptide from the resin. The first method to selectively monoalkylate the N terminus was described by Miller and Scanlan and later further developed by Kessler and coworkers (Fig. 2) [7, 10].

On-resin methylation of the N terminus is based on selective mono-protection and activation of the amine, followed by methylation via either direct alkylation or Mitsunobu chemistry and subsequent deprotection to afford the mono-methylated secondary N-terminal amine [7, 11]. Several methods, in which the base, solvent, and reaction time were varied, have been reported for the first step [8, 10, 12]. However, all methods have the transient protecting group in common, the *o*-NBS group. This group is chosen due to its ability to activate the N terminus for alkylation, not to form sulfenes and to be readily removed by 2-mercaptoethanol and DBU. For the second step, two methods are available, direct alkylation with alkyl halides or dimethyl sulfate or alkylation via Mitsunobu chemistry using MeOH [8, 12]. Direct alkylation is straightforward, but the alkylating reagents are generally carcinogenic and methylation of the N terminus is troublesome for Cys and His, as these residues are prone to side-chain alkylation. For residues prone to side-chain alkylation by Mitsunobu chemistry, using MeOH is preferred [8]. Even though dry reaction conditions are required, Mitsunobu chemistry can also be advised for the synthesis of peptoids due to the high number of suitable commercially available primary alcohols [12]. Finally, after alkylation, the *o*-NBS group is removed by treatment with 2-mercaptoethanol and base [8, 10, 12]. Initially the use of thiophenol was recommended, as it would cleave the sulfonamide faster and is less toxic than 2-mercaptoethanol [10]. However, thiophenol might cause ester hydrolysis (in case of anchoring of the peptide through an ester bond) and release of the peptide from a resin affording peptide acids [12]. The sulfonamide hydrolysis can be monitored by the release of 2-(2-nitrophenylthio)ethanol, which has a strong yellow color.

Coupling onto, i.e., N-acylation of, N-methylated peptides can be difficult and may require harsh or expensive reagents combined with long coupling time to afford satisfactory incorporation [13–16]. Microwave heating has emerged as a powerful tool in peptide synthesis and has also proved its value in the synthesis of N-methylated peptides [15, 17, 18]. Additionally, a wide variety of coupling reagents, like DIC/HOAt, HATU, and COMU, have been reported in combination with microwave-assisted couplings [19, 20]. We have recently established that the coupling yield strongly depends both on the side chain of the N-methylated residue and on the amino acid that is coupled onto the methylated N terminus [21]. For example, we found that in contrast to coupling onto sterically nondemanding residues such as ^{Me}Gly, which proceeds with full conversion, coupling onto ^{Me}Ile is very difficult at room temperature [17, 18]. Fortunately, coupling yields are greatly improved when using microwave heating with either HATU or COMU as coupling reagent. Additionally, microwave heating can not only increase yields and purity but will also reduce the synthesis time both for non-methylated and N-methylated peptides [17, 20, 21].

2 Materials

All solutions have to be freshly prepared from the following materials using peptide-grade NMP and the highest grade available for the other reagents:

2.1 Resin

1. TentaGel with Rink amide linker, loading: 0.24 mmol/g.

2.2 Organic Solvents

1. 1-Methyl-2-pyrrolidine (NMP).
2. *N,N*-Dimethylformamide (DMF).
3. Dichloromethane (DCM).

2.3 Other Reagents

1. *N,N*-Diisopropylethylamine (DIEA).
2. Trifluoroacetic acid (TFA).
3. Triethylsilane (TES).
4. 1-[(1-(Cyano-2-ethoxy-2-oxoethylideneaminooxy)-dimethylamino-morpholino-methylene)] methanaminium hexafluorophosphate (COMU).

2.4 Standard Solutions for On-Resin N-Methylation by Direct Alkylation (See Notes 1 and 2)

1. *Solution A* (4 equiv. *o*-NBS-Cl + 10 equiv. collidine in NMP) to prepare 20 mL: weigh out 1,182 g of *o*-NBS-Cl in a 50 mL Falcon tube, add NMP to a volume of 10 mL, and dissolve by agitation. Add 1.76 mL collidine and fill up to 20 mL with NMP.
2. *Solution B* (3 equiv. DBU in NMP) to prepare 20 mL: place 598 μL DBU in a 50 mL Falcon tube and fill to 20 mL with NMP.
3. *Solution C* (10 equiv. dimethyl sulfate (CAUTION) in NMP) to prepare 20 mL: place 1.26 mL dimethyl sulfate (*see* **Notes 1** and **3**) in a 50 mL Falcon tube and fill to 20 mL with NMP.
4. *Solution D* (10 equiv. 2-mercaptoethanol and 5 equiv. DBU in NMP) to prepare 20 mL: place 938 μL mercaptoethanol (*see* **Note 4**) in a 50 mL Falcon tube, add NMP to 15 mL. Then add 996 μL DBU and fill to 20 mL with NMP.

3 Methods

3.1 Procedure for Manual N-Methylation with Dimethyl Sulfate, Scale: 0.1 mmol, Resin-Loading 0.24 mmol/g (TentaGel) [16]

With this procedure the N terminus will be protected and simultaneously activated by reacting it with *o*-NBS-Cl. After this step, the methylation can take place by treating the resin with DBU and dimethyl sulfate. Direct alkylation of the N terminus is preferred for all amino acids, except Cys and His, as the process is faster and does not require a change of solvent. The direct alkylation method is therefore suitable for automated synthesis on a peptide synthesizer. Finally, the *o*-NBS protecting group is removed with

2-mercaptoethanol and the mono-methylated N terminus is obtained (schematic representation in Fig. 2, the alkylation step is written as method B in bold).

All steps are performed at room temperature.

1. Place the resin in a 10 mL syringe fitted with a polyethylene filter.
2. Swell resin for 15 min in twice the resin bed volume NMP (*see* **Note 5**).
3. Drain the solution and wash once with twice the resin bed volume NMP.
4. Add 1.5 mL of solution A to the resin and agitate for 15 min.
5. Drain the solution.
6. Wash resin five times with twice the resin bed volume NMP.
7. If desired a Kaiser test (*see* **Note 6**) can be performed to check if the protection with *o*-NBS is complete.
8. Add 1.5 mL of solution B to the resin and agitate for 3 min.
9. Add 1.5 mL of solution C (*see* **Notes 3** and 7) and agitate for another 2 min.
10. Drain the solution (*see* **Note** 7) and wash the resin once with 4 mL NMP (approximately twice the resin bed volume).
11. Repeat steps 8–10.
12. Wash the resin five times with twice the resin bed volume NMP.
13. If desired a test cleavage could be performed to check if the alkylation is complete (*see* **Note 8**).
14. Add 1.5 mL of solution D to the resin (*see* **Note 4**) and agitate for 5 min.
15. Drain the solution and wash the resin once with twice the resin bed volume NMP.
16. Repeat **steps 14** and **15**.
17. Wash resin five times with twice the resin bed volume NMP.
18. Dispose of all filtrates from reactions and washing steps appropriately (*see* **Note** 7).

3.2 Coupling onto N-Methylated Amino Acids, Scale: 0.025 mmol, Resin Loading 0.24 mmol/g (TentaGel)

We use a Biotage Initiator™ for microwave heating, or alternatively the Biotage SyroWave™ (*see* **Note 9**).

Coupling onto the methylated N terminus, which is a sterically hindered secondary amine, can be difficult and will generally be accompanied by low coupling yields. Here, we describe the use of microwave heating in combination with the coupling reagent COMU to obtain high coupling yields, while the reaction time is only 20 min (Fig. 3).

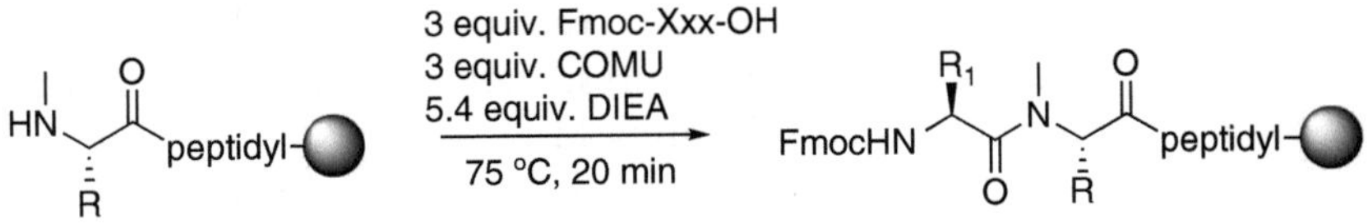

Fig. 3 Schematic representation of the coupling of an Fmoc-amino acid onto an N-methylated peptidyl resin. The use of 3 equiv. of amino acid, coupling reagent COMU, and 5.4 equiv. of the base DIEA, for 20 min at 75 °C, offers satisfactory incorporation for most couplings

1. The resin is placed in a 0.5 or 2.0 mL glass Biotage microwave vial (*see* **Note 10**).
2. An appropriate magnetic stirring bar is added (*see* **Note 11**).
3. The amino acid (3 equiv.) and coupling reagent (COMU, 3 equiv.) are weighed and placed into a 1.5 mL Eppendorf tube.
4. DMF (400 μL) is added and the reagents are dissolved by agitation, to a final concentration of approximately 0.18 M of amino acid and coupling reagent.
5. 5.4 equiv. DIEA is added and mixed by gently shaking the Eppendorf tube.
6. The mixture is added directly to the resin and the microwave vial is tightly sealed with the appropriate cap.
7. The vial is placed in the microwave cavity and irradiated without external cooling. *Settings microwave* (*see* **Notes 9** and **12**):

Temperature	75 °C
Reaction time	20 min
Pre-stirring	5 s
Absorption level	Very high
Stirring speed	600 rpm

The sample will be irradiated with approximately 60 W.

1. After reacting for 20 min, the cap is removed and 2 mL DCM is added to the resin solution. The solution is then transferred to a 5 mL syringe fitted with a polyethylene filter (*see* **Note 13**).
2. Wash resin four times with NMP (2 mL).
3. Perform a test cleavage to test if the incorporation is complete (*see* **Note 14**).
4. If required, repeat **steps 1–10**.

4 Notes

1. Due to the highly hazardous nature of some of the reagents, we suggest that the solutions are prepared in disposable containers, e.g., Falcon tubes.

2. The amount of solution needed depends on the scale of the reaction and the resin loading. The volume mentioned here is sufficient for 1.3 mmol (or 13 × 0.1 mmol) of TentaGel resin with 0.24 mmol/g loading; this volume is required to allow the resin to swell and agitate properly. Therefore, modify the concentrations of the reagents and volumes of solution A–D depending on the type of resin and its loading.
3. Dimethyl sulfate is a very hazardous reagent and a known carcinogen. Only handle this reagent, solutions containing this reagent, and all waste inside a fume hood. Avoid inhalation and skin contact at all costs. Wear full protective clothing, safety glasses, and gloves. In case of a spill of any of the solutions containing dimethyl sulfate, dispose the gloves directly. Consult the MSDS datasheet for more information.
4. 2-Mercaptoethanol has a very unpleasant odor. Have a beaker with a solution of household chlorine (bleach) in water (1:1) ready in case of a spill. Wash all disposables and glassware with the chlorine solution before disposal (Falcon tubes, pipettes, etc.).
5. The volume of the resin bed is approximately 3 mL for 0.1 mmol TentaGel resin. Therefore, when washing the resin with twice the resin bed volume, use 6 mL NMP. Adjust the volume depending on the type and amount of resin used.
6. The procedure of a Kaiser test is described in Chapter 2.
7. The total volume is now 3 mL and the final concentrations of the reagents in solutions B and D are halved. However, this is done to allow a proper swelling and agitation in **step 9**.
8. The filtrate will contain the excess of dimethyl sulfate; dispose the waste in an appropriate manner. Rinse the container in which you collect the filtrate three times with acetone to remove the last traces of dimethyl sulfate.
9. For the test cleavage, remove a small amount of resin from the reactor and wash with DCM (3×) and dry thoroughly before treatment with TFA:TES:H_2O (95:2.5:2.5) for 2 h at room temperature. The TFA is removed under a stream of nitrogen and the peptide is triturated from ether. After centrifugation and removal of the ether, the peptide can be dissolved in a mixture of MeCN/H_2O and analyzed by LC-MS.
10. Other microwave reactors can be used; however, make sure to adjust the settings so they correspond to the settings described in Subheading 3.2 (**steps 1**, **2,** and 7). Domestic microwave ovens are not suitable and may be dangerous to use here. If a laboratory microwave instrument is not available, heating with a heater-shaker can be considered.
11. The resin is easily transferred after it has been washed with DCM twice and dried by suction for 5 min.

12. We use a small and triangular magnet, as we find that this type of magnet bounces through the resin solution, rather than grinding the resin particles.
13. We set the absorption level to "very high" when using microwave heating on peptides. We do this to avoid that the temperature increases above the set value during the initial heating.
14. Some resin might have attached itself to the glass surface of the vial during the transfer of the resin. These resin particles may not have reacted and should be removed with a spatula or cotton swab before adding the DCM to avoid mixing reacted and unreacted resin beads.
15. As the N terminus is methylated, and thus is a secondary amine, the Kaiser test cannot be used to test that the incorporation of the following amino acid is complete.

References

1. Fusetani N, Matsunaga S (1993) Bioactive sponge peptides. Chem Rev 93:1771–1791
2. Wipf P (1995) Synthetic studies of biologically active marine cyclopeptides. Chem Rev 95:2115–2134
3. Gilon C, Dechantsreiter MA, Burkhart F, Friedler A, Kessler H (2002) Synthesis of peptides and peptidomimetics. In: Goodman M, Felix A, Moroder L, Tonolio C (eds) Houben-Weyl methods of organic chemistry. Georg Thieme, Stuttgart, pp 215–291
4. Chatterjee J, Laufer B, Beck JG, Helyes Z, Pinter E, Szolcsanyi J, Horvath A, Mandl J, Reubi JC, Keri G, Kessler H (2011) N-methylated sst_2 selective somatostatin cyclic peptide analogue as a potent candidate for treating neurogenic inflammation. ACS Med Chem Lett 2:509–514
5. Mas-Moruno C, Rechenmacher F, Kessler H (2010) Cilengitide: the first anti-angiogenic small molecule drug candidate. Design, synthesis and clinical evaluation. Anticancer Agents Med Chem 10:753–768
6. Desgrosellier JS, Cheresh DA (2010) Integrins in cancer: biological implications and therapeutic opportunities. Nat Rev Cancer 9:9–22
7. Biron E, Chatterjee J, Kessler H (2006) Optimized selective *N*-methylation of peptides on solid support. J Pept Sci 12:213–219
8. Biron E, Kessler H (2005) Convenient synthesis of N-methylamino acids compatible with Fmoc solid-phase peptide synthesis. J Org Chem 70:5183–5189
9. Govender T, Arvidsson PI (2006) Facile synthesis of Fmoc-N-methylated alpha- and beta-amino acids. Tetrahedron Lett 47:1691–1694
10. Miller SC, Scanlan TS (1997) Site-selective N-methylation of peptides on solid support. J Am Chem Soc 119:2301–2303
11. Chatterjee J, Gilon G, Hoffman A, Kessler H (2008) N-methylation of peptides: a new perspective in medicinal chemistry. Acc Chem Res 41(10):1331–1342
12. Reichwein JF, Liskamp RMJ (1998) Site-specific N-alkylation of peptides on the solid-phase. Tetrahedron Lett 39:1243–1246
13. Sabatino G, Mulinacci B, Alcaro M, Chelli M, Rovero P, Papini P (2002) Assessment of new 6-Cl-HOBt based coupling reagents for peptide synthesis. Part 1: coupling efficiency study. Int J Pept Res Ther 9:119–123
14. Kaminski ZJ, Kolesinska B, Kolesinska J, Sabatino G, Chelli M, Rovero P, Blaszczyk M, Glowka ML, Papini AM (2005) *N*-triazinylammonium tetrafluoroborates. A new generation of efficient coupling reagents useful for peptide synthesis. J Am Chem Soc 127:16912–16920
15. Teixido M, Albericio F, Giralt E (2005) Solid-phase synthesis and characterization of *N*-methyl rich-peptides. J Pept Res 65:153–166
16. Chatterjee J, Laufer B, Kessler H (2012) Synthesis of *N*-methylated cyclic peptides. Nat Protoc 7:432–444
17. Rodriguez H, Suarez M, Albericio F (2010) A convenient microwave-enhanced solid-phase synthesis of short chain *N*-methyl-rich peptides. J Pept Sci 16:136–140
18. Pedersen SL, Mehrotra A, Jensen KJ (2010) In: Lebl M, Meldal M, Jensen KJ, Hoeg-Jensen T (eds) Proceedings of the 31st European peptide symposium, San Diego, USA. Prompt Scientific Publishing

19. Pedersen SL, Tofteng AP, Malik L, Jensen KJ (2011) Microwave heating in solid-phase peptide synthesis. Chem Soc Rev 41:1826–1844
20. Malik L, Tofteng AP, Pedersen SL, Sørensen KK, Jensen KJ (2010) Automated 'X-Y' robot for peptide synthesis with microwave heating: application to difficult peptide sequences and protein domains. J Pept Sci 16:506–512
21. Roodbeen R, Pedersen SL, Hosseini M, Jensen KJ (2012) Microwave heating in solid-phase synthesis of *N*-methylated peptides: when is room temperature better? Eur J Org Chem (36):7106–7111.

Chapter 11

Synthesis of Antimicrobial Peptoids

Paul R. Hansen and Jens K. Munk

Abstract

Peptoids (*N*-substituted glycines) are mimics of α-peptides in which the side chains are attached to the backbone N^{α}-amide nitrogen instead of the C^{α}-atom. Peptoids hold promise as therapeutics since they often retain the biological activity of the parent peptide and are stable to proteases. In recent years, peptoids have attracted attention as new potential antibiotics against multiresistant bacteria. Here we describe the submonomer solid-phase synthesis of an antimicrobial peptoid, H-*N*mbn-*N*lys-*N*lys-*N*nap-*N*but-*N*mbn-*N*lys-NH_2.

Key words Peptoids, *N*-substituted glycines, Submonomer solid-phase synthesis, Antimicrobial peptides

1 Introduction

Peptoids are oligomers of *N*-substituted glycines [1]. They are mimics of α-peptides in which the side chains are attached to the backbone N^{α}-amide nitrogen instead of the C^{α}-atom (Fig. 1). Peptoids are achiral, and adopt different conformations from peptides; however, they retain the same overall properties of peptides [2]. Peptoids hold promise as therapeutics since they often retain the biological activity of the parent peptide and are stable to proteases [3].

Peptoids have been used to study a number of biological processes, including cellular uptake and delivery [4], ligands for GPCRs [5], protein–protein interactions [6], protease inhibitors [7], and calcium channel blockers [8]. Recently, Lee and Zuckermann were able to introduce peptoid residues at defined positions into bovine pancreatic ribonuclease A using a combination of genetic engineering and chemical synthesis [9]. Furthermore, Chen and coworkers reported on an expanded polyglutamine-binding peptoid as a novel therapeutic agent for treatment of Huntington's disease [10]. Finally, several peptoid conjugates with interesting properties have been reported. This group includes,

Knud J. Jensen et al. (eds.), *Peptide Synthesis and Applications: Second Edition*, Methods in Molecular Biology, vol. 1047, DOI 10.1007/978-1-62703-544-6_11, © Springer Science+Business Media New York 2013

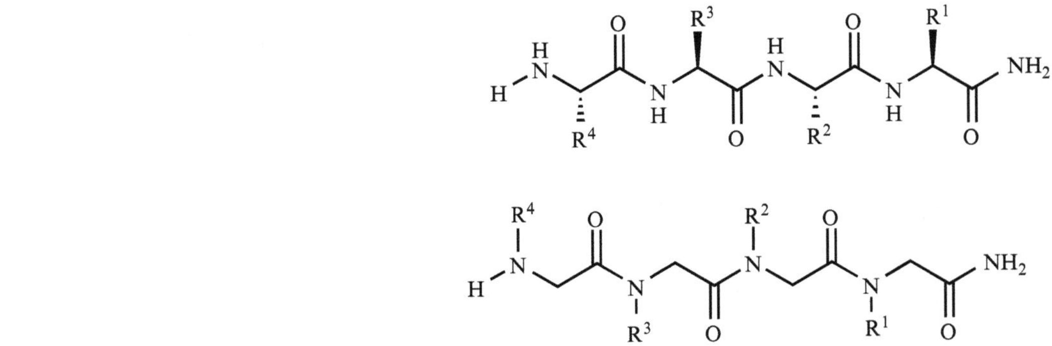

Fig. 1 Comparison of the general structure of peptides and peptoids

e.g., lipitoids [11], glycopeptoids [12], and carboxyalkyl peptoid PNAs [13]. For recent reviews of peptoids, *see* refs. 14–18.

Antimicrobial peptides are a major part of innate immunity. They act as first line of defense against invading pathogens, killing a wide range of pathogenic microorganisms, including Gram-positive bacteria, Gram-negative bacteria, and fungi [19]. Unlike conventional antibiotics which target cell wall synthesis, protein synthesis, metabolism, or transcription/replication, most antimicrobial peptides work by disrupting the bacterial membrane. However, the exact mechanism has not been completely elucidated.

In recent years, peptoid research has focused on the development of new peptoid-based antimicrobial agents either from de novo design [20–24], cyclic peptoids [25], libraries [26, 27], or by transforming naturally occurring antimicrobial peptides into peptoid analogs [28].

Four methods for the solid-phase synthesis of peptoids have been reported: (i) the monomer approach in which *N*-substituted glycine building blocks are synthesized and characterized and then coupled sequentially [1]. (ii) reductive alkylation of glycine with the appropriate aldehyde or ketone to obtain the desired N^{α}-*alkylated* glycine derivative [29], (iii) the photolithographic approach [30], and (iv) the submonomer approach [31]. Only the latter, which by far is the most popular, will be described in detail below.

In the submonomer approach, peptoids are synthesized from the C to N terminus on a Rink amide linker resin using a two-step coupling cycle (Fig. 2). First step involves a DIC-mediated coupling of bromoacetic acid to the resin followed by a S_N2-nucleophilic displacement by a primary amine. This cycle is repeated until the desired compound is obtained. The peptoid is then cleaved from the resin with 95 % aqueous TFA. Following removal of TFA by evaporation, the peptoid is precipitated and lyophilized. Finally, the identity of the product is verified by chromatography combined

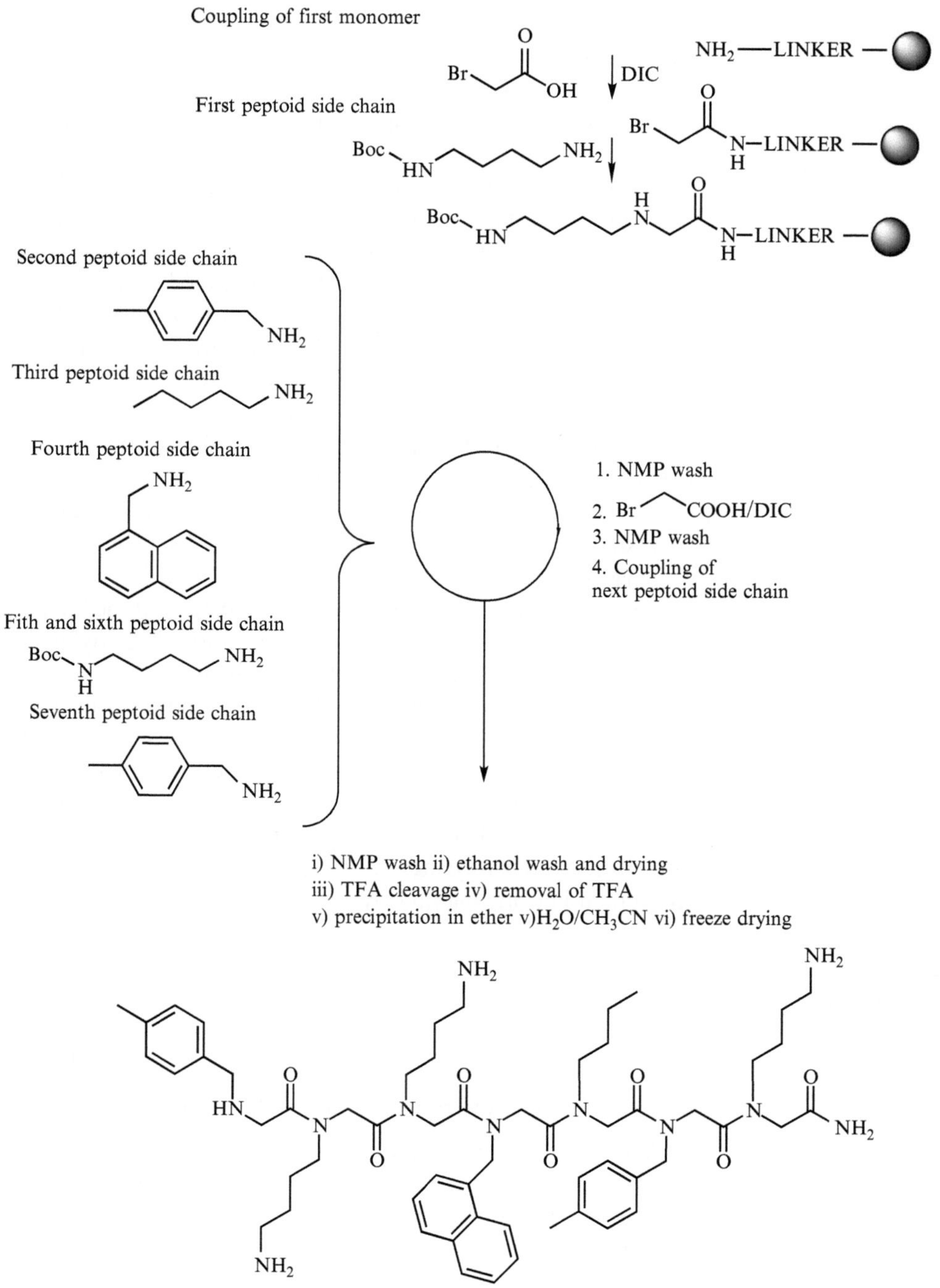

H-Nmbn-Nlys-Nlys-Nnap-Nbut-Nmbn-Nlys-NH_2

Fig. 2 Submonomer synthesis of H-*N*mbn-*N*Lys-*N*Lys-*N*nap-*N*but-*N*mbn-*N*ys-NH_2 described in this paper: *Bu* butyl, *lys* 4-aminobutyl, *mbn* 4-methylbenzyl, *nap* 1-napthylmethyl, *DIC N′*,*N*-diisopropylcarbodiimide, and *NMP N*-Methyl-2-pyrrolidone

with mass spectrometry, e.g., LC-MS or analytical HPLC and MALDI-TOF-MS.

Peptoids have also been characterized by NMR spectroscopy [32]. However, they are complicated systems to study at longer chain lengths, since both the *cis* and *trans* conformations of the backbone amide bonds of peptoids can be significantly populated [33].

Peptoids containing up to 50 residues have been synthesized in reasonable yield using the submonomer approach [17]. Different instrumentations have been used for peptoid synthesis such as manual synthesis in syringes equipped with a fritted filter [21], automated peptide synthesizers [28], automated microwave peptide synthesizers [12], and parallel synthesizers [34].

Most primary amines are suitable for peptoid synthesis. Protection is required if the amine carries a side-chain functionality such as carboxylic acid, amine, thiol, and/or heterocyclic compound. The review by Culf and Ouellette [15] lists several amines which successfully have been used for peptoid synthesis.

Three side reactions have been reported. The first one is the formation of *N*-substituted diketopiperazines instead of the linear dimer. The compounds 2-methoxyethylamine and 2-phenylethylamine belong to this group [27]. Secondly, *N*-substituted glycines carrying a nucleophile three or four atoms from the amino group may react with the main-chain bromoacetyl group. This group includes 4-(2-aminoethyl)morpholine, 4-(2-aminoethyl)morpholine, and N^{Im}-tritylhistamine [35]. Finally, some submonomers are unstable to TFA cleavage [15] including *p*-methoxy α-methylbenzylamine, 2,4,6-trimethoxybenzylamine, *p*-guanidino α-methylbenzylamine, and *p*-guanidinophenylethylamine.

2 Materials

1. Syringe (5 mL) with a stopper and PTFE filter (*see* **Notes 1–3**).
2. TentaGel R RAM resin (loading = 0.18 mmol/g).
3. A vacuum device (pump) with a liquid collection reservoir which fits onto the syringe tip, preferably equipped with a valve to allow swift connection and disconnection between reservoir and syringe.
4. *N*-methyl-2-pyrrolidone (NMP).
5. Piperidine.
6. Bromoacetic acid.
7. Diisopropylcarbodiimide (DIC).
8. Lab shaker.
9. Butylamine.
10. *N*-Boc-1,4-butandiamine.

11. 4-Methylbenzylamine.
12. 1-Napthylmethylamine.
13. Acetic acid.
14. Ethanol.
15. Diethyl ether.
16. Trifluoroacetic acid (TFA).
17. TFA:H_2O 95:5 (*see* **Note 4**).
18. Preparation of 0.6 M bromoacetic acid in NMP and 3.2 M DIC in NMP. Prepare fresh solutions daily: Dissolve 0.83 g of bromoacetic acid in 10 mL NMP. Dissolve 500 μL of DIC in 500 μL NMP.

3 Methods

To overcome some of the poor pharmaceutical properties of peptides discussed previously, researchers have focused on chemical modification of the peptide backbone. By incorporation of nonproteinogenic amino acids, such as *N*-substituted glycines, improvement of the biological lifetime and stabilization of secondary structures can be achieved.

Here we describe the design and synthesis of a peptoid, H-*N*mbn-*N*lys-*N*lys-*N*nap-*N*but-*N*mbn-*N*lys-NH_2 (Fig. 2), which displays antimicrobial activity against Gram-positive and Gram-negative bacteria. Since most antimicrobial peptides are cationic and amphipathic, we designed the peptoid to have an even distribution of cationic and hydrophobic *N*-substituted residues. Butyl 1,4-diamine was chosen as the cationic peptoid residue for several reasons. First, once incorporated into the peptoid, the resulting *N*-(4-aminobutyl) glycyl residue (*N*lys) resembles a lysine residue. Second, it has been shown that introduction of *N*lys in peptoids may reduce hemolytic activity [2]. Third, it is easier to synthesize *N*lys containing peptoids than the corresponding *N*arg derivatives, which requires an additional synthesis step [36]. Finally, butylamine, 4-methylbenzylamine, and 1-napthylmethylamine were chosen as hydrophobic building blocks since we previously have shown that peptoids containing these building blocks may display antimicrobial activity [21].

3.1 Preparation of Resin TentaGel R RAM (250 mg, Loading = 0.18 mmol/g)

1. Remove the plunger (piston) and insert the PTFE filter. Use the piston to push the filter all the way to the syringe bottom. Make sure that the filter is tight and remove the piston again.
2. Add the resin to the syringe and connect the vacuum device, leaving the syringe in an upright position.
3. Wash briefly with NMP (1 × 4 mL) using the vacuum device to drain the NMP.

4. Disconnect the vacuum, leaving the syringe in an upright position, and add NMP (approx. 4 mL).
5. Allow the resin to swell for at least 30 min, preferably overnight.

3.2 Removing the Fmoc Group

1. Drain the NMP and disconnect the vacuum.
2. Remove the Fmoc group by adding 20 % piperidine in NMP (v/v) (2 mL).
3. After 3 min drain the piperidine solution and disconnect the vacuum.
4. Wash briefly with NMP (1 × 2 mL) using vacuum.
5. Add 20 % piperidine in NMP (v/v) (2 mL), leave for 7 min, and then drain.
6. Wash thoroughly with NMP (2 × 2 mL, then 8 × 4 mL) using vacuum.

3.3 Coupling Synthesis

1. Reinsert the plunger into the syringe just above the resin.
2. Acylation step: For 250 mg TentaGel R RAM (loading = 0.18 mmol/g), mix 750 μL of premade 0.6 M bromoacetic acid in NMP and 150 μL of premade 3.2 M DIC in NMP in an Eppendorf tube. Allow the mixture to react for 3 min.
3. Suck the acylation mixture into the syringe by dipping the syringe tip in the acylation mixture and pulling the plunger. Make sure the resin is completely soaked. If needed, add a minimum volume of NMP.
4. Seal the syringe tip with a stopper and agitate on a shaking table for 30 min at room temperature.
5. Remove the stopper and plunger and reconnect the vacuum. Rinse the plunger with NMP and drain the syringe using vacuum.
6. Wash briefly with NMP (2 × 4 mL) using vacuum.
7. Repeat **steps 1–6**.
8. Reinsert the plunger just above the resin.
9. Displace bromide by adding the primary amine. For 250 mg TentaGel R RAM, use 1 mmol amine dissolved in 1 mL of NMP. Transfer the amine solution to the syringe using the plunger. Make sure that the resin is completely soaked. If needed, add a minimum volume of NMP.
10. Seal the syringe tip with a stopper and agitate on a lab shaker for 2 h at room temperature.
11. Remove the stopper and plunger, rinse the plunger with NMP, reconnect the vacuum, and then drain.
12. Wash briefly with NMP (6 × 4 mL) using vacuum.

13. Repeat **steps 1–12** to elongate peptoid.
14. After the final displacement reaction, wash thoroughly with NMP (3×4 mL) and then ethanol (3×4 mL).
15. Reinsert the plunger into the syringe.
16. Dry off any remaining ethanol by, e.g., lyophilization or by placing the syringe in vacuum for at least 1 h, preferably overnight.

3.4 Cleavage

1. Place the syringe in an upright position and seal with a stopper.
2. Cleave off the peptoid (as well as side-chain protecting groups) by transferring TFA:H_2O (95:5) (3 mL) to the resin and wrap with parafilm.
3. After 2 h, place the syringe in a cryotube and carefully remove parafilm and stopper.
4. Allow liquid to elute from the syringe and collect eluate.
5. Wash resin with TFA:H_2O (95:5) (3–4 mL) and collect eluate.
6. Evaporate most of the combined eluate with a gentle stream of N_2.
7. When a few hundred microliters of liquid remains, add diethyl ether (5 mL) to precipitate peptoid.
8. Shake briefly and leave standing for 10 min.
9. Centrifuge sample at 1,400×*g* for 5 min and discard supernatant.
10. Add diethyl ether (5 mL) to wash peptoid.
11. Repeat **steps 8–10**.
12. Leave samples open in the fume hood overnight to allow remaining diethyl ether to evaporate.
13. Redissolve crude product in CH_3CN/H_2O (1:1).
14. Lyophilize.
15. Repeat **steps 13** and **14**, which generally results in a white fluffy powder. Notice the product will be obtained as the TFA salt.
16. Analyze the product by LC-MS or MALDI-TOF-MS and HPLC.
17. For the described synthesis, analytical HPLC was performed using a Phenomenex C_{12} reversed phase column on a Waters 600E system equipped with Empower software (Waters, Milford, MA, USA). Samples were chromatographed at a flow rate of 1.5 mL/min starting with 0.1 % aqueous trifluoroacetic acid (TFA) (buffer A) and increasing over 20 min to 0.1 % TFA in CH_3CN/H_2O (7:1) followed by 0.1 % TFA in

CH_3CN/H_2O (9:1) (buffer B) over 10 min, detection at 220 nm.

Preparative HPLC was performed as described above on a Vydac C_{18} reversed phase column. Samples were initially chromatographed at a flow rate of 1 mL/min starting with buffer A/buffer B (9:1) for 10 min and increasing over 60 min to buffer A/buffer B (4:6) and finally to buffer B over an additional 8 min. Flow rate was 4 mL/min, detection at 220 nm.

Matrix-assisted laser desorption/ionization time-of-flight mass spectrometry was carried out on a Bruker Flex, using α-cyano-*p-hydroxycinnamic* acid as matrix and peptide standard calibration.

4 Notes

1. One syringe per synthesis, preferably with rubber gasket. For approx. 250 mg resin, use a 5-mL syringe.
2. We recommend a 200-μL pipette tip cut to fit the syringe tip and melted closed.
3. The filter is prepared with a punch and should fit snugly into the syringe, preventing resin from escaping when draining solvent.
4. TFA is highly toxic, volatile, etc.

Acknowledgements

The present work is performed as a part of the Danish Center for Antibiotic Research and Development (DanCARD) financed by The Danish Council for Strategic Research (grant no. 09-067075).

References

1. Simon RJ, Kania RS, Zuckermann RN et al (1992) Peptoids: a modular approach to drug discovery. Proc Natl Acad Sci USA 89:9367–9371
2. Patch JA, Kirshenbaum K, Seurynck S et al (2004) Versatile oligo (*N*-substituted) glycines: the many roles of peptoids in drug discovery. In: Neilsen PE (ed) Pseudo-peptides in drug development. Wiley-VCH Verlag BmbH & Co. KGaA, Weinheim, pp 1–31
3. Zuckermann RN, Kodadek T (2009) Peptoids as potential therapeutics. Curr Opin Mol Ther 11:299–307
4. Murphy JE, Uno T, Hamer JD et al (1998) A combinatorial approach to the discovery of efficient cationic peptoid reagents for gene delivery. Proc Natl Acad Sci USA 95:1517–1522
5. Zuckermann RN, Martin EJ, Spellmeyer DC et al (1994) Discovery of nanomolar ligands for 7-transmembrane G-protein-coupled receptors from a diverse *N*-(substituted) glycine peptoid library. J Med Chem 37: 2678–2685
6. Nguyen JT, Turck CW, Cohen FE et al (1998) Exploiting the basis of proline recognition by SH3 and WW domains: design of N-substituted inhibitors. Science 282:2088–2092
7. Mazaleyrat JP, Rage I, Mouna AM et al (1994) Peptoid mimics of a C2-symmetric inhibitor of the HIV-1 protease. Bioorg Med Chem Lett 4:1281–1284

8. Garcia-Martinez C, Humet M, Planells-Cases R et al (2002) Attenuation of thermal nociception and hyperalgesia by VR1 blockers. Proc Natl Acad Sci USA 99:2374–2379
9. Lee B-C, Zuckermann RN (2011) Protein side-chain translocation mutagenesis via incorporation of peptoid residues. ACS Chem Biol 6:1367–1374
10. Chen X, Wu J, Luo Y et al (2011) Expanded polyglutamine-binding peptoid as a novel therapeutic agent for treatment of Huntington's disease. Chem Biol 18:1113–1125
11. Huang C, Uno T, Murphy J et al (1998) Lipitoids-novel cationic lipids for cellular delivery of plasmid DNA in vitro. Chem Biol 5:345–354
12. Seo J, Michaelian N, Owens SC et al (2009) Chemoselective and microwave-assisted synthesis of glycopeptoids. Org Lett 11: 5210–5213
13. De Cola C, Manicardi A, Corradini R et al (2012) Carboxyalkyl peptoid PNAs: synthesis and hybridization properties. Tetrahedron 68:499–506
14. Fowler SA, Blackwell HE (2009) Structure–function relationships in peptoids: recent advances toward deciphering the structural requirements for biological function. Org Biomol Chem 7:1508–1524
15. Culf AS, Ouellette RJ (2010) Solid-phase synthesis of N-substituted glycine oligomers (α-peptoids) and derivatives. Molecules 15:5282–5335
16. Godballe T, Nilsson LL, Petersen PD et al (2011) Antimicrobial α-peptides and α-peptoids. Chem Biol Drug Des 77:107–116
17. Zuckermann RN (2011) Peptoid origins. Biopolymers 96:545–555
18. Olsen CA (2010) Peptoid-peptide hybrid backbone architectures. Chembiochem 11:152–160
19. Brogden NK, Brogden KA (2011) Will new generations of modified antimicrobial peptides improve their potential as pharmaceuticals? Int J Antimicrob Agents 38:217–225
20. Ryge TS, Frimodt-Moller N, Hansen PR (2008) Antimicrobial activities of twenty lysine-peptoid hybrids against clinically relevant bacteria and fungi. Chemotherapy 54: 152–156
21. Ryge T, Hansen P (2005) Novel lysine-peptoid hybrids with antibacterial properties. J Pept Sci 11:727–734
22. Goodson B, Ehrhardt A, Ng S, Nuss J et al (1999) Characterization of novel antimicrobial peptoids. Antimicrob Agents Chemother 43:1429–1434
23. Kapoor R, Eimerman PR, Hardy JW et al (2011) Efficacy of antimicrobial peptoids against *mycobacterium tuberculosis*. Antimicrob Agents Chemother 55:3058–3062
24. Kapoor R, Wadman MW, Dohm MT et al (2011) Antimicrobial peptoids Are effective against *pseudomonas aeruginosa* biofilms. Antimicrob Agents Chemother 55:3054–3057
25. Huang ML, Shin SBY, Benson MA et al (2012) A comparison of linear and cyclic peptoid oligomers as potent antimicrobial agents. ChemMedChem 7:114–122
26. Ryge TS, Hansen PR (2006) Potent antibacterial lysine-peptoid hybrids identified from a positional scanning combinatorial library. Bioorg Med Chem 14:4444–4451
27. Humet M, Carbonell T, Masip I et al (2003) A positional scanning combinatorial library of peptoids as a source of biological active molecules: identification of antimicrobials. J Comb Chem 5:597–605
28. Chongsiriwatana NP, Patch JA, Czyzewski AM et al (2008) Peptoids that mimic the structure, function, and mechanism of helical antimicrobial peptides. Proc Natl Acad Sci USA 105:2794–2799
29. Tal-Gan Y, Freeman NS, Klein S et al (2010) Synthesis and structure-activity relationship studies of peptidomimetic PKB/Akt inhibitors: the significance of backbone interactions. Bioorg Med Chem 18:2976–2985
30. Li S, Bowerman D, Marthandan N et al (2004) Photolithographic synthesis of peptoids. J Am Chem Soc 126:4088–4089
31. Zuckermann RN, Kerr JM, Kent SBH et al (1992) Efficient method for the preparation of peptoids (oligo(N-substituted glycines)) by submonomer solid-phase synthesis. J Am Chem Soc 114:10646–10647
32. Armand P, Kirshenbaum K, Goldsmith RA et al (1998) NMR determination of the major solution conformation of a peptoid pentamer with chiral side chains. Proc Natl Acad Sci USA 95:4309–4314
33. Sui Q, Borchardt D, Rabenstein DL (2007) Kinetics and equilibria of cis/trans isomerization of backbone amide bonds in peptoids. J Am Chem Soc 129:12042–12048
34. Shin SBY, Kirshenbaum K (2007) Conformational rearrangements by water-soluble peptoid foldamers. Org Lett 9:5003–5006
35. Figliozzi GM, Goldsmith R, Ng S, Banville SC, Zuckermann RN (1996) Synthesis of N-substituted glycine peptoid libraries. Methods Enzymol 267:437–447
36. Zhu WL, Hahm KS, Shin SY (2007) Cathelicidin-derived Trp/Pro-rich antimicrobial peptides with lysine peptoid residue (nlys): therapeutic index and plausible mode of action. J Pept Sci 13:529–535

Chapter 12

Synthesis of Lipidated Peptides

Federica Rosi and Gemma Triola

Abstract

One of the main reasons of the high diversity and complexity of the human proteome compared to the human genome is the extensive work performed by the posttranslational machinery to incorporate numerous different functionalities on proteins. The covalent attachment of chemical moieties in proteins after translation is known as posttranslational modification (PTM) and has a crucial role in controlling protein localization and activity. Relevant modifications include phosphorylation, carboxymethylation, glycosylation, acetylation, or lipidation. Despite their essential role on protein function, the synthesis of fully posttranslationally modified proteins has been challenging. However, important advances on chemical biology have enabled the synthesis of fully posttranslationally modified peptides and proteins. As a result of this, peptides bearing, i.e., phosphorylated amino acids, C-terminal methylations, lipid modifications, or non-natural tags have become accessible. These peptides, as well as the corresponding proteins obtained using ligation techniques, have been invaluable tools in biochemical and biophysical studies. As an example of these advances, this chapter describes the methods developed for the synthesis of lipidated peptides from the Ras and Rab families.

Key words Lipidated peptides, Farnesyl, Palmitoyl, Ras, Rab

1 Introduction

Proteins can be posttranslationally modified by covalent attachment of lipids such as cholesterol, myristic acid, palmitic acid, or isoprenoid groups [1, 2]. Relevant examples of lipidated proteins are the members of the Ras and Rab families in which it has been clearly shown that lipidation has a pivotal role in controlling not only protein localization but also their functioning [3, 4].

Ras superfamily of small GTPases comprises more than 100 different proteins that are involved in many cellular signalling processes regulating among others cell growth and proliferation, cell motility, or vesicle transport. Ras members are expressed with a C-terminal sequence known as CAAX box (where C is a cysteine, A can be any aliphatic amino acid, and X is any amino acid and determines the isoprenyl group to be attached to the cysteine) [5]. This CAAX box is recognized by prenyl transferases (farnesyltransferase

Knud J. Jensen et al. (eds.), *Peptide Synthesis and Applications: Second Edition*, Methods in Molecular Biology, vol. 1047, DOI 10.1007/978-1-62703-544-6_12, © Springer Science+Business Media New York 2013

(FTase) or geranylgeranyltransferase (GGTase)) which attach a 15-carbon farnesyl (Far) or a 20-carbon geranylgeranyl (GerGer), respectively [6]. If X is Phe or Leu, a GerGer group is linked to the cysteine, in all the other cases a Far moiety is attached [7]. Proteins containing this CAAX box undergo additional modifications. The three last amino acids are then cleaved by the Ras converting enzyme (Rce1), and the resulting C-terminal cysteine is then methylated by isoprenylcysteine carboxymethyl transferase (Icmt) [5]. Alternatively, Rab proteins lack the C-terminal CAAX box and are doubly geranylgeranylated by Rab geranylgeranyltransferase (RabGGTase) which requires previous association of Rab proteins with the Rab escorting protein (REP) [8].

Although the isoprenoid moiety is essential for localization, it is usually not enough for ensuring stable membrane association. In the case of the K-Ras isoform, this is achieved by a polybasic sequence formed by eight lysines, which specifically interact with the acidic phospholipids present at the membrane. In some other cases, such as the N-Ras or H-Ras isoforms, a second lipidation motif is necessary, and these proteins are subsequently palmitoylated at cysteine residues located near the prenylated cysteine. The palmitoylation of cysteines is a reversible modification catalyzed by protein acyltransferases (PATs) and by acyl protein thioesterases (APTs) (Fig. 1) [9].

Preparative amounts of lipidated proteins would be a prime tool for the study of processes were such proteins are involved. However, these fully modified proteins are not easily accessible using standard techniques. In this context, important advances in chemical biology have enabled the synthesis of lipidated proteins via ligation of truncated expressed proteins and lipidated peptides [10, 11].

Important aspects have to be considered for the synthesis of these lipidated peptides. The strategy to be adopted depends on many factors, such as the number and the nature of the lipid groups, their position on the peptide, the length of the peptide chain, the nature of C-terminus, and the presence of nonnatural groups (fluorophores or tags for ligation). Many examples have been reported that rely on different synthetic approaches, such as the use of solution or solid-phase strategies or the different methods applied for the incorporation of the lipid groups (either by lipidation of a selectively deprotected amino acid or via the use of lipidated building blocks).

In our group, much work has been done to establish efficient methodologies that allow the synthesis of lipidated peptides incorporating all the desired functionalities. As a result of this, many advances have been done in the last decade in this direction [10, 12–21]. Current methods are mainly based on the synthesis of peptides using solid-phase strategies and previously synthesized lipidated amino acids as building blocks. Particularly, efficient

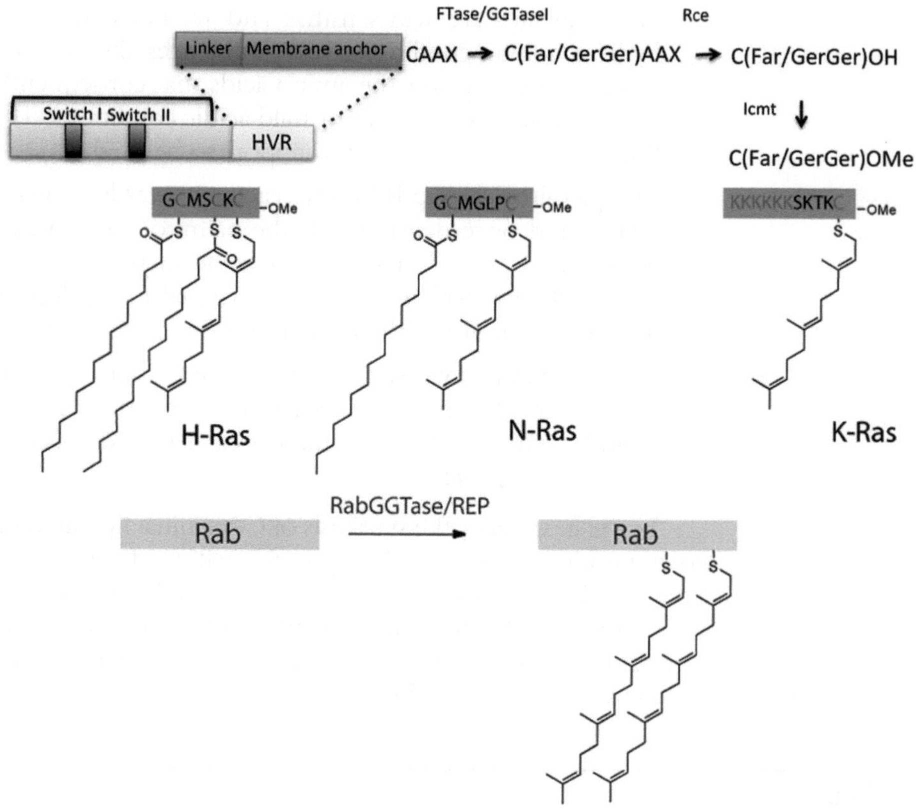

Fig. 1 C-terminal processing of CAAX box containing proteins and C-terminal lipidation pattern of Ras (H-Ras, N-Ras, and K-Ras) and Rab proteins

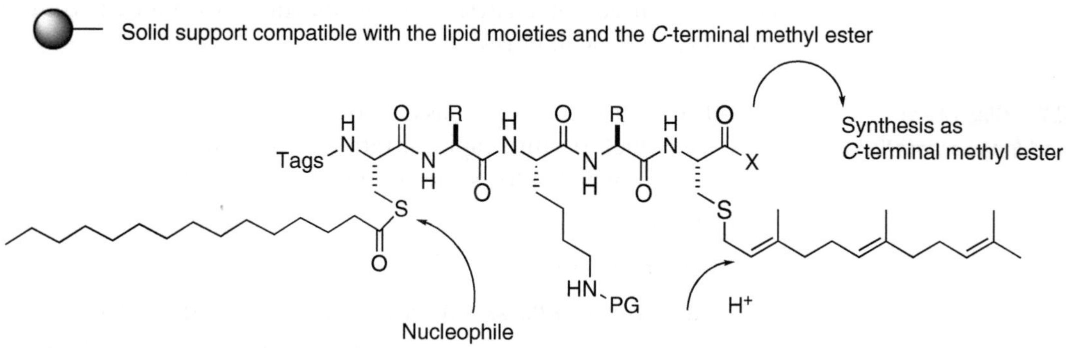

Fig. 2 General considerations for the synthesis of lipidated peptides on solid support

strategies for the synthesis of fully posttranslationally modified peptides corresponding to the C-terminus of Ras and Rab proteins have been developed. In this specific case, several considerations need to be taken into account (Fig. 2).

- The required lipidated building blocks have to be synthesized first. In some cases, new methods for the synthesis of these lipidated cysteines need to be developed.

- Prenyl groups are acid sensitive and are not compatible with hydrogenolytic conditions. This influences the choice of the protecting groups for the amino acids that consequently need to be removed under basic or mild acidic conditions. The same limitation concerns the solid support to be used.
- The palmitoyl group is highly reactive towards a nucleophilic attack, and once deprotected, the palmitoyl group can easily undergo an *S*- to *N*-acyl shift. Specific methods are required for coupling as well as for the N-terminal Fmoc deprotection in order to minimize the formation of side products.
- An appropriate solid support has to be chosen which is compatible with the lipid moieties and that can afford the required peptides with a C-terminal methyl ester in case this is present in the native protein.

In the next pages the synthesis of C-terminal lipidated peptides corresponding to Ras and Rab proteins will be described in detail. First of all, the synthesis of the required building blocks will be shown followed by different solid-phase strategies for the synthesis of a large variety of monolipidated, double lipidated, labeled, or differently functionalized peptides.

2 Materials

Amino acids, resins, all reagents, and solvents are commercially available and, if not specified, used without further purification. The water used for analysis and purification is obtained using a MilliQ system (Millipore).

2.1 Glassware and Instrumentation

1. 50 or 100 mL reactor flasks with a frit at the bottom, a stopcock, which permits rapid filtration and washing of the resin, a side arm, which allows the supply of nitrogen or argon.
2. Orbital shaker.
3. Magnetic stirrer.
4. Round-bottom flasks equipped with a magnetic stirring bar.
5. Polypropylene syringe reactors for solid-phase synthesis (pore size 25 μm) (Fig. 3).
6. Syringes equipped with a steel needle.
7. Rotary evaporator.
8. Glass column for chromatography (silica gel for chromatography—0.035–0.070 mm, 60 A).
9. TLC plates (silica gel).
10. Nuclear magnetic resonance spectrometer.
11. UV/Vis spectrophotometer.

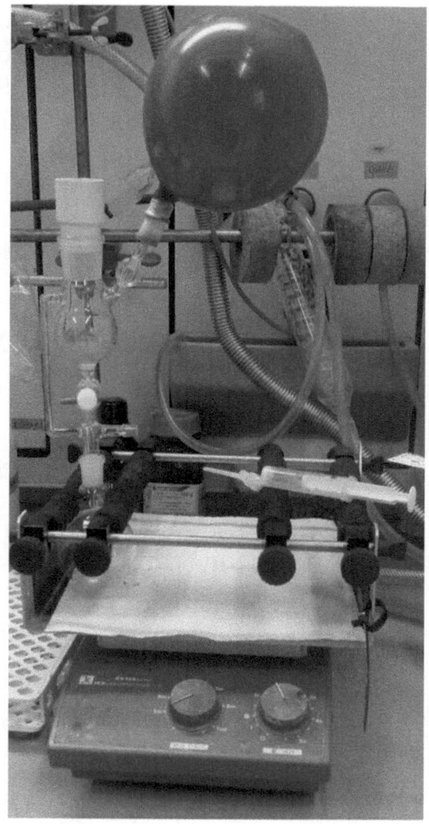

Fig. 3 Glass reactor and syringe reactor employed for solid-phase synthesis on an orbital shaker

12. Quartz cuvettes.
13. Microwave reaction vessel (reactions are carried out under stirring in a 10-mL closed reaction vessel, and temperature is controlled by an IR temperature sensor).
14. Microwave reactor (microwave irradiation is performed in a single-mode Discovery system coupled to an Explorer system (CEM)).
15. Sinter funnel.
16. Water reflux condenser.
17. Analytical reversed phase liquid chromatography–mass spectrometer (LC–MS) equipped with C4 and C18 column, coupled to an ESI spectrometer. Standard conditions for the analysis: detection at 210 or 260 nm; flow rate: 1.0 mL/min; total time: 15 min; solvents: A—0.1 % HCOOH in water, B—0.1 % HCOOH in acetonitrile; gradient: 10 % B for 1 min, from 10 to 100 % B in 10 min, 100 % B for 1 min, 10 % B for 3 min.

18. Preparative reversed phase HPLC equipped with C4 and C18 column—standard conditions for the purification: detection at 210 or 260 nm; flow 25 mL/min; total time 30 min; solvents: A—0.1 % TFA in water, B—0.1 % TFA in acetonitrile; gradient: 10 % B for 1 min, from 10 % B to 100 % B over 23 min, 100 % B for 2 min, 10 % B for 4 min.

2.2 Chemicals and Solvents

1. Methanol.
2. *n*-Pentane.
3. Dichloromethane (DCM).
4. Acetonitrile.
5. Toluene.
6. Dimethylformamide (DMF).
7. *N*-Methylpyrrolidone (NMP).
8. Diethyl ether.
9. Phosphomolybdic acid (staining solution for thin-layer chromatography, 5 g in 100 mL absolute ethanol).
10. Aluminum oxide (alumina).
11. Fmoc *N*-hydroxysuccinimide ester (Fmoc-OSu).
12. Triethylsilane (TES).
13. Trimethylsilyl chloride.
14. Palmitoyl chloride, farnesyl chloride, geranylgeranyl chloride.
15. Hexadecene.
16. 2,2′-Azobis(2-methylpropionitrile) (AIBN).
17. Bromophenol.
18. Tetrakis(triphenylphosphine)palladium(0) ($Pd(PPh_3)_4$).
19. Phenylsilane ($PhSiH_3$).
20. Copper(II)acetate ($Cu(OAc)_2$).
21. Magnesium sulfate ($MgSO_4$).
22. 1-Methylimidazole (1-MeIm).
23. Iodoacetonitrile.
24. Tetrahydrofuran (THF).
25. *N*-Methylmorpholine (NMM).
26. Dichloroethane (DCE).

2.3 Amino Acids, Resins, and Activators

1. Cysteine hydrochloride monohydrate.
2. H-Cys-OMe.
3. Fmoc-Cys(Trt)-OH, Fmoc-Cys(S*t*Bu)-OH, Fmoc-Ser(Trt)-OH, Fmoc-Glu(OAll)-OH, Fmoc-Pro-OH, Fmoc-Gly-OH, Fmoc-Lys(Boc)-OAll, Fmoc-Leu-OH, Fmoc-Met-OH.

4. Maleimide caproic acid.
5. HOBt (Hydroxybenzotriazole).
6. HBTU (2-(1*H*-Benzotriazole-1-yl)-1,1,3,3-tetramethylaminium hexafluorophosphate).
7. HATU (2-(1*H*-7-Azabenzotriazol-1-yl)-1,1,3,3-tetramethylaminium hexafluorophosphate).
8. HCTU (2-(6-Chloro-1-*H*-Benzotriazole-1-yl)-1,1,3,3-tetramethylaminium hexafluorophosphate).
9. PyBOP (Benzotriazole-1-yl-oxy-tris-pyrrolidino-phosphonium hexafluorophosphate).
10. DIC (*N,N′*-diisopropylcarbodiimide).
11. Fmoc-4-hydrazinobenzoyl AM NovaGel resin (Novabiochem).
12. 4-Sulfamylbutyryl-AM resin (Novabiochem).
13. Chlorotrityl chloride resin (Novabiochem).

2.4 Bases and Acids

1. Trifluoroacetic acid (TFA).
2. Triethylamine.
3. Piperidine.
4. Pyridine.
5. Acetic acid.
6. Ammonia 7N in methanol.
7. Collidine (2,4,6-trimethylpyridine (TMP)).
8. 1,8-Diazabicyclo[5.4.0]undec-7-ene (DBU).
9. *N,N′*-Diisopropylethylamine (DIEA).

3 Methods

We have developed several methods for the synthesis of lipidated peptides corresponding to the C-termini of Ras and Rab proteins. The most straightforward strategy relies on the use of solid-phase peptide synthesis methods and Fmoc-protected lipidated cysteines. These approaches have allowed us to obtain the desired peptides including both natural and nonnatural modifications with high purity and good yields in a short time. Different linkers have been investigated that match the requirements of the desired peptides. We will here give an overview of the different methods applied using as examples peptides of the Rab7, N-Ras, and K-Ras C-termini.

Prior to starting the synthesis of the lipidated peptides on solid phase, it is necessary to prepare the Fmoc-protected cysteines equipped with the different lipid modifications. The prenylated cysteines (either with a farnesyl or a geranylgeranyl group) can be

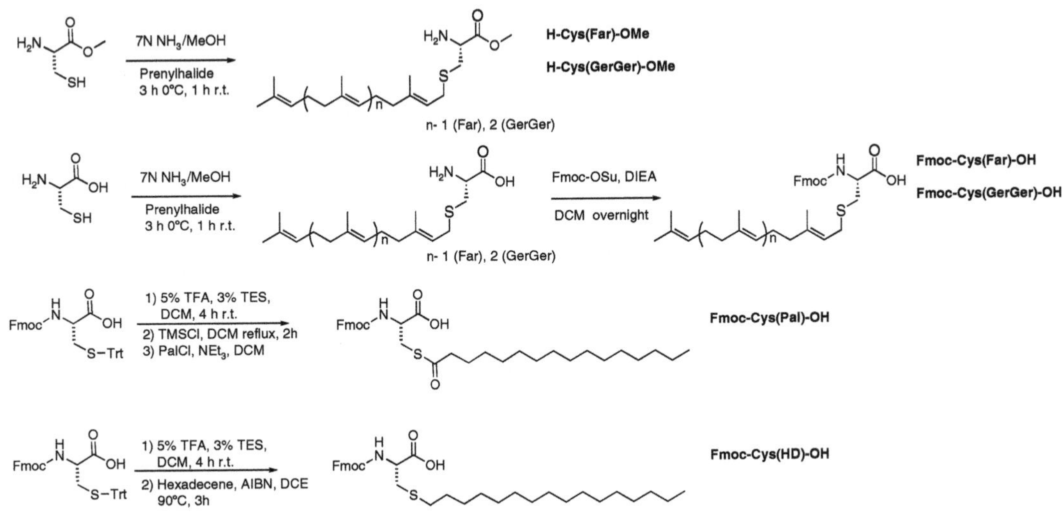

Fig. 4 Synthesis of the required lipidated cysteine building blocks

prepared in two different versions, the Fmoc-protected cysteine or the cysteine methyl ester with the free amino group [19]. The use of one or another depends on the design of the synthesis, as it will be shown below. The synthesis of the other necessary building blocks, the palmitoylated cysteine as well as its non-hydrolyzable analog Fmoc-Cys(HD)-OH where the labile thioester has been replaced by a more stable thioether, is performed in a different manner. The Fmoc-Cys(Pal)-OH is synthesized from Fmoc-Cys(Trt)-OH after removal of the trityl group and reaction with the corresponding palmitoyl chloride [19]. The Fmoc-Cys(HD)-OH is alternatively obtained applying a thiol-ene reaction to the corresponding thiol and a long chain alkene and using 2,2′-azobis(2-methylpropionitrile) (AIBN) as a radical initiator (Fig. 4) [22].

3.1 Synthesis of Lipidated Cysteines

3.1.1 H-Cys(Far)-OMe and H-Cys(GerGer)- OMe

1. In a flask (*see* **Note 1**) prepare a solution of the cysteine hydrochloride monohydrate or the cysteine methyl ester (5.7 mmol) in methanol (11 mL), cool it to 0 °C, and keep under inert atmosphere (*see* **Note 1**). Add slowly and under stirring a 7 N solution of ammonia in methanol (15 mL, *see* **Notes 2** and **3**).
2. Add to the prepared solution the prenyl halide (1 equiv.) and stir at 0 °C for 3 h and at room temperature for 1 h.
3. Evaporate the solvent under reduced pressure.
4. Wash the solid residue with *n*-pentane (3 × 50 mL).
5. Purify the crude by column chromatography on silica gel using a gradient of 0–3 % methanol in DCM, detect the purified fractions by TLC using UV lamp or a staining solution.
6. Evaporate the fractions containing the final product to obtain a pale yellow oil, which identity will be confirmed by ^{1}H-NMR and LC–MS.

3.1.2 Fmoc-Cys(Far)-OH and Fmoc-Cys(GerGer)-OH

1. In a flask (*see* **Note 1**) prepare a solution of the cysteine hydrochloride monohydrate or the cysteine methyl ester (5.7 mmol) in methanol (11 mL), cool it to 0 °C, and keep under inert atmosphere (*see* **Note 1**). Add slowly and under stirring a 7 N solution of ammonia in methanol (15 mL, *see* **Notes 2** and **3**).
2. Add to the prepared solution the prenyl halide (1 equiv.) and stir at 0 °C for 3 h and at room temperature for 1 h.
3. Evaporate the solvent under reduced pressure.
4. Wash the solid residue with *n*-pentane (3 × 50 mL).
5. Place the residue in a round-bottom flask and add DCM (50 mL).
6. Cool the resulting suspension to 0 °C, and add triethylamine (0.88 mL, 6.28 mmol) and Fmoc *N*-hydroxysuccinimide ester (Fmoc-OSu) (2.12 g, 6.28 mmol).
7. Stir the reaction mixture overnight at room temperature. Check the completeness of the reaction by TLC (use as an eluent a solution of methanol in DCM 1:9).
8. Evaporate the solvent under reduced pressure.
9. Purify the crude by column chromatography on silica gel using a gradient of 0–3 % methanol in DCM (Fmoc-Cys(Far)-OH) or 0–10 % methanol in DCM (Fmoc-Cys(GerGer)-OH), detect the purified fractions by TLC using a UV lamp or a staining solution.
10. Evaporate the fractions containing the final product to obtain a pale yellow oil, which identity will be confirmed by ^{1}H-NMR and LC–MS.

3.1.3 Palmitoylated Cysteine (Fmoc-Cys(Pal)-OH)

1. In a round-bottom flask (*see* **Note 1**), dissolve the Fmoc-Cys(Trt)-OH (2 g, 3.4 mmol) in DCM (50 mL) and keep under inert atmosphere (*see* **Note 2**).
2. Add to the stirred solution trifluoroacetic acid (TFA) (*see* **Note 4**) (2.5 mL) and triethylsilane (TES) (1.5 mL).
3. Stir at room temperature for 4 h under inert atmosphere.
4. Add toluene (15 mL) and co-evaporate under reduced pressure (*see* **Note 5**). Repeat the operation three times, evaporating each time to dryness.
5. Transfer the solid residue into a sinter funnel (*see* **Note 6**), and wash it five times with pentane (5 mL for 1 mmol of starting cysteine) to remove the released triphenylmethane.
6. Transfer the remaining solid to a round-bottom flask (*see* **Note 7**) and dissolve it in DCM (25 mL for 1 g of residue). Equip the reaction with a water reflux condenser.
7. Add trimethylsilyl chloride (0.48 mL, 3.7 mmol) and heat the solution to reflux for 2 h.

8. Cool the mixture to room temperature.
9. Add palmitoyl chloride (3.1 mL, 10.2 mmol).
10. Remove the reflux condenser and place an addition funnel. Add dropwise over 3 h a solution of triethylamine (0.78 mL, 5.6 mmol) in DCM (15 mL).
11. Stir under inert atmosphere at room temperature for 1 h. Check the completeness of the reaction by TLC (use as an eluent a solution of methanol in DCM 1:9).
12. Evaporate the solvent under reduced pressure, and purify the crude mixture by flash column chromatography on silica gel using a gradient of 0–5 % methanol in DCM. Detect the purified fractions by TLC using UV lamp or staining solution.
13. Evaporate the fractions containing the final product to obtain a pale yellow oil, which identity will be confirmed by ^{1}H-NMR and LC–MS.

3.1.4 Hexadecylated Cysteine (Fmoc-Cys(HD)-OH)

1. In a round-bottom flask (*see* **Note 1**), dissolve the Fmoc-Cys(Trt)-OH (2 g, 3.4 mmol) in DCM (50 mL) and keep under inert atmosphere (*see* **Note 2**).
2. Add to the stirred solution TFA (2.5 mL, *see* **Note 4**) and TES (1.5 mL).
3. Stir at room temperature for 4 h under inert atmosphere.
4. Add toluene (15 mL) and co-evaporate under reduced pressure (*see* **Note 5**). Repeat the operation three times, evaporating each time to dryness.
5. Transfer the solid residue into a sinter funnel (*see* **Note 6**), and wash it five times with *n*-pentane (5 mL for 1 mmol of starting cysteine) to remove the released triphenylmethane.
6. Degas the dichloroethane under an argon stream in a sonicator bath for 15 min.
7. Transfer the remaining solid to a round-bottom flask and dissolve it in the degassed dichloroethane. Equip the reaction with a water reflux condenser.
8. Add the hexadecene (3 equiv.) and the AIBN (0.5 equiv.) and heat it to reflux at 90 °C for 3 h (*see* **Note 8**). Check the completeness of the reaction by TLC (eluent mixture 10 % methanol in DCM).
9. Cool the mixture to room temperature and remove the solvent under reduced pressure.
10. Purify the crude by flash column chromatography on silica gel using a gradient of 0–4 % methanol in DCM. Detect the purified fractions by TLC using UV lamp or staining solution.

3.2 Solid-Phase Peptide Synthesis: General Methods

As mentioned above, the linker chosen for the solid-phase synthesis of lipidated peptides is of utmost importance. High concentrations of acid during the synthesis or for the release of the peptide from the solid support should be avoided to keep the isoprenoid group intact. In case a palmitoyl group is present, different conditions for the Fmoc deprotection and the coupling of amino acids should be used to minimize a nucleophilic attack to the thioester as well as an *S*- to *N*-acyl shift. Finally, the linker should be able to afford the desired peptide as a C-terminal methyl ester in case this functionality is present at the native sequence.

Not so many linkers meet all these requirements. Among them we can find the hydrazide linker and the sulfonamide linker. Both linkers are stable under acid and basic conditions allowing the synthesis of the peptides and their orthogonal release from the solid support. In the case of the hydrazide linker, the peptide is obtained after an oxidative cleavage [18]. The first step is the oxidation of the hydrazine to hydrazide, and then, methanol is used as a nucleophile to release the peptide with a C-terminal methyl ester. In the case of the sulfonamide linker, the coupling of the first amino acid generates an acyl sulfonamide stable under acid or basic conditions until it is selectively alkylated with haloacetonitriles, and then it becomes susceptible to nucleophilic attack thus releasing the peptide from the solid support. Unfortunately, the use of methanol as a nucleophile is discarded in this case due to racemization of the cysteine residue in the presence of the necessary catalytic amounts of DMAP. Alternatively, peptides can be obtained by immobilization of the second C-terminal amino acid and subsequent release from the solid support by using the amino group of H-Cys(Far)-OMe as a nucleophile [21].

Another linker successfully applied for the synthesis of lipidated peptides is the trityl linker, which affords the desired peptides after treatment with low concentrations of acid (1 % TFA) without affecting the integrity of the prenyl group. The main disadvantage of this linker is that it affords the peptides as free acid. In order to obtain the C-terminal methyl ester functionality, the second C-terminal amino acid can be immobilized on the solid support via the side-chain functionality. After incorporation of the required prenylated cysteine methyl ester, the peptide chain can be normally elongated and released from the solid support (Fig. 5) [23].

Some examples will now be described in more detail. Peptides corresponding to the C-terminal sequence of Rab7 [24] or N-Ras are described using the hydrazide linker. Similarly, a double lipidated N-Ras peptide is described using the Ellman sulfonamide linker. Finally, the synthesis of a peptide corresponding to the C-terminus of K-Ras will be described using the trityl linker. Please note that although the palmitoylated version of N-Ras is described using the hydrazide linker, both linkers, hydrazide and sulfonamide, are suitable for the synthesis of peptides bearing palmitoyl groups, fluorophores, biotin tags, or any other modification. Some

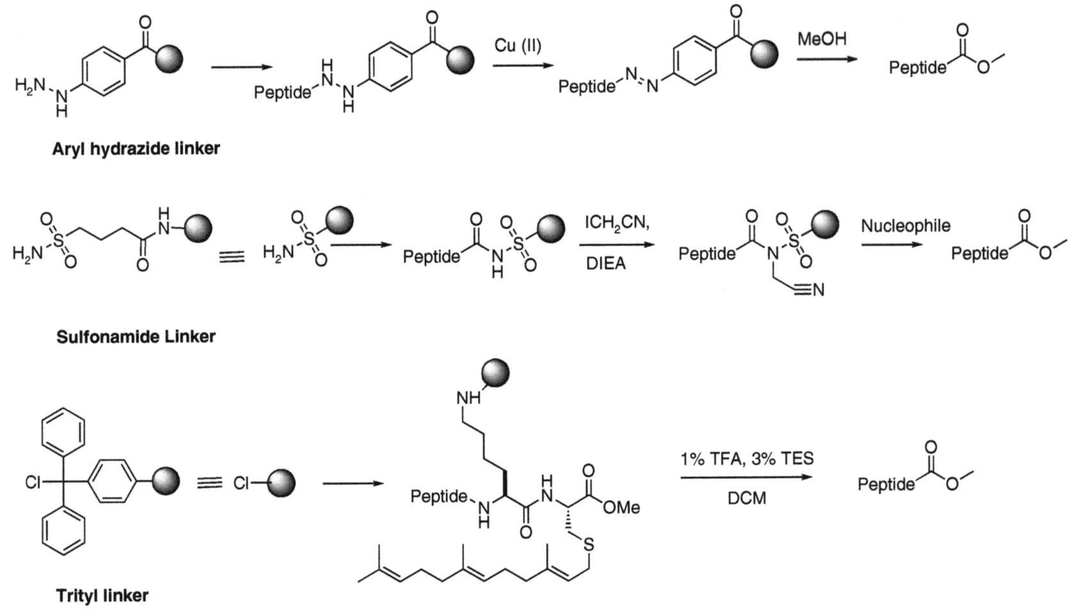

Fig. 5 General approaches for the synthesis of lipidated peptides using the hydrazide, sulfonamide, and trityl linkers

minor differences between both linkers should be taken into account. The hydrazide linker requires the use of degassed solvents and reagents to avoid oxidation before the release of the peptide, which would result in lower yields. In the case of the sulfonamide linkers, loading of the first amino acid without racemization may be an issue in some particular cases, and the appropriate conditions should be carefully chosen [25]. In addition, the release of the peptide is a two-step procedure that may take more than 24 h, and the deprotection of the side-chain protecting groups should be made after release of the peptide.

3.3 Fmoc Deprotection

3.3.1 Fmoc Removal with 20 % Piperidine/Dimethylformamide (DMF) (v/v)—Solution A

1. Treat the resin (*see* **Note 9**) with the Fmoc deprotecting solution A (2 mL for 100 mg of resin) and shake gently for 10 min.
2. Remove the solution by filtration and repeat the operation a second time.
3. Remove the solution and wash the resin with DMF (add 2 mL for 100 mg of resin, shake for 5 min, remove the solvent, and repeat the whole procedure five times).

3.3.2 Fmoc Removal with 1 % 1,8-Diazabicyclo[5.4.0]undec-7-ene (DBU)/DMF—Solution B (see **Note 10**)

1. Treat the resin (*see* **Notes 9** and **10**) with the Fmoc deprotecting solution B (2 mL for 100 mg of resin) and shake for 30 s.
2. Remove the solution and wash the resin with DMF (add 2 mL for 100 mg of resin, shake for 1 min, remove the solvent, and repeat the whole procedure three times).

3.3.3 Fmoc Removal with 2 % DBU and 2 % Piperidine/DMF (v/v/v)—Solution C

1. Treat the resin (*see* **Note 9**) twice with the Fmoc removal solution C (2 mL for 100 mg of resin) and shake for 10 min.
2. Remove the solution and wash the resin with DMF (add 2 mL for 100 mg of resin, shake for 5 min, remove the solvent, and repeat the whole procedure five times).

3.4 Coupling

3.4.1 Coupling with HBTU/HOBt/DIEA

1. Prepare a 0.4 M solution of the amino acid (4 equiv. with respect to the resin loading) in DMF (*see* **Note 1**) in a round-bottom flask.
2. Add in this order: HBTU (4 equiv.), HOBt (4 equiv.), and DIEA (8 equiv.) under stirring at room temperature.
3. Add this solution to the resin (*see* **Note 9**) and shake for 2 h at room temperature.
4. Remove the solution by filtration and wash the resin consecutively with DMF, DCM, and again with DMF (add 2 mL for 100 mg of resin, shake for 2 min, remove the solvent, and repeat the whole procedure three times with each solvent).

3.4.2 Coupling with HATU (See ***Note 10****)*

1. Prepare a 0.4 M solution of the amino acid (5 equiv. with respect to the resin loading) in a 7:1 (v/v) mixture of DCM and DMF (*see* **Note 1**).
2. Add in this order: HATU (5 equiv.) and DIEA (20 equiv.) at room temperature, and stir for 5 min to preactivate the amino acid.
3. Add this solution to the resin and shake for 2 h.
4. Remove the solution by filtration and wash the resin with DMF (add 2 mL for 100 mg of resin, shake for 2 min, remove the solvent, and repeat the whole procedure five times).

3.4.3 Coupling with HCTU/DIEA

1. Prepare a 0.4 M solution of the amino acid (4 equiv. with respect to the resin loading) in DMF (*see* **Note 1**) in a round-bottom flask.
2. Add in this order: HCTU (4 equiv.) and DIEA (8 equiv.) under stirring at room temperature.
3. Add this solution to the resin and shake for 2 h at room temperature.
4. Remove the solution by filtration and wash the resin five times with DMF and with DCM (add 2 mL for 100 mg of resin, shake for 2 min, remove the solvent, and repeat the whole procedure five times for each solvent).

3.4.4 Coupling with DIC/ HOBt

1. Prepare a 0.4 M solution of the amino acid (4 equiv. with respect to the resin loading) in *N*-Methylpyrrolidone (NMP) (*see* **Note 1**) in a round-bottom flask.
2. Add in this order: *N,N'*-diisopropylcarbodiimide (DIC) (4 equiv.) and HOBt (4 equiv.) under stirring at room temperature.

3. Add this solution to the resin and shake for 2.5 h at room temperature.
4. Remove the solution by filtration and wash the resin five times with DMF and with DCM (add 2 mL for 100 mg of resin, shake for 2 min, remove the solvent, and repeat the whole procedure five times for each solvent).

3.4.5 Coupling of the Cysteine with HBTU/HOBt/TMP (See **Note 11**) [26]

1. Prepare a solution of the Fmoc-protected cysteine (*see* Table 1 for the quantity, *see* **Note 1**) in a mixture of DCM and DMF 1:1 (4 mL for 100 mg of resin).
2. Add in this order: HBTU, HOBt, and 2,4,6-trimethylpyridine (TMP) (*see* Table 1 for the quantity) and stir at room temperature for 2 min.
3. Add this solution to the resin and shake for the right reaction time at room temperature.
4. Remove the solution by filtration and wash the resin five times with DMF and with DCM (2 mL for 100 mg of resin) for 2 min.

3.5 Fmoc Loading Determination

1. Weight two times 3 mg of dried loaded resin, write the exact weight in each case.
2. *See* Chapter 2 loading test protocol.

3.6 Completeness of the Coupling Reaction

1. The completeness of the coupling step can be monitored using the bromophenol blue test. Please note that all traces of DIEA and TMP need to be removed by extensive washing to obtain a reliable result (*see* **Note 12**).
2. In case the coupling reaction is not complete, another coupling step is recommended. Double coupling is specially recommended after the incorporation of lipidated cysteines.

3.7 Test Cleavage

1. Weigh between 10 and 20 mg of the dried resin (*see* **Note 13**).
2. Put the resin in a syringe reactor and swell it with 2 mL of DCM for 20 min.

Table 1
Coupling of lipidated cysteine building blocks with HBTU/HOBt/TMP

AA	Coupling	Solvent	Time
Fmoc-Cys(Far)-OH (4 equiv.)	HBTU/HOBt/TMP (4:4:4)	DCM/DMF (1:1)	5 h
Fmoc-Cys(Pal)-OH (4 equiv.)	HBTU/HOBt/TMP (4:4:4)	DCM/DMF (1:1)	Overnight
Fmoc-Cys(S*t*Bu)-OH (4 equiv.)	HBTU/HOBt/TMP (4:4:4)	DCM/DMF (1:1)	4 h
Fmoc-Cys(HD)-OH (4 equiv.)	HBTU/HOBt/TMP (4:4:4)	DCM/DMF (1:1)	Overnight

3. Remove the solvent and treat the resin with the appropriate cleavage mixture (*see* detailed conditions in the next sections).
4. Collect the solution in a flask and evaporate the solvent (if TFA is present, always co-evaporate with toluene, *see* **Note 5**).
5. Analyze the released peptide by analytical reversed phase LC–MS.

3.8 Synthesis of the Rab7 C-Terminus Using the Fmoc-Hydrazinobenzoyl Resin (Fig. 6)

3.8.1 Loading of the Fmoc-Cys(GerGer)-OH on Fmoc-4-Hydrazinobenzoyl Resin

Fmoc-4-hydrazinobenzoyl AM NovaGel resin with an initial loading 0.40–0.70 mmol/g is used (*see* **Note 14**).

1. In a reactor flask swell the resin with dry DCM (2–3 mL for 100 mg of resin) for 10 min.
2. Remove the DCM by filtration and wash the resin two times with dry DCM and two times with dry DMF (add 2 mL for 100 mg of resin, shake for 5 min, remove the solvent, and repeat the whole procedure).
3. Remove the Fmoc group by treatment with the Fmoc deprotecting solution A (2 mL for 100 mg of resin).
4. Couple the Fmoc-Cys(GerGer)-OH using HBTU/HOBT/TMP (*see* Table 1 and Subheading 3.4).
5. After the right time remove the solvent and wash the resin with DMF, DCM, and DMF (add 2 mL for 100 mg of resin, shake for 2 min, remove the solvent, and repeat the whole procedure five times for each solvent).
6. Determine the loading by Fmoc UV test (*see* Subheading 3.5).

3.8.2 Elongation of the Peptide Chain

1. Remove the Fmoc group by treatment with the Fmoc deprotecting solution A (2 mL for 100 mg of resin, synthesis is usually performed at 0.1 mmol scale).
2. Couple the Fmoc-Ser(Trt)-OH following using HBTU/HOBt/DIEA (*see* Subheading 3.4).
3. After Fmoc deprotection (solution A), couple the Fmoc-Cys(GerGer)-OH using HBTU/HOBT/TMP (*see* Table 1 and Subheading 3.4).
4. After Fmoc deprotection (solution A), couple the Fmoc-Ser(Trt)-OH and following the same procedure the Fmoc-Glu(OAll)-OH using HBTU/HOBt/DIPEA (*see* Sub-heading 3.4).
5. After Fmoc deprotection, couple the Fmoc-Cys(S*t*Bu)-OH using HBTU/HOBT/TMP (*see* Table 1 and Subheading 3.4).
6. For deprotection of the trityl group (Ser(Trt)), treat the resin containing the protected peptide with a DCM solution containing 1 % TFA (*see* **Note 4**) and 2 % TES and shake for 1 h (*see* **Note 15**).
7. Remove the solvent and wash with DCM, DMF, and DCM (add 2 mL for 100 mg of resin, shake for 5 min, remove the solvent, and repeat the whole procedure five times for each solvent).

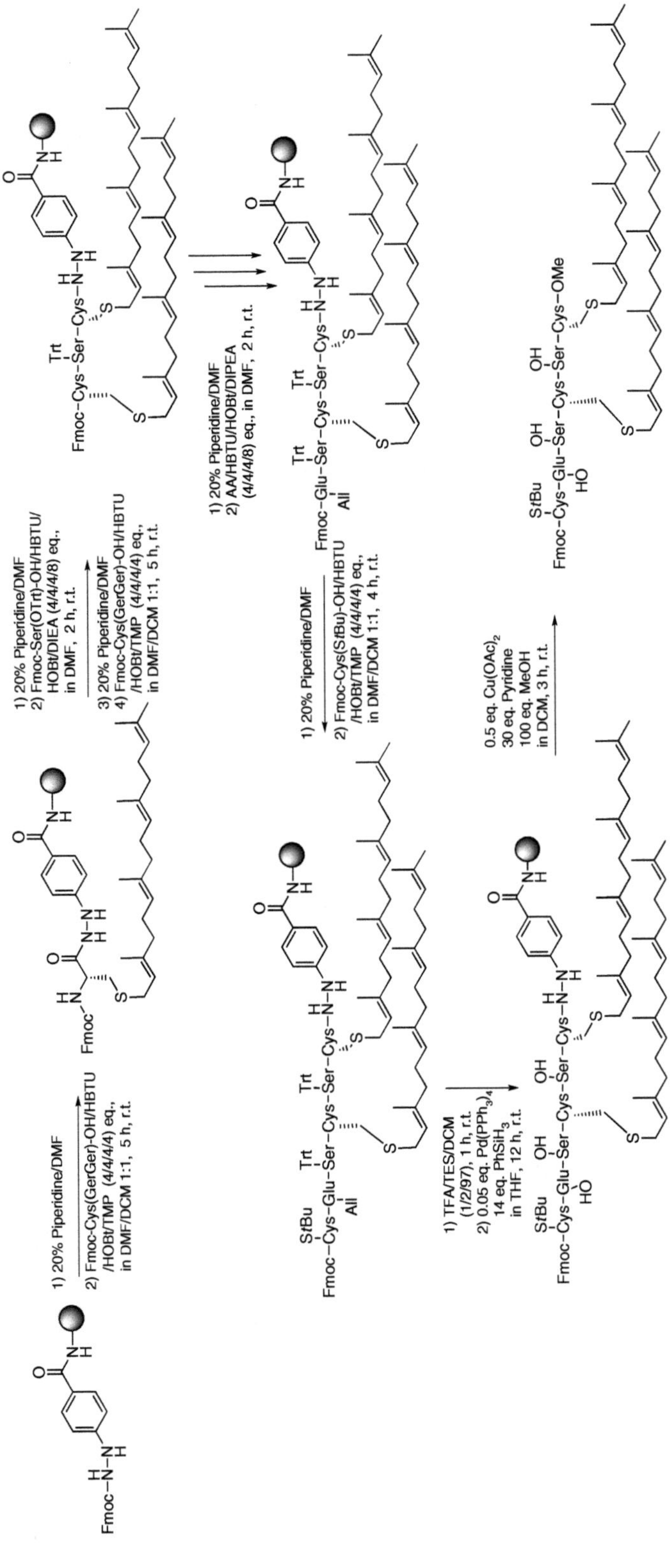

Fig. 6 Synthesis of Rab7 C-terminus using the hydrazide linker

8. For the deprotection of the allyl group (Glu(OAll)), prepare a solution containing phenylsilane ($PhSiH_3$) (14 equiv. with respect to the loaded amino acid), Tetrakis(triphenylphosphine) palladium(0) ($Pd(PPh_3)_4$) (0.05 equiv. with respect to the loaded amino acid—*see* **Note 16**) in THF (30 mL for 1 mmol of loaded amino acid).
9. Add the prepared solution to the resin and shake for 12 h at room temperature under inert atmosphere.
10. Wash the obtained brown resin with THF (*see* **Note 17**), DCM, DMF, and DCM (add 2 mL for 100 mg of resin, shake for 5 min, remove the solvent, and repeat the whole procedure five times for each solvent).
11. Dry the resin under high vacuum for some hours and perform a test cleavage (*see* Subheading 3.7).

3.8.3 Oxidative Cleavage

1. Prepare a solution of copper acetate ($Cu(OAc)_2$) (0.5 equiv. with respect to the initial loading of the resin) and pyridine (30 equiv.) in DCM, containing the nucleophile (methanol—100 equiv.).
2. Treat the resin with the prepared solution and shake for 3 h under oxygen atmosphere (*see* **Note 18**).
3. Collect the solution by filtration and co-evaporate with toluene under reduced pressure.
4. Wash the residue with toluene evaporating to dryness after each washing.
5. Solve the residue in DCM and wash with a 0.1 M solution of HCl in water. The organic phase is dried under $MgSO_4$, filtered, and the solvent is eliminated under reduced pressure.
6. Purify the crude on short silica gel column using a 0–5 % gradient of methanol in DCM or by preparative HPLC using a C4 column and the identity will be confirmed by LC–MS (*see* Subheading 2.1 for HPLC and LC–MS gradient and flow information).

3.9 Synthesis of the N-Ras C-Terminus Using the Fmoc-Hydrazinobenzoyl Resin(Fig. 7)

Fmoc-4-hydrazinobenzoyl AM NovaGel resin with an initial loading 0.40–0.70 mmol/g (*see* **Note 14**).

3.9.1 Loading of the Fmoc-Cys(Far)-OH on the Fmoc-4-Hydrazinobenzoyl Resin

1. In a solid-phase reactor, put the Fmoc-4-hydrazinobenzoyl resin (1 mmol) and swell it with DCM (2–3 mL for 100 mg of resin) for 10 min.
2. Remove the DCM by filtration and wash the resin two times with DCM and DMF (add 2 mL for 100 mg of resin, shake for 5 min, remove the solvent, and repeat the whole procedure five times for each solvent).
3. Remove the Fmoc group from the resin by treatment with the Fmoc deprotecting solution A.

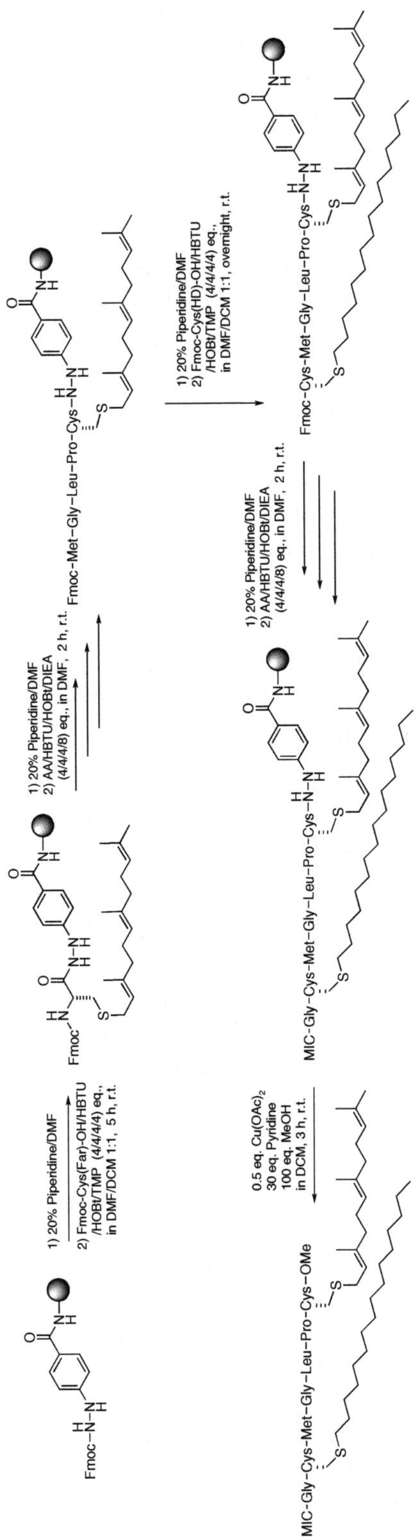

Fig. 7 Synthesis of farnesylated and hexadecylated N-Ras C-terminus using the hydrazide linker

4. Couple the Fmoc-Cys(Far)-OH using HBTU/HOBt/TMP (*see* Table 1 and Subheading 3.4).
5. After the right reaction time, remove the solution by filtration and wash with DMF, DCM, and DMF (add 2 mL for 100 mg of resin, shake for 5 min, remove the solvent, and repeat the whole procedure five times for each solvent).

3.9.2 Elongation of the Peptide Chain

1. Remove the Fmoc group on the cysteine by treatment with the Fmoc deprotecting solution A.
2. Couple the corresponding amino acids using in HBTU/HOBt/DIEA (*see* Subheading 3.4).
3. Follow the deprotection/coupling cycles for peptide chain elongation.
4. Couple the Fmoc-Cys(HD)-OH using HBTU/HOBt/TMP (*see* Table 1 and Subheading 3.4).
5. After the right reaction time, remove the solvent and wash more times with DMF, DCM, and DMF (add 2 mL for 100 mg of resin, shake for 5 min, remove the solvent, and repeat the whole procedure five times for each solvent).
6. Dry the resin under high vacuum for some hours and perform the test cleavage (*see* Subheading 3.7).

3.9.3 Oxidative Cleavage

1. Prepare a solution of $Cu(OAc)_2$ (0.5 equiv. with respect to the initial loading of the resin), pyridine (30 equiv.), acetic acid (50 equiv.) and methanol as a nucleophile (215 equiv.—5 mL for 100 mg of resin).
2. Add the solution to the resin and shake for 3 h under oxygen atmosphere (*see* **Note 18**).
3. Collect the solution by filtration and co-evaporate with toluene under reduced pressure.
4. Solve the residue in DCM and wash with a 0.1 % solution of HCl in water. The organic phase is dried under $MgSO_4$, filtered, and the solvent is eliminated under reduced pressure.
5. Purify the crude on short silica gel column using a 0–5 % gradient of methanol in DCM or by preparative HPLC using a C4 column, and the identity will be confirmed by LC–MS (*see* Subheading 2.1 for HPLC and LC–MS gradient and flow information).

3.10 Synthesis of N-Ras C-Terminus Using the Sulfonamide Linker (Fig. 8)

4-Sulfamylbutyryl-AM resin with an initial loading 1.1 mmol/g is used.

3.10.1 Loading of the First Amino Acid to the 4-Sulfamylbutyryl-Aminomethyl Resin

1. Weight the amount of resin corresponding to 0.1 mmol into a syringe reactor. For resin swelling add DCM (2–3 mL for 100 mg of resin) and shake gently for 15 min.
2. Remove the solvent by filtration and wash the resin two times with DCM (2 mL for 100 mg of resin).

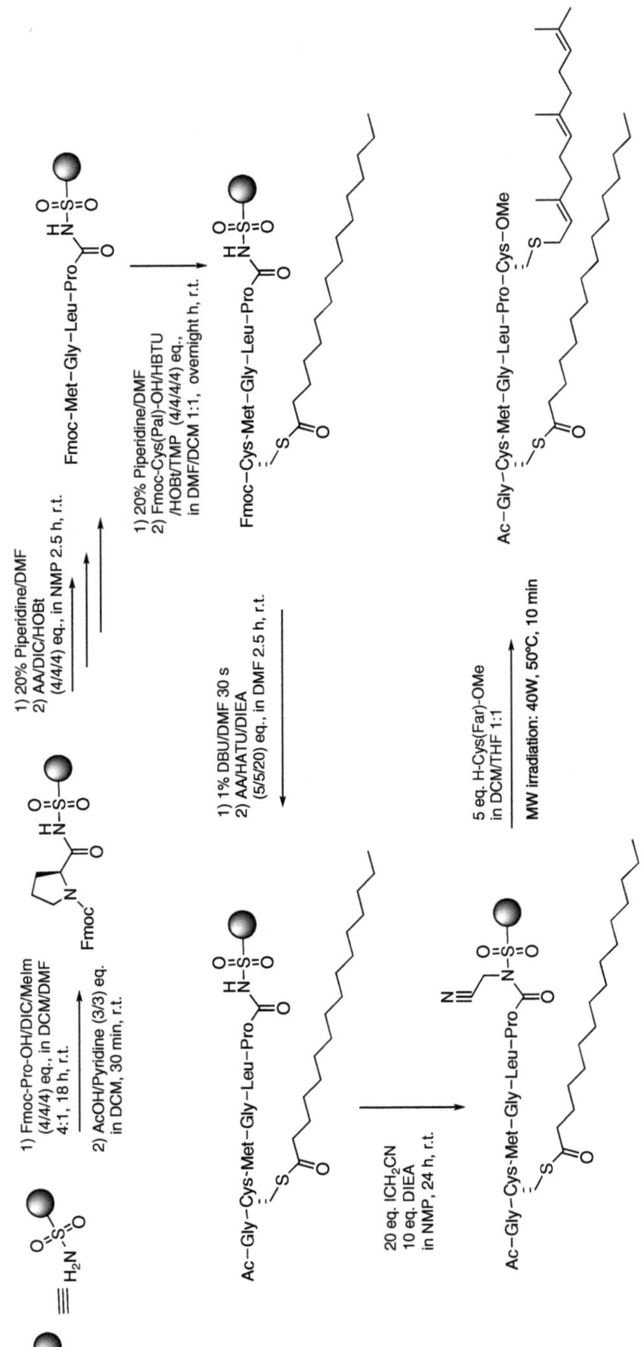

Fig. 8 Synthesis of farnesylated and palmitoylated N-Ras C-terminus using the sulfonamide linker

3. In a flask (*see* **Note 1**) equipped with a magnetic stirring bar, dissolve the Fmoc-Pro-OH (0.4 mmol) in a 4:1 (v/v) mixture of DCM and DMF to afford a final concentration of 0.4 M.
4. Add the 1-methylimidazole (1-MeIm) (0.4 mmol) and DIC (0.4 mmol) and stir at room temperature under inert atmosphere for 2 min (*see* **Note 19**).
5. Add the prepared solution to the resin and shake for 18 h.
6. Remove the solution by filtration and wash the resin with DMF and DCM (add 2 mL for 100 mg of resin, shake for 5 min, remove the solvent, and repeat the whole procedure five times for each solvent).

3.10.2 Capping of the Unreacted Amino Groups

1. Treat the resin with a DCM solution containing acetic acid (3 equiv.) and pyridine (3 equiv.) and shake for 30 min.
2. Wash the resin with DCM and DMF (add 2 mL for 100 mg of resin, shake for 5 min, remove the solvent, and repeat the whole procedure five times for each solvent).
3. Determine the loading by Fmoc UV test with the use of Fmoc deprotecting solution A (*see* **Note 20**, *see* Subheading 3.5 and Chapter 2).

3.10.3 Elongation of the Peptide (See ***Notes 21–23****)*

1. Treat the resin with Fmoc deprotecting solution A (2–3 mL for 100 mg of resin) and shake for 30 min.
2. Remove the solution and wash the resin for 5 min with DMF, repeat it five times.
3. In stirred flask dissolve the amino acid (4 equiv. with respect to the loading of the first amino acid) in NMP.
4. Add in this order: DIC (4 equiv. with respect to the loading of the first amino acid) and HOBt (4 equiv. with respect to the loading of the first amino acid) and stir for 2 min (*see* Subheading 3.4).
5. Add this solution to the resin and shake for 2.5 h at room temperature.
6. Remove the solution and wash the resin with DMF, DCM, and DMF (add 2 mL for 100 mg of resin, shake for 2 min, remove the solvent, and repeat the whole procedure five times for each solvent).
7. Repeat bullet 1–6 for each amino acid.
8. Couple the Fmoc-Cys(Pal)-OH with HBTU/HOBt/TMP (*see* Table 1, Subheading 3.4).
9. Wash the resin extensively with DMF.
10. Prepare a solution of Fmoc-Gly-OH (5 equiv.), HATU (5 equiv.), and DIEA (20 equiv.) in DMF and stir for 5 min (*see* **Note 10** and Subheading 3.4).

11. The deprotection of the Fmoc group of Fmoc-Cys(Pal)-OH is carried out using a DBU solution in DMF (Fmoc deprotection solution B).
12. Couple quickly the next amino acid by adding the preactivated solution to the resin and shake for 3 h.
13. Fmoc deprotection will be again carried out with solution B.
14. Prepare a solution of acetic acid (5 equiv.), HATU (5 equiv.), and DIEA (20 equiv.) in DMF and stir for 5 min (*see* **Note 10**). After this time add to the resin and shake for 3 h.

*3.10.4 N-Alkylation (See **Note 24**)*

1. In a round-bottom flask equipped with a magnetic stirring bar, dissolve the iodoacetonitrile (20 equiv. with respect to the loading of the first AA) and DIEA (10 equiv.) in anhydrous NMP.
2. Filter the solution through an alumina basic plug (*see* **Note 25**).
3. Add NMP to the resin and shake for 2 min. Remove the solvent and repeat this procedure three times (3 mL for 100 mg of resin).
4. Add the prepared and filtered solution of iodoacetonitrile/DIEA in NMP to the resin.
5. Resin is shaken under inert atmosphere for 24 h protected from light (i.e., covered with aluminum foil).
6. Remove the solvent by filtration and wash the resin consecutively with dry NMP and dry DCM (add 3 mL for 100 mg of resin, shake for 10 min, remove the solvent, repeat the whole procedure five times for each solvent) and with anhydrous THF (add 3 mL for 100 mg of resin, shake for 5 min, remove the solvent, repeat the whole procedure three times for each solvent—*see* **Note 26**).
7. Dry the resin under high vacuum for several hours.

3.10.5 Cleavage

1. Prepare a solution of the Cys derivative (H-Cys(Far)-OMe, 5 equiv. with respect to the loading of the first amino acid) in a 1:1 (v/v) mixture of DCM and THF (3 mL for 100 mg of resin).
2. Put the resin in a microwave vessel reactor and add the cysteine solution.
3. Irradiate the vessel with an initial power of 40W, at 50 °C for 10 min.
4. Remove the solvent containing the desired peptide by filtration, and wash the resin three times with THF (3 mL for 100 mg of resin) and three times with DCM (3 mL for 100 mg of resin).
5. Collect all the solvents and evaporate them under reduced pressure.

6. The resulting crude peptide can be purified by preparative reverse phase HPLC using a C4 column and the identity will be confirmed by LC–MS (*see* Subheading 2.1 for HPLC and LC–MS gradient and flow information).

3.11 Synthesis of the K-Ras4B C-Terminus on the 2-Chlorotrityl Linker (Fig. 9)

2-Chlorotrithyl chloride resin with an initial loading of 0.7 mmol/g is used.

3.11.1 Loading of the Fmoc-Lys-OAll on 2-Chlorotrityl Resin

1. In a flask (*see* **Note 1**) dissolve the Fmoc-Lys(Boc)-OAll with DCM (3 mL for 1 mmol amino acid).
2. For the removal of the Boc protecting group, add TFA (3 mL for 1 mmol amino acid) and stir at room temperature for 3 h (*see* **Note 4**).
3. Co-evaporate with toluene three times under reduced pressure and dry the residue in high vacuum for 15 h.
4. In a solid-phase reactor put the 2-chlorotrityl resin (1 mmol) and dry it under high vacuum for 12 h (*see* **Note 27**).
5. Dissolve the deprotected Fmoc-Lys-OAll (0.5 equiv.) in dry DCM (7 mL) and add dry DIEA (2 equiv. with respect to the Lys).

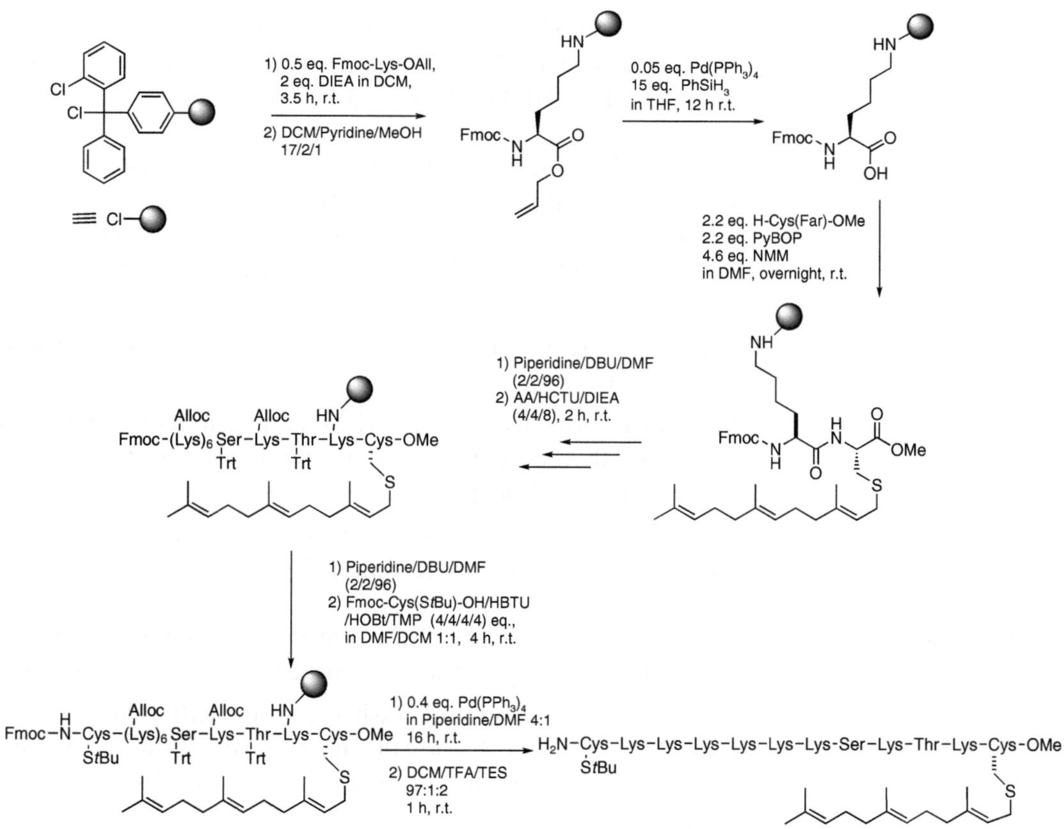

Fig. 9 Synthesis of the C-terminal peptide from K-Ras using the 2-chlorotrityl linker

6. For the resin swelling, add dry DCM to the resin and gently shake for 20 min under inert atmosphere.
7. Remove the solvent by filtration, add the solution containing the Fmoc-Lys-OAll to the resin, and shake under argon for 3.5 h.
8. Filter off the solution and treat the resin with a mixture of DCM/MeOH/DIEA (17/2/1—3 min, three times, anhydrous solvents) (*see* **Note 28**) for the blocking of unreactive groups.
9. Remove the solution and wash the resin with DCM, DMF, and again DCM (add 2 mL for 100 mg of resin, shake for 5 min, remove the solvent, and repeat the whole procedure five times—*see* **Note 28**).
10. Determine the loading of the resin with the UV Fmoc test (*see* Subheading 3.5). The average loading for this step is 0.3 mmol/g (*see* **Note 29**).

3.11.2 Allyl Deprotection

1. Prepare a solution containing $PhSiH_3$ (14 equiv. with respect to the loaded AA), $Pd(PPh_3)_4$ (0.05 equiv. with respect to the loaded amino acid—*see* **Note 16**) in THF (30 mL for 1 mmol of loaded amino acid).
2. Add the prepared solution to the loaded resin and shake for 12 h at room temperature under inert atmosphere.
3. Wash the obtained brown resin several times with THF (*see* **Note 17**), DCM, DMF, and DCM.

3.11.3 H-Cys(Far)-OMe Coupling

1. Dissolve the H-Cys(Far)-OMe (2.2 equiv. with respect to the last loading estimated) and the PyBOP (2.2 equiv. with respect to the last loading estimated) in a mixture of DMF and *N*-methylmorpholine (NMM, 4.6 equiv. with respect to the last loading estimated).
2. Add the mixture to the resin and shake overnight at room temperature.
3. Wash extensively with DMF, DCM, and again with DMF (add 2 mL for 100 mg of resin, shake for 5 min, remove the solvent, and repeat the whole procedure five times).
4. Dry the resin under high vacuum for some hours and perform the test cleavage (*see* Subheading 3.7).

3.11.4 Peptide Elongation

1. Treat the resin two times with Fmoc deprotecting solution C (2–3 mL for 100 mg of resin) and shake for 10 min (*see* **Note 30**).
2. Remove the solution and wash the resin with DMF, DCM, and DMF (add 2 mL for 100 mg of resin, shake for 5 min, remove the solvent, and repeat the whole procedure five times for each solvent).

3. Couple all the amino acids following the general conditions for the coupling using HCTU/DIEA (*see* Subheading 3.4), repeating the coupling twice for each amino acid (*see* **Note 31**).
4. Wash the resin with DCM and dry it overnight.
5. For the deprotection of the Lys(Alloc) as well as the N-terminal Fmoc group, prepare a solution of $Pd(PPh_3)_4$ (0.4 equiv.) in DMF/piperidine 1:4.
6. Treat the resin with the prepared solution and shake under inert atmosphere for 16 h.
7. Remove the solvent by filtration, and wash with DMF, DCM, and methanol (add 2 mL for 100 mg of resin, shake for 5 min, remove the solvent, and repeat the whole procedure five times for each solvent) (*see* **Note 16**).
8. Dry the resin under high vacuum for some hours and perform the test cleavage.

3.11.5 Cleavage

1. Treat the resin with a 1 % TFA and 2 % TES solution in DCM (7 mL for 100 mg of resin) and shake for 1 h (*see* **Note 15**).
2. Collect the solvent in a flask and wash the resin with DCM, DCM/methanol 1:1, and methanol.
3. Co-evaporate three times the solvent with toluene under reduced pressure (*see* **Note 15**).
4. Dissolve the crude in a small amount of methanol and add cold diethyl ether.
5. Centrifuge the precipitate, dissolve it in water, freeze it on liquid nitrogen, and lyophilize it.
6. Purify the peptide by preparative HPLC on C18 column (*see* Subheading 2.1 for HPLC and LC–MS gradient and flow information).

4 Notes

1. It is recommended to add the solid slowly to the flask containing the solvent to obtain a better dissolution.
2. In order to keep the reaction under inert atmosphere, fill a balloon with argon or nitrogen and connect it to the flask using a syringe and a septum.
3. The solution of ammonia in methanol is commercially available and can be diluted by addition of more methanol. To achieve this take the required amount of ammonia/methanol solution carefully with a syringe and add it to the methanol in a round-bottom flask. Ammonia solution in methanol is flammable and toxic, it should be always handled very carefully. Work always under inert atmosphere and do not get in contact with the toxic vapors.

4. TFA should be carefully handled; it is a toxic and corrosive acid.
5. Toluene forms an azeotrope with TFA facilitating its removal.
6. This solid may be difficult to handle or too fine to be filtered. In that case put the solid into a Falcon tube, add *n*-pentane, and mix it (a sonicator bath can be used). Remove the *n*-pentane by centrifugation and repeat the operation five times.
7. In this case it is not necessary to dry the solid before the next reaction.
8. AIBN is required as a radical initiator. Handle it carefully as it may be explosive.
9. A dry resin should be first treated with DCM for 5 min and DMF for 5 min for a proper swelling of the resin.
10. After the coupling of Fmoc-Cys(Pal)-OH, Fmoc deprotection is quickly performed using DBU as a base (Fmoc deprotection solution B), and next coupling is performed using HATU as activating agent to avoid S- to N-acyl shift once the amino group is free and the nucleophilic attack to the thioester moiety. A solution containing amino acid/HATU/DIEA should be preactivated for 5 min and added directly to the resin after Fmoc deprotection.
11. Cysteine is one of the amino acids more prone to racemization. Coupling conditions that minimize this racemization has been extensively studied. In this context, the use of equimolar conditions of HBTU/HOBt and 2,4,6-trimethylpyridine (Collidine) as base has given excellent results [26].
12. Bromophenol blue test is an alternative to the more toxic Kaiser test. A change from deep blue to yellow indicates disappearance of aminomethyl groups [27, 28].
13. Prior drying, resin is placed in a syringe reactor and washed with DCM and methanol. Then the syringe with the resin is dried overnight under high vacuum.
14. When using the Fmoc-4-hydrazinobenzoyl resin, all the solvents (DCM, DMF) and reagents (piperidine) need to be dry and degassed. To degas solvents/reagents place them for 10 min in a sonicator bath (located in a hood) under a stream of argon or nitrogen. Fmoc deprotection and coupling steps should be performed always under argon. For that reason a solid-phase reactor is preferred than a syringe reactor when using this resin.
15. When an isoprenoid moiety is present, the concentration of TFA cannot exceed 2 % to avoid side reactions with the farnesyl group. Special care has also to be taken when eliminating the solvent under reducing conditions; always add toluene as a cosolvent and the temperature of the water bath should not exceed 40 °C.

16. The $Pd(PPh_3)_4$ can be easily degraded. In that chase it loses its yellow color and turns brown. Keep always the reagent under argon. A good quality reagent is essential for the completeness of the reaction. Cleavage test is recommended after this step to check the completeness of the reaction. If it is not finished, this step should be repeated.
17. The resin should be extensively washed with THF until the filtered off solution is colorless.
18. Oxidative cleavage can be performed under oxygen by filling a balloon with oxygen and connecting it to the flask. Alternatively, reaction can be performed in an open vessel under air atmosphere.
19. Special care has to be taken with the loading of the sulfonamide resin to minimize racemization. Because of that, the conditions for the loading can vary depending on the amino acid to be loaded [21, 25].
20. The average loading obtained following this protocol is 0.6 mmol/g.
21. The general coupling conditions (HBTU/HOBt/DIEA) are not used here since incomplete couplings have been observed (probably because of a side reaction between the sulfonamide group and the HBTU under basic conditions). Alternatively, the use of DIC/HOBt has shown in this case to be a better option for the elongation of the peptide chain.
22. Double coupling is strongly recommended after introducing the lipidated cysteines. Deletion of sequences resulting from incomplete couplings cannot be easily separated from the desired peptide due to the high hydrophobicity of the double lipidated peptides, thus co-eluting in the HPLC.
23. The final acetylation of the N-terminal Gly is achieved by coupling the acetic acid in the same conditions than any amino acid.
24. The *N*-acyl sulfonamide is highly stable under acidic and basic conditions. This stability is lost when it is *N*-alkylated. In these conditions, the C-terminal acyl functionality of the peptide chain becomes more susceptible to nucleophilic attack, allowing its release from the solid support. Thus, the cleavage is performed in two steps: the alkylation of the resin and the nucleophilic attack of the peptide at the C-terminal position. The most used alkylating agent is iodoacetonitrile, while different nucleophilic agents can be used to afford different functionalities at the C-terminus (amide, thioesters).
25. Prepare a cartridge of alumina by filling a syringe reactor for solid-phase synthesis with a 2–3 cm layer of alumina, and filter the solution through this cartridge.
26. Extensive washing is required at this point to remove any traces of the reagents.

27. The 2-chlorotrityl resin is highly sensitive to moisture. For this reason, it must be always stored under argon and handled carefully. Until the first amino acid has been loaded, all the solvent used should be dry and all the reactions have to be performed under argon. Better results are obtained if the amino acid is also dried overnight at high vacuum before being used.
28. Unreacted groups are capped by the use of a DCM/methanol/pyridine solution. After this, there is no need to use dry solvents anymore and reactions can be performed under air atmosphere.
29. Low loadings are in some cases preferred because they result in higher yields of the final peptide.
30. The combination of DBU and piperidine for the Fmoc removal has shown to be more efficient in the case of polybasic peptides such as K-Ras.
31. Although single coupling protocols generally also afford the peptides, better yields and more pure products have been obtained when performing double couplings during the synthesis.

References

1. Walsh CT, Garneau-Tsodikova S, Gatto GJ (2005) Protein posttranslational modifications: the chemistry of proteome diversifications. Angew Chem Int Ed Engl 44:7342–7372
2. Triola G (2011) The protein lipidation and its analysis. J Glycom Lipidom S:2. doi: 10.4172/2153-0637.S4172-4001
3. Brunsveld L, Kuhlmann K, Alexandrov K et al (2006) Lipidated Ras and Rab peptides and proteins—synthesis, structure, and function. Angew Chem Int Ed Engl 45:6622–6646
4. Levental I, Grzybek M, Simons K (2010) Greasing their way: lipid modifications determine protein association with membrane rafts. Biochemistry (Mosc) 49:6305–6316
5. Konstantinopoulos PA, Karamouzis MV, Papavassiliou AG (2007) Post-translational modifications and regulation of the RAS superfamily of GTPases as anticancer targets. Nat Rev Drug Discov 6:540–555
6. Omerovic J, Hammond DE, Clague MJ et al (2008) Ras isoform abundance and signalling in human cancer cell lines. Oncogene 27: 2754–2762
7. Beese LS, Reid TS, Terry KL et al (2004) Crystallographic analysis of CaaX prenyltransferases complexed with substrates defines rules of protein substrate selectivity. J Mol Biol 343:417–433
8. Itzen A, Goody RS (2011) GTPases involved in vesicular trafficking: structures and mechanisms. Semin Cell Dev Biol 22:48–56
9. Triola G, Waldmann H, Hedberg C (2012) Chemical biology of lipidated proteins. ACS Chem Biol 7:87–99
10. Brunsveld L, Kuhlmann J, Waldmann H (2006) Synthesis of palmitoylated Ras-peptides and -proteins. Methods 40:151–165
11. Gerauer M, Koch S, Brunsveld L, Waldmann H (2008) Lipidated peptide synthesis. Wiley Encyclopedia of Chem Biol 1–11
12. Nagele E, Schelhaas M, Kuder N et al (1998) Chemoenzymatic synthesis of N-Ras lipopeptides. J Am Chem Soc 120:6889–6902
13. Bader B, Kuhn K, Owen DJ et al (2000) Bioorganic synthesis of lipid-modified proteins for the study of signal transduction. Nature 403:223–226
14. Machauer R, Waldmann H (2000) Synthesis of the N-terminal N-myristoylated and S-palmitoylated undetrigintapeptide of endothelial NO-synthase. Angew Chem Int Ed Engl 39:1449–1453
15. Kuhn K, Owen DJ, Bader B et al (2001) Synthesis of functional Ras lipoproteins and fluorescent derivatives. J Am Chem Soc 123:1023–1035
16. Ludolph B, Eisele F, Waldmann H (2002) Solid-phase synthesis of lipidated peptides. J Am Chem Soc 124:5954–5955
17. Peters C, Waldmann H (2003) Solid-phase synthesis of peptide esters employing the hydrazide linker. J Org Chem 68:6053–6055
18. Kragol G, Lumbierres M, Palomo J et al (2004) Solid phase synthesis of lipidated peptides. Angew Chem Int Ed Engl 43:5839–5842
19. Lumbierres M, Palomo J, Kragol G et al (2005) Solid phase synthesis of lipidated peptides. Chem Eur J 11:7405–7415

20. Brunsveld L, Kuhlmann J, Alexandrov K et al (2006) Lipidated Ras and Rab peptides and proteins-synthesis, structure, and function. Angew Chem Int Ed 45: 6622–6646
21. Triola G, Gerauer M, Gormer K et al (2010) Solid-phase synthesis of lipidated ras peptides employing the Ellman sulfonamide linker. Chem Eur J 16:9585–9591
22. Triola G, Brunsveld L, Waldmann H (2008) Racemization-free synthesis of S-alkylated cysteines via thiol-ene reaction. J Org Chem 73:3646–3649
23. Chen YX, Koch S, Uhlenbrock K et al (2010) Synthesis of the Rheb and K-Ras4B GTPases. Angew Chem Int Ed Engl 49:6090–6095
24. Brunsveld L, Watzke A, Durek T et al (2005) Synthesis of functionalized Rab GTPases by a combination of solution- or solid-phase lipopeptide synthesis with expressed protein ligation. Chem Eur J 11:2756–2772
25. Ingenito R, Dreznjak D, Guffler S et al (2002) Efficient loading of sulfonamide safety-catch linkers by Fmoc amino acid fluorides. Org Lett 4:1187–1188
26. Han YX, Albericio F, Barany G (1997) Occurrence and minimization of cysteine racemization during stepwise solid-phase peptide synthesis. J Org Chem 62:4307–4312
27. Backes BJ, Ellman JA (1999) An alkanesulfonamide "safety-catch" linker for solid-phase synthesis. J Org Chem 64:2322–2330
28. Krchnak V, Vagner J, Lebl M (1988) Noninvasive continuous monitoring of solid-phase peptide-synthesis by acid-base indicator. Int J Pept Protein Res 32:415–416

Chapter 13

Solid-Phase Synthesis of Phosphopeptides

Kim B. Højlys-Larsen and Knud J. Jensen

Abstract

Phosphopeptides are generally prepared by incorporation of suitable, protected phosphoamino acid derivatives during peptide synthesis using routine coupling protocols. The feasibility of chemical synthesis of phosphorylated peptides by Fmoc-SPPS was greatly enhanced by the introduction of the monobenzyl protecting group for the phosphate group. This minimized β-elimination of the phosphate group and made Fmoc-based synthesis of phosphopeptides the preferred synthesis strategy. Described here is our strategy for the synthesis of phosphopeptides attached to the solid support PEGA via a backbone amide linker type. This linker allows removal of side-chain protection groups without releasing the phosphopeptide from the solid support, thus enabling solid-phase-based pull-down reactions and peptide-protein interaction studies.

Key words Phosphopeptide, Monobenzyl protection, Fmoc-based protocol, PEGA, Backbone amide linker

1 Introduction

Posttranslational modification (PTM) such as phosphorylation, glycosylation, or acetylation often controls whether a protein is in an active or inactive form. Phosphorylation comes in different flavors, as not only Ser and Thr but also Tyr and His can be side-chain phosphorylated. Protein phosphorylation alters the conformation of the protein, as well as its activity, binding properties, and subcellular distribution. Phosphorylation is regulated by the interplay between protein kinases and protein phosphatases. This process controls many diverse processes, such as metabolism, membrane transport, gene transcription, and kinase cascade activation [1]. The human genome contains 518 protein kinase domains, thus representing 2 % of the total human genome [2], and roughly 204 phosphatases [3]. A PubMed search using the term "protein phosphorylation" gives more than 186,645 citations [4], whereas a similar search on "phosphopeptide synthesis" only gave 4,611 hits (of these are 130 reviews). The study of phosphorylation and

Knud J. Jensen et al. (eds.), *Peptide Synthesis and Applications: Second Edition*, Methods in Molecular Biology, vol. 1047, DOI 10.1007/978-1-62703-544-6_13,

phosphoproteomics has been extensively reviewed elsewhere including methods for chemical tagging strategies [5], affinity methods utilized specifically for phosphoprotein and phosphopeptide enrichment prior to MS analysis [6], and proteomic technologies for the characterization of the cell surface proteins and their modifications [7]. Phosphopeptides are very useful tools for the study of protein-protein interactions, protein phosphorylation, and dephosphorylation [8]. Using phosphopeptides to raise antibodies against defined motives in proteins can provide information on phosphorylation status, thus deriving information on the roles that proteins play in the cell [9–11]. Our own research on Ser/Thr-phosphorylated peptides has focused on the study of sequence specificity of phosphatases [12, 13], as well as on the affinity properties of phosphopeptide binding proteins [14, 15].

In this chapter, we will focus on Ser/Thr phosphorylation. Tyr phosphorylation is described in Ottinger et al. [15, 16]. The chapter describes a novel and general methodology for the solid-phase synthesis of phosphopeptides, featuring direct incorporation of *N*α-Fmoc-phospho-l-Tyr, with an unprotected side chain. This technique obviated the formation of peptide by-products containing Tyr H-phosphonate.

Phosphopeptides can be synthesized by either of two general methods, global phosphorylation or the building block approach. Global phosphorylation (post-assembly phosphorylation) involves the synthesis of a peptide in which Ser, Thr, or Tyr is incorporated without side-chain protection. Phosphorylation is then accomplished by treating the peptide with a phosphorochloridate or phosphoramidite followed by oxidation. In the more convenient building block approach, the phosphate is incorporated into a *Nα-protected* amino acid prior to peptide synthesis. The phosphoamino acid building block is then incorporated into the growing peptide chain [8]. Prior to the mid-1990s, most phosphopeptides were synthesized via Boc-SPPS [17, 18] or via peptide assembly followed by global phosphorylation by treatment of the peptide with a phosphorochloridate or phosphoramidite, followed by oxidation [19–21]. One challenge using Fmoc-SPPS for phosphopeptide synthesis was the use of bis-protected Ser phosphate, where treatment with piperidine in the Fmoc removal step resulted in partial β-elimination with loss of the phosphate group and formation of the corresponding dehydroalanine residue. Introduction of the monobenzyl protection group for the phosphate group [22, 23] minimized β-elimination thus facilitating Fmoc-SPPS of phosphopeptides. Fmoc-SPPS involves the incorporation of protected phosphoamino acid derivatives in peptide synthesis using routine HBTU-HOBt coupling protocols. Upon completion of peptide assembly, release of the peptide with concomitant deprotection is affected using a standard TFA/scavenger mixture. This allows a quite reliable preparation of phosphopeptides.

A second focus of this chapter is the preparation of resin-bound phosphopeptides, which can be used for affinity pull-down. On-bead screenings necessitate the use of a handle that is acid stable so that phosphopeptides remain support bound after removal of side-chain protecting groups. One such handle is 4-formylphenoxyacetic acid (or "monoBAL"), which requires trifluoromethane sulfonic acid (TFMSA) for release of peptides when needed for analytical or quality control purposes [24, 25], *see* Chapter 9. However, other linkers such as the HMBA handle can also be used. Another important factor is the property of the polymeric support, as the resin has to be compatible with the biological assay. A useful support is the poly[acryloyl-bis(aminopropyl)polyethylene glycol] (PEGA) resin developed by Meldal [26, 27]. The PEGA resin has excellent swelling in both aqueous systems and in most organic solvents used for peptide synthesis. The resin has a molecular cutoff at ~35–70 kDa.

On the other hand, if the phosphopeptides are intended to be used in solution-phase assays, the synthesis can be carried out on a standard resin with common linkers. This includes the Rink amide or Wang-type linkers on polystyrene, TentaGel, or ChemMatrix supports. Upon completion of the peptide chain assembly, the peptides are released from the solid support by treatment with TFA-containing cocktails with concomitant removal of acid-labile side-chain protecting groups. Using these conditions robust protocols for rapid and efficient synthesis of phosphopeptides have been established.

Here we describe a protocol for the use of the monoBAL linker on PEGA for the Fmoc-SPPS of phosphopeptides that are to be used in on-resin display, e.g., for pull-down proteomics. However, this protocol can also be used for general Fmoc-SPPS of phosphopeptides, provided a TFA-cleavable handle is used together with the appropriate chemistry for anchoring the first amino acid.

2 Materials

1. Resins for peptide synthesis: Poly[acryloyl-bis(aminopropyl) polyethylene glycol] (PEGA) [26, 27]. The PEGA resin was obtained from Polymer Laboratories Ltd (Now Agilent, England).
2. Building blocks: Amino acid derivatives were from Iris Biotech or Novabiochem, except for Fmoc-Ser[PO(OBn)OH]-OH, Fmoc-Thr[PO(OBn)OH]-OH, and Fmoc-Tyr[PO(OBn) OH]-OH which were obtained from Bachem (Switzerland).

Amino acids were generally $N\alpha$-Fmoc protected and with the following side-chain protecting groups: *tert*-butyl for Ser and Tyr; trityl for Asn; 2,2,4,6,7-pentamethyl-dihydrobenzofuran-5-sulfonyl (Pbf) for Arg.

3. Coupling reagents: HBTU (*N*-[(1H-benzotriazol-1-yl)-(dimethylamino)-methylene]-*N*-methylmethanaminium hexafluorophosphate *N*-oxide), DIC (diisopropylcarbodiimide), HOBt (1-hydroxybenzotriazole), and DIEA (*N,N-diisopropylethylamine*) were from Iris Biotech GmbH or Novabiochem. All other chemicals were obtained from Sigma-Aldrich.
4. Instruments: Syro II peptide synthesizer (MultiSynTech GmbH, Germany; now Biotage, Sweden), Biotage Initiator™ microwave instrument, and a solid-phase reaction vessel (Biotage AB, Uppsala, Sweden) [15].

3 Methods

3.1 Standard Procedure for the Synthesis of Phosphopeptides [28, 29]

1. Preloading of monoBAL handle (*see* **Note 1**): 4-Formylphenoxy acetic acid (4 equiv.), HBTU (3.9 equiv.), HOBt (4 equiv.), and DIEA (10 equiv.) in NMP are mixed for 2 min and added to the resin. The mixture is microwave heated at 60 °C for 20 min. The solution was drained and washed (*see* **Note 2**).

   A negative Kaiser test indicates completion of the coupling. The resin is acetylated with Ac_2O in DCM (1:3) for 2 × 5 min.
2. Attachment of the first amino acid via reductive amination: H-Aaa-NH_2·HCl (20 equiv.; typically H-Gly-NH_2·HCl) and $NaBH_3CN$ (20 equiv.) are suspended in AcOH–MeOH (1:99), added to the preloaded resin from **step 1**, and microwave heated at 60 °C for 10 min. The solution is drained, and the resin is washed. The coupling and washing steps are repeated once.
3. Incorporation of the second amino acid by acylation with symmetrical anhydride: Fmoc-Aaa-OH (10 equiv., typically Fmoc-Gly-OH) and DIC (5 equiv.) are suspended in NMP–DCM (1:9) and activated for 10–15 min. The mixture is then added to the resin and microwave heated at 60 °C for 10 min. The solution is drained, and the resin is washed. The coupling and washing steps are repeated once.
4. Peptide elongation (*see* **Note 3**): Peptide synthesis starting from Fmoc-Gly(monoBAL-PEGA)Gly-NH_2 resin (500 mg, 0.1 mmol) is performed using Fmoc amino acid (15 equiv. to activated PEGA resin loading, *see* **Note 4**) under standard HBTU, HOBt, and DIEA (*see* Chapter 2), coupling times of

45 min and 10 min deprotection with piperidine-DMF (1:4) repeated twice.

5. End-capping: The *N*α-Fmoc protection group is removed by treatment with piperidine-DMF (1:4) for 1 min, followed by draining, and additional piperidine-DMF (1:4) treatment (2 × 5 min) at room temperature. Next, the solution is drained, and the resin is washed. The resin is acetylated using Ac_2O-DCM (1:3) for 2 × 5 min. The solution is drained, and the resin is washed.
6. Removal of side-chain protecting groups (*see* **Note 5**): These are removed by treatment with Reagent B* [Triethylsilane/TFA/H_2O (95:2.5:2.5)] for 2 h. The resin is washed with TFA, DCM twice, and finally swelled in MeOH.
7. Release of phosphopeptides from the monoBAL handle for analytical purposes: A small sample of resin can be removed to analyze the resin-bound peptide. Treatment with a 9:1 mixture of trifluoroacetic acid (TFA) and trifluoromethane sulfonic acid (TFMSA) for 2 h at room temperature released the peptide from the monoBAL handle for LC-MS analysis. The solvents were drained, and the resin was washed with TFA and DCM (2×). The combined phases were concentrated, redissolved in DCM, and the peptide was extracted with H_2O. The water phase was concentrated for analysis by LC-MS using a Dionex Ultimate 3000 coupled to a Thermo MSQ+ electrospray mass spectrometer under the control of Dionex Chromeleon software.

3.2 Example of Synthesized Phosphopeptides

Using the method in Section 3.1, which are summarized in Fig. 1, the following peptides were synthesized (Table 1).

4 Notes

1. The following procedure is described for 4-formylphenoxy acetic acid, which is the precursor for the "monoBAL" handle, but it can equally be used for the *o*-BAL handle, *see* Chapter 9. However, treatment with conc. TFA will release the peptide from the support, thus providing the phosphopeptide directly in solution. Other linkers, such as Rink amide or Wang-type handles, can also be used for the Fmoc-SPPS of peptides in solution; however, the first amino acid is then anchored by conventional N-acylation.
2. After each step the resin is washed using the following standard conditions: NMP (3×), DCM (3×), and NMP (3×).

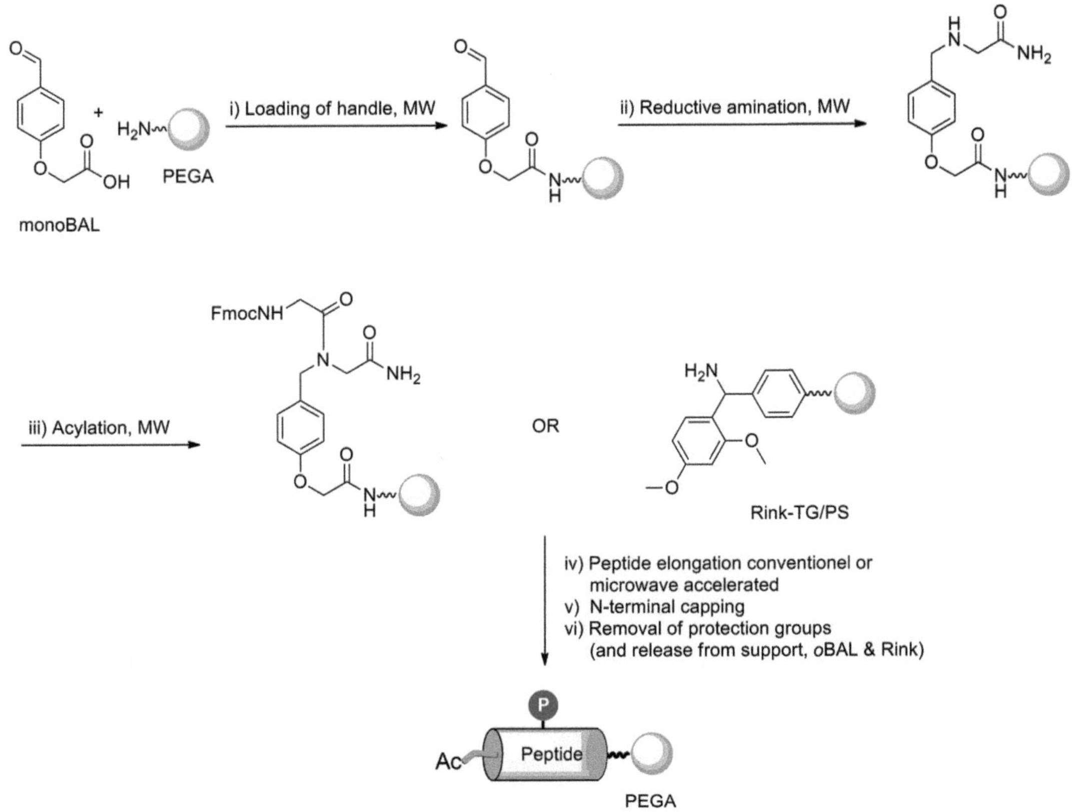

Fig. 1 Phosphopeptide synthesis on solid support. *General conditions*: (*i*) monoBAL, HBTU, HOBt, DIEA, NMP; (*ii*) H-Gly-NH_2, $NaBH_3CN$, AcOH–MeOH (1:99); (*iii*) Fmoc-Gly-OH, DIC, DCM-NMP (9:1); (*iv*) Fmoc-Aaa-OH, HBTU, HOBt, DIEA, NMP; (*v*) Ac_2O, DCM; (*vi*) TFA-TES-H_2O (95:2.5:2.5). *MW* microwave heating, 60 °C; *P* phosphate group on Ser, Thr, or Tyr; *PEGA* poly[acryloyl-bis(aminopropyl)polyethylene glycol]; *TG* TentaGel; *PS* polystyrene [14, 15, 29]

3. For synthesis of phosphopeptides in solution, Rink-TG or Rink-PS supports can be used skipping **steps 1–3**.
4. The 15 equiv. Fmoc amino acid to resin loading ratio used ensures proper swelling of the PEGA resin without compromising the concentration of the loaded amino acids. This also ensures a ~100 % coupling efficiency, which is important when the synthesized peptides are not cleaved and purified prior to biological testing.
5. These conditions will cleave off the peptides from the solid support when Rink or *o*BAL linkers are used.

Table 1
Synthesized peptides

Peptide no.	Target protein and phosphor residue	Sequence	Elemental composition	Calculated MS	Found MS	Assigned
1	p38 (180pT)	Ac-TDDEM**pT**GYVATGG-NH_2	$C_{55}H_{85}N_{14}O_{27}PS$	1436.52	1458.7	$[M+Na]^+$
2	UNG2 (31pT)	Ac-PAVQG**pT**GVAGVPGGG-NH_2	$C_{54}H_{90}N_{17}O_{21}P$	1343.62	1344.8	$[M+H]^+$
3	ERK5 (218pT, 220pY)	Ac-HQYFM**pT**E**pY**VATRGGG-NH_2	$C_{78}H_{114}N_{22}O_{29}P_2S$	1916.73	1917.2	$[M+H]^+$
4	JNK1 (183pT, 185pY)	Ac-TSFMM**pT**P**pY**VVTRGGG-NH_2	$C_{72}H_{115}N_{19}O_{27}P_2S_2$	1803.71	1804.5	$[M+H]^+$

The bold letters indicates the phosphorylated residues

References

1. Hunter T (2000) Signaling—2000 and beyond. Cell 100:113–127
2. Manning G, Whyte DB, Martinez R, Hunter T, Sudarsanam S (2002) The protein kinase complement of the human genome. Science 298:1912–1918
3. Alonso A, Sasin J, Bottini N, Friedberg I, Friedberg I, Osterman A, Godzik A, Hunter T, Dixon J, Mustelin T (2004) Protein tyrosine phosphatases in the human genome. Cell 117:699–711
4. Search performed 2012/12/12
5. Leitner A, Lindner W (2009) Chemical tagging strategies for mass spectrometry-based phospho-proteomics. Methods Mol Biol 527:229–243
6. Thingholm TE, Jensen ON, Larsen MR (2009) Analytical strategies for phosphoproteomics. Proteomics 9:1451–1468
7. Wisniewski JR (2011) Tools for phospho- and glycoproteomics of plasma membranes. Amino Acids 41:223–233
8. McMurray JS, Coleman DR IV, Wang W, Campbell ML (2001) The synthesis of phosphopeptides. Biopolymers 60:3–31
9. Sun T, Campbell M, Gordon W, Arlinghaus RB (2001) Preparation and application of antibodies to phosphoamino acid sequences. Biopolymers 60:61–75
10. Brumbaugh K, Johnson W, Liao W-C, Lin M-S, Houchins JP, Cooper J, Stoesz S, Campos-Gonzalez R (2011) Overview of the generation, validation, and application of phosphosite-specific antibodies. Methods Mol Biol 717:3–43
11. Teraishi T, Miura K (2009) Toward an in situ phospho-protein atlas: phospho- and site-specific antibody-based spatio-temporally systematized detection of phosphorylated proteins in vivo. Bioessays 31:831–842
12. Yamaguchi H, Minopoli G, Demidov ON, Chatterjee DK, Anderson CW, Durell SR, Appella E (2005) Substrate specificity of the human protein phosphatase 2C delta Wip1. Biochemistry 44:5285–5294
13. Hojlys-Larsen KB, Sorensen KK, Jensen KJ, Gammeltoft S (2012) Probing protein phosphatase substrate binding: affinity pull-down of ILKAP phosphatase 2C with phosphopeptides. Mol Biosyst 8:1452–1460
14. Brandt M, Madsen JC, Bunkenborg J, Jensen ON, Gammeltoft S, Jensen KJ (2006) On-bead chemical synthesis and display of phosphopeptides for affinity pull-down proteomics. Chembiochem 7:623–630
15. Larsen K, Brandt M, Sørensen KK, Gammeltoft S, Jensen KJ (2007) Solid-phase peptide synthesis as an enabling tool for proteomics. Chim Oggi 25:34–36
16. Ottinger EA, Shekels LL, Bernlohr DA, Barany G (1993) Synthesis of phosphotyrosine-containing peptides and their use as substrates for protein tyrosine phosphatases. Biochemistry 32:4354–4361
17. Otaka A, Miyoshi K, Kaneko M, Tamamura H, Fujii N, Nomizu M, Burke TR, Roller PP (1995) Development of efficient 2-step deprotection methodology for dimethyl-protected phosphoamino acid-containing peptide resins and its application to the practical synthesis of phosphopeptides. J Org Chem 60:3967–3974
18. Ueno Y, Suda F, Taya Y, Noyori R, Hayakawa Y, Hata T (1995) Solid-phase synthesis of peptides containing o-phosphoryl serine and o-phosphoryl threonine using allyl group for phosphate protection. Bioorg Med Chem Lett 5:823–826
19. Hubalek F, Edmondson DE, Pohl J (2002) Synthesis and characterization of a collagen model delta-O-phosphohydroxylysine-containing peptide. Anal Biochem 306:124–134
20. Kupihar Z, Varadi G, Monostori E, Toth GK (2000) Preparation of an asymmetrically protected phosphoramidite and its application in solid-phase synthesis of phosphopeptides. Tetrahedron Lett 41:4457–4461
21. Johnson T, Packman LC, Hyde CB, Owen D, Quibell M (1996) Backbone protection and its application to the synthesis of a difficult phosphopeptide sequence. J Chem Soc 1: 719–728
22. Wakamiya T, Saruta K, Yasuoka J-i, Kusumoto S (1994) An efficient procedure for solid-phase synthesis of phosphopeptides by the Fmoc strategy. Chem Lett 23:1099–1102
23. Wakamiya T, Togashi R, Nishida T, Saruta K, Yasuoka J, Kusumoto S, Aimoto S, Kumagaye KY, Nakajima K, Nagata K (1997) Synthetic study of phosphopeptides related to heat shock protein HSP27. Bioorg Med Chem 5: 135–145
24. Bourne GT, Meutermans WDF, Alewood PF, McGeary RP, Scanlon M, Watson AA, Smythe ML (1999) A backbone linker for BOC-based peptide synthesis and on-resin cyclization: synthesis of stylostatin 1. J Org Chem 64:3095–3101
25. Boas U, Brask J, Jensen KJ (2009) Backbone amide linker in solid-phase synthesis. Chem Rev 109:2092–2118
26. Meldal M (1992) PEGA—a flow stable polyethylene-glycol dimethyl acrylamide

copolymer for solid-phase synthesis. Tetrahedron Lett 33:3077
27. Auzanneau FI, Christensen MK, Harris SL, Meldal M, Pinto BM (1998) Synthesis and characterization of polyethylene glycol polyacrylamide copolymer (PEGA) resins containing carbohydrate ligands. Evaluation as supports for affinity chromatography. Can J Chem 76:1109
28. Pedersen SL, Tofteng AP, Malik L, Jensen KJ (2012) Microwave heating in solid-phase peptide synthesis. Chem Soc Rev 41:1826–1844
29. Brandt M, Gammeltoft S, Jensen KJ (2006) Microwave heating for solid-phase peptide synthesis: general evaluation and application to 15-mer phosphopeptides. Int J Pept Res Ther 12:349–357

Chapter 14

Synthesis of *O*-Glycopeptides and Construction of Glycopeptide Microarrays

Ola Blixt and Emiliano Cló

Abstract

O-Glycosylation of proteins is an important modification which affects biological function and immunity. In this chapter, we provide protocols for efficient solid-phase *O*-glycopeptide synthesis (SPGPS) and protocols for the construction of glycopeptide microarray chips for screening applications. This will be exemplified for mucin-type glycopeptides and the construction of glycopeptide microarrays. To this end, the protocols provided are particularly suited for small-scale robotic parallel synthesis. N-Terminal amine capping of deletion peptides during synthesis stands out as vital to this strategy. It allows for direct on-slide enrichment of the full-length target product and thereby bypasses tedious isolation and purification procedures.

Key words Glycosylation, *O*-glycans, Glycopeptides, Posttranslational modifications (O-PTM), Microarrays, Glycan-binding proteins, Antibodies, Posttranslational glycosylation (PTG), Specificities

1 Introduction

O-Glycosylation of proteins is one of the most complex posttranslational modifications (O-PTM) of proteins in nature [1]. The two major forms of protein glycosylation are *N*- and *O*-glycosylation. From a biosynthetical standpoint, there are several fundamental differences between them. *N*-Glycan biosynthesis is initiated by the transfer of a preformed oligosaccharide precursor of a set structure to asparagine of nascent proteins at a recognition motif (N-X-S/T). *O*-Glycosylation, on the other hand, occurs at the hydroxyl groups of serine and threonine and is much more variable, because (a) the set of monosaccharides initiating the glycosylation event is very diverse including 2-*N*-acetylglucosamine (GlcNAc), mannose (Man), fucose (Fuc), and 2-*N*-acetylgalactosamine (GalNAc) [2, 3], and (b) there is no clear consensus sequence of the underlying protein scaffold that guides to the sites of glycosylation. There are six main types of *O*-glycosylation, which are highlighted in Table 1.

Knud J. Jensen et al. (eds.), *Peptide Synthesis and Applications: Second Edition*, Methods in Molecular Biology, vol. 1047, DOI 10.1007/978-1-62703-544-6_14, © Springer Science+Business Media New York 2013

Table 1
Major types of *O*-glycosylation. Hydroxylic sites used for core structure extension are highlighted with dashed boxes

Structure	Description	Reference
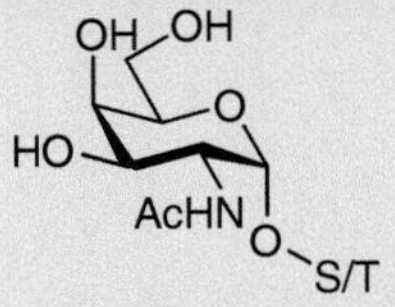 α-O-GalNAc	Most common form of *O*-glycosylation in eukaryotes, known as mucin-type glycosylation, as mucin displays dense clusters of highly extended glycans. GalNAc as α-anomer is the first sugar unit attached to S/T (giving the "Tn antigen"). It is the foundation for the formation of eight core structures resulting from extension to C-3 and/or C-6	[33]
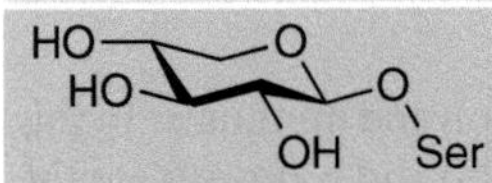 β-O-xylose	The tetrasaccharide Glc(*β*1-3)Gal(*β*1-3) Gal(*β*1-4)Xyl*β*-Ser. Constitutes the core structure of proteoglycans, like heparan and chondroitin sulfate, which are the principal components of extracellular matrix. They are implicated in many biological processes like cell adhesion, enzyme inhibition, embryonic development, and blood coagulation	[34]
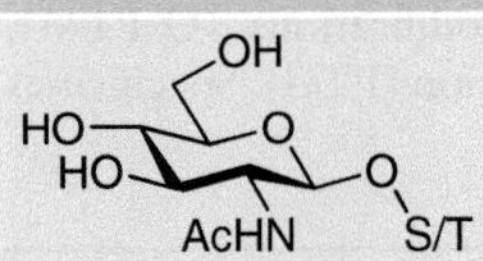 β-O-GlcNAc	Intracellular glycosylation type (cytosol and nucleus) characterized by the attachment of single GlcNAc-β units. It is highly dynamic and it often acts synergically with protein phosphorylation to control the activity of diverse cellular processes. It plays a role in cancer, neurodegeneration, and diabetes	[35]
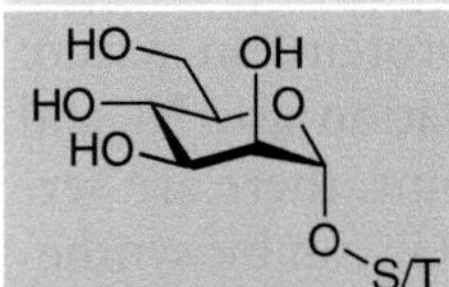 α-O-mannose	Essential to allow correct interaction of α-dystroglycan subunit (α-DG) of the muscular cell cytoskeleton with the extracellular matrix. In case of interaction with laminin, full glycan hapten requires Neu5Ac(*α*2–3)Gal(*β*1-4)GlcNAc(*β*1–2) Man-*α*-S/T. Genetic defects leading to hypoglycosylation are responsible for various forms of muscular dystrophy	[36]
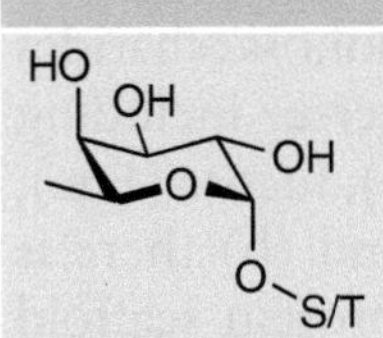 α-O-fucose	*O*-Fucosylation is a rare type of glycosylation that occurs at EGF-like repeats, like in the transmembrane Notch factor, which controls cell fate determination in many organisms. *O*-Fucosylation is involved in intracellular signal transduction	[37]
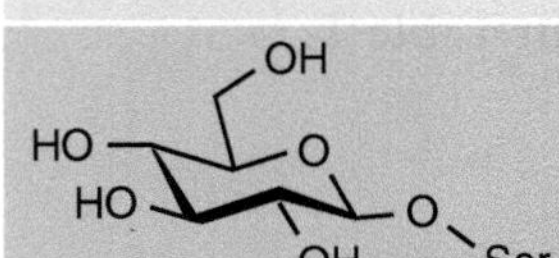 β-O-glucose	Another rare type of glycosylation present, like fucosylation, in the EGF-like repeats of Notch factor with implication in signal transduction	[37]

One of the most common kinds of *O*-glycosylation is *O*-GalNAc mucin-type glycosylation. Biosynthesis is initiated by 20 unique ppGalNAc transferases with different specificities but also with partial redundancies [4]. Major synthetic efforts in preparation of diverse glycan libraries and presenting them on various high-throughput screening platforms, such as microarray chips, have dramatically improved studies between proteins and complex carbohydrates in the past decade [5, 6]. Examples of binding interactions including lectin, antibody, viral, cell, and enzyme interaction have been studied and increased the impact of glycomics [7]. However, it has become increasingly evident that glycans presented in close proximity to its underlying peptide/protein structure, such as *O*-glycans linked to Ser or Thr, have a profound effect on their interactions. For example, many anti-carbohydrate tumor antibodies towards STn, T, and Tn *O*-glycans do require appropriate presentation of the underlying peptide backbone for optimal binding [8]. In addition, aberrant *O*-glycosylation can break immune tolerance eliciting glycopeptide-specific antibodies [9, 10]. Such immune response underpins efforts towards development of serodiagnostics [11–13] and vaccines [14].

There are two different ways to synthesize *O*-glycopeptides: (1) enzymatic and (2) chemical synthesis. Enzymatic glycosylation of peptides by recombinant GalNAcTs is feasible but is generally complicated by incomplete reactions and uncertainties of site-specific attachments, which generates heterogeneous products. Preparation of *O*-glycopeptides by means of solid-phase peptide synthesis (SPPS) provides chemical control over the sites of glycosylation and structural homogeneity. It relies on preformed *O*-glycosylated and suitably protected amino acids. *N*-Glycosylated amino acids can be incorporated in a similar manner. The Fmoc methodology is most suited to accomplish glycopeptide synthesis over the Boc methodology, as the latter requires the exposure of the growing peptide to strong acid after each elongation step, in order to remove the Boc group from N-terminal, as well as HF for final cleavage from the resin. These conditions will ultimately compromise the integrity of the acid-sensitive glycosidic linkage, and acid-induced beta elimination will occur (Fig. 1a). Fmoc-SPPS, on the other hand, is conducive to *O*-glycopeptide preparation since exposure to acid occurs only for release of the peptide from the resin with TFA-based cocktails, yet reaction protocols must be carefully selected as still many common reagents for this methodology are incompatible with the labile glycosidic bond [15]. The main hurdle is to select protecting group strategies for the glycans that not only allow them to survive SPPS but that can also be removed with ease without detrimental effect for the target structure. To this end, the choice is rather limited, and acetylation of the glycan hydroxyls is by far the most widespread; benzoylation is also reported, but to a lesser extent as it requires somewhat more forceful conditions for its removal. Acetyl groups can be removed

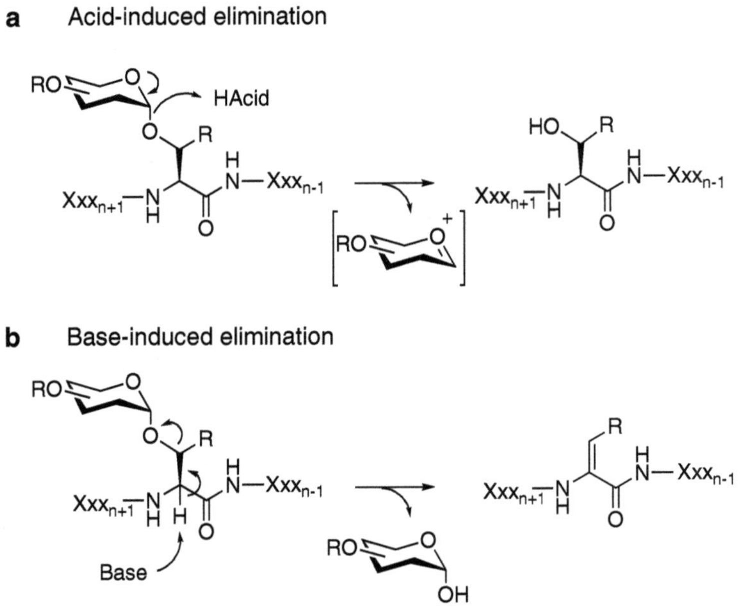

Fig. 1 Mechanism of (**a**) acid- and (**b**) base-induced glycan beta elimination

with either dilute sodium methoxide in methanol, hydrazine hydrate [16], or saturated ammonia in methanol [17], the former being the most widespread. Still, the operation is not trouble free as base-catalyzed β-elimination of the sugar moiety can occur extensively if the pH is not tightly controlled and kept below 11 (Fig. 1b). Importantly, piperidine and morpholine, the commonly used hindered bases for Fmoc removal, neither affect the acetyl groups nor trigger the onset of β-elimination [18], [19]. Additionally, acetyl groups, by virtue of their withdrawing effect, are able to enhance the stability of the glycosidic bond towards acid-catalyzed elimination [20]. It is important to be aware that the basic conditions needed for acetyl removal can negatively affect cysteine [21] and asparagine [22] side chains. A more generalized problem during the coupling of glycosylated amino acids concerns the risks related to acyl migration, when the reactions are sluggish, from the sugar to the N-terminal of the growing peptide chain, thus bringing about permanent capping that stalls the synthesis [23].

Fmoc-based solid-phase glycopeptide synthesis is compatible with the standard linkers (handles) widely used in peptide synthesis, such as Rink amide, Wang, SASRIN (mild acid cleavage), 2-chlorotrityl chloride (protected peptide acid), and Sieber (protected peptide amides). For combinatorial one-bead one-compound studies, both the safety-catch amide linker (SCAL, stable to both acid and base until sulfone reduction) [24] and the photocleavable linker 4-[4-(1-aminoethyl)-2-methoxy-5-nitrophenoxy] butyric acid [25] have been successfully applied. These linkers are suitable for displaying the fully deprotected glycopeptides on bead while allowing their nondestructive recovery after bead assay. Concerning the coupling conditions, the most common combinations of coupling reagents and activators found in Fmoc-SPPS are compatible with glycopeptides synthesis. Popular among these are DIC/HOBt, HBTU/HOBt, and HATU/HOAt.

The main limitation in achieving complex *O*-glycopeptides via Fmoc-SPPS is related to the accessibility of glycosylated building blocks. Ser and Thr building blocks carrying short glycan haptens like GalNAc*α* (Tn antigen) and di- and trisaccharides like Gal*β*(1-3)GalNAc*α* (T antigen) and Neu5Ac*α*(2-3)Gal*β*(1-3)GalNAc*α* (ST antigen), respectively, are commercially available. However, more complex oligosaccharides are difficult to obtain and require custom synthesis [26]. In order to shortcut these difficulties, enzymatic oligosaccharide extension with soluble recombinantly expressed glycosyltransferases has proven to be a valuable alternative [27, 28].

In our laboratories, we have optimized the synthesis of *O*-glycopeptides for the production of *O*-glycopeptide microarrays, in order to dramatically simplify the purification requirements for each synthetic target. Peptide microarrays [29] have evolved during the last decade as a powerful tool for high-throughput screening of a host of biological events. We [30] and others [31] have shown that microarray format can be harnessed to display up to hundreds of different synthetic *O*-glycopeptides, anchored in a covalent fashion, for Seromic profiling. Here, we provide first a detailed protocol based on our latest developments for the high-throughput production of different *O*-glycopeptide structures exploiting glycosylated amino acid building blocks. It combines parallel and automated Fmoc-based solid-phase glycopeptides synthesis at a scale of 2 μmol [30]. This entails capping of unreacted N-terminal amines after each coupling cycle, so that ultimately, only full-length target structures retain the free N-terminal amine. Next, the glycopeptides are used directly for the construction of glycopeptide microarrays. The fact that only full-length glycopeptides have an N-terminal amine enables an on-chip purification technique as these amine-functionalized glycopeptides can be covalently captured by amine-reactive glass slides (NHS-ester present throughout the surface, Fig. 2). The slide support we apply, by virtue of its hydrogel surface, has proven to be particularly suited for sera screening giving very low unspecific binding and background in the off-feature areas [6]. Moreover, enzymatic elongation of glycan haptens is also well tolerated, thus giving access to a multilayered screening platform [32]. Detailed explanation for the preparation of glycopeptide microarrays is also provided.

2 Materials

1. TentaGel S with Fmoc-Rink amide linker, loading 0.23 mmol/g (Rapp Polymere GmbH, Tübingen, Germany).
2. *N*α-Fmoc amino acids (Fmoc-AA-OH), containing the following side-chain protecting groups: *tert*-butyl (O*t*Bu, for Glu, Asp, Ser,

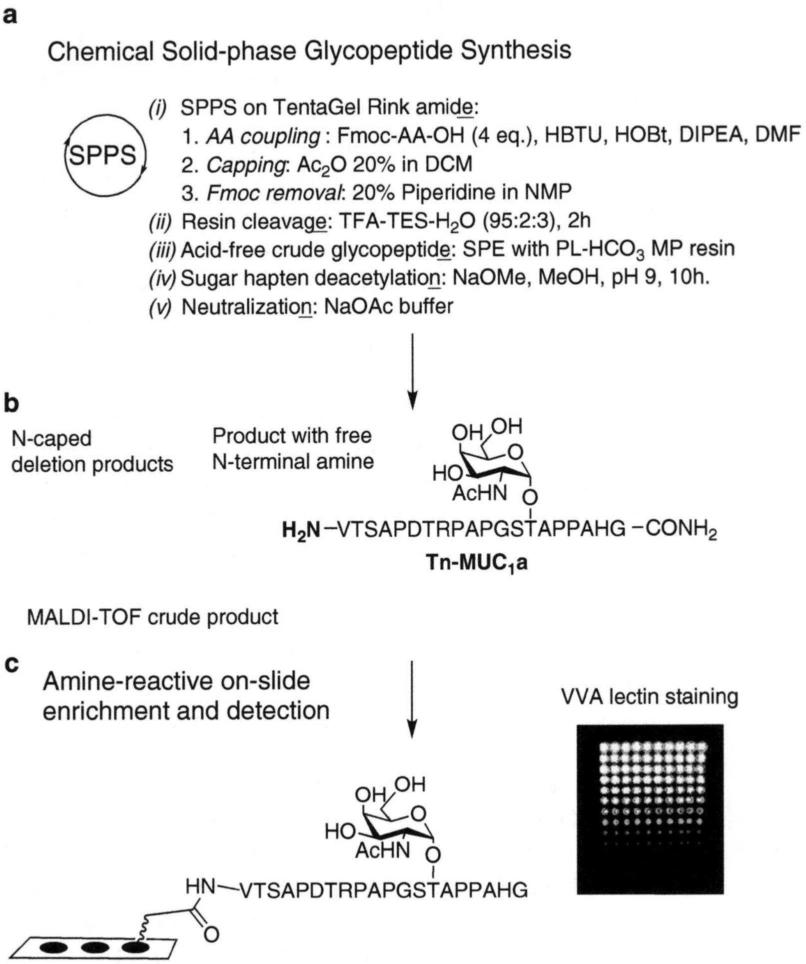

Fig. 2 Chemical synthesis and on-slide enrichment of glycopeptides. (**a**) Solid-phase glycopeptide synthesis with *N*-acetylation capping step. (**b**) MALDI-TOF spectra of crude Tn-MUC1 20mer glycopeptide. (**c**) On-slide microarray enrichment of crude Tn-MUC1 20mer glycopeptide product via amine-reactive glass surface and detection of GalNAc-specific lectin (VVL)

Thr, Tyr), 2,2,4,6,7-pentamethyl-dihydrobenzofuran-5-sulfonyl (Pbf, for Arg), trityl (Trt, for Asn, Gln, His), and *tert*-butyloxycarbonyl (Boc for Lys) are commercially available from different sources (e.g., Novabiochem; Advanced ChemTech; Bachem Bioscience; Iris Biotech).

3. Fmoc-protected Tn (GalNAcα)-Ser/Thr, Core1 (T antigen) (Galβ1-3GalNAcα)-Ser/Thr, Core3 (GlcNAcβ1-3GalNAcα)-Thr, and Core4 (GlcNAcβ1-3[GlcNAcβ1-6]GalNAcα)-Thr were purchased from Sussex Research (OT, Canada).
4. Coupling reagent: HBTU (*N*-[(1*H*-benzotriazol-1-yl) (dimethylamino)methylene]-*N*-methylmethanaminium hexafluorophosphate *N*-oxide).
5. Auxiliary nucleophiles: HOBt (1-hydroxybenzotriazole), HOAt (*N*-hydroxybenzotriazole).

6. Capping reagent: acetic anhydride.
7. Solvents: NMP (*N*-methyl-2-pyrrolidone), DMF (*N,N-dimethylformamide*), DCM (dichloromethane), diethyl ether (Et_2O), methanol (MeOH).
8. Bases: piperidine, DIEA (*N,N*-diisopropylethylamine), MeONa 0.5 M (sodium methoxide 0.5 M solution in methanol).
9. Acids: trifluoroacetic acid (TFA), acetic acid (HOAc).
10. Scavengers: triethylsilane (TES), deionized water.
11. Plastic syringe (2 mL) fitted with polypropylene filter frit and Teflon tap.
12. Syro II, fully automatic peptide synthesizer (MultiSynTech GmbH, Germany; now from Biotage AB, Sweden).
13. Multi-syringe suction block.
14. PL-HCO_3 (hydrogen carbonate) MP SPE resin (Agilent Technologies, # PL3540).
15. 96-Well filter plates (1 mL deep) (Pall, #5054, Ann Arbor, MI, USA).
16. Vacuum manifold (Pall, #5017, Ann Arbor, MI, USA).
17. 96-Well polypropylene round-bottomed microtiter plates (2 mL deep) (Ritter, #43001-0020, Schwabmünchen, Germany).
18. Schott NEXTERION® Slide H (Schott AG, #1070936, Jena, Germany).
19. SpeedVac concentrator (Thermo Fischer Scientific #SPD131 DDA-230) fitted with ultralow temperature refrigerator (Thermo Fischer Scientific #RVT4104-230) and oil-free Teflon vacuum pump (Thermo Fischer Scientific, #OFP400-230).
20. Lyophilizer.
21. MicroGrid II automated microarrayer (Digilab, Holliston, MA).
22. Print buffer (PB) contains 150 mM phosphate, 0.005 % CHAPS, 0.03 % NaN_3, pH 8.5.
23. Blocking buffer (BB) containing 50 mM ethanolamine in 50 mM borate buffer, pH 9.2.
24. 384-Well source plates (BD Falcon MicrotestTM 384-well 30 μL assay plates from BD Biosciences, Le Pont-de-Claix, France).

3 Methods

3.1 General Considerations

The protocol for *O*-glycopeptide synthesis described below has been optimized in our labs to fulfill the requirements for our covalent microarray platform. Key features in this respect are (1) small-scale synthesis, in which only few milligrams of resin are used per peptide, in order to allow for higher throughput by parallel robotic automation; (2) capping of unreacted N-terminal amines with acetic anhydride after each elongation step is undertaken after each coupling step. This prevents the formation of deletion peptides and allows for direct on-slide enrichment of the full-length target product. The protocol also entails direct mixing of Fmoc-amino acid-HOBt/HOAt solution with the resin followed by addition of coupling reagent and base, thus avoiding pre-activation. This feature aids the smooth transfer of the protocol to robotic synthesizer as well.

3.2 Fmoc-Based Solid-Phase O-Glycopeptide Synthesis

Ahead of embarking on the synthetic effort, working solutions of (a) Fmoc-AA-OH (4.0 equiv., 0.5 M) dissolved in a 9:1 mixture of the auxiliaries nucleophiles HOBt:HOAt (4.0 equiv., 0.5 M) in DMF (*see* **Notes 1** and **2**), (b) HBTU (3.8 equiv., 0.48 M) in DMF, and (c) DIPEA 2 M in NMP to be used during peptide chain elongation are prepared as well as the piperidine (2:3 v/v) solution in NMP for Fmoc removal and acetic anhydride solution (1:4) in DMF containing also 1 % v/v DIPEA and few crystals of HOBt for capping failure sequences. All these solutions are bench stable at room temperature for several days.

The synthesis procedure is optimized for the preparation of 2 μmol scale of glycopeptides. The standard solid support employed is TentaGel S RAM (0.24 mmol/g). The resin is equipped with a Rink amide handle, which is cleaved in TFA-based cleavage cocktails alongside side-chain protection groups generating C-terminal amides. The handle is Fmoc protected; thus, removal is necessary before coupling the first amino acid. Polypropylene syringe barrels, fitted with a filter frit, are used as reactors (2 mL size).

The typical protocol to be adopted per each elongation cycle entails Fmoc removal at the N-terminal amine, Fmoc-AA-OH coupling to the growing peptide chain, and capping of the unreacted structures.

1. Fmoc removal from N-terminal amine: add 100 μL piperidine in NMP (2:3 v/v) solution to the resin and incubate under agitation for 15 min. Flush away the liquid phase and wash (3× 1 min, 1 mL each) with NMP. Add 50 μL NMP followed by 50 μL piperidine in NMP (1:4 v/v) solution and incubate under agitation for 15 min. Flush away the liquid phase and wash (3× 1 min, 1 mL each) with NMP.
2. Chain elongation: add 4.0 equiv. of Fmoc-AA-OH/HOBt:HOAt solution followed by 3.8 equiv. of HBTU solution and finally 8.0 equiv. of DIEA 2 M solution to the resin. Allow to gently stir for 120 min, whereupon flush the coupling solution away and wash the drained resin (4× 1 min, 1 mL each) with NMP (*see* **Note 3**).
3. Capping unreacted structures: add 100 μL of acetic anhydride capping solution to the resin and allow to gently vortex for 10 min. Flush away the liquid phase and wash (3× 1 min, 1 mL each). Repeat capping a total of three times. Finally wash the resin 5× 2 min, 1.5 mL each.

The synthesis ends after Fmoc removal from the last amino acid in order to leave the N-terminal amine ready for coupling with NHS-ester in microarray slide. Wash the resin with 5× 1 min, 1.0 mL each, and proceed to *O*-glycopeptide cleavage. Alternatively, the peptide on resin can be stored for long periods at −18 °C provided the resin had been previously dried under vacuum for a few hours.

We recommend performing the Kaiser test (*see* Chapter 2) only after the addition of *O*-glycoamino acids or when the operator is aware that a given coupling to be particularly difficult.

3.3 O-Glycopeptide Cleavage from Solid Support

1. Freshly prepare the cleavage cocktail in an Erlenmeyer flask as follows: TFA:TES:water (94:3:3 v/v). Swirl the flask around regularly to make sure that TES droplets remain micro-dispersed in the cocktail (*see* **Note 4**).
2. Carefully add approximately 1 mL of cleavage cocktail per peptidyl-resin; seal the syringe barrel with the plunger and allow gentle agitation for 120 min.
3. Collect the liquid phase by suction in a snap cap tube and wash the beads with 300 μL fresh TFA collecting it with the first TFA filtrate.
4. Reduce the volume of the liquid TFA phase down to 1–200 μL with a SpeedVac centrifuge (*see* **Note 5**).
5. Add approximately 1 mL of cold Et_2O to the cleavage cocktail residue and vortex vigorously. The white amorphous crude glycopeptide precipitate will appear.
6. Spin down the precipitate in the tube in a high-speed centrifuge cooled at 4 °C at 3,000 × *g* or more for 5 min and discard the Et_2O which can be easily decanted off.
7. Repeat **steps 5** and **6** twice.
8. Let the residual Et_2O evaporate from the pellets inside a fume hood for 2–4 h.
9. Dissolve the pellet in 500 μL 2 M acetic acid in water and lyophilize.

3.4 Deprotection of Glycan Hapten for O-Glycopeptide Sequences Devoid of Carboxylic Acid Side Chains

Removal of the acetyl esters masking the hydroxyl groups of the sugar moieties of the glycopeptides is carried out in Zémplen conditions with diluted MeONa in MeOH. It is of the uttermost importance to carefully control the pH at this stage. The main hazard is that if the pH is too high, >12, then β-elimination of the glycan hapten ensues, thus spoiling the whole synthetic effort. On the other hand, if the pH is too low, <8.5, the deacetylation becomes very sluggish and will fail to reach completion even after prolonged incubation. When the glycopeptides have no carboxylic acid side chain (Glu or Asp), we find it convenient to use hydrogen carbonate beads (PL-HCO_3 MP SPE, Agilent) in order to obtain the glycopeptides in free-base form. The beads neutralize the excess of acid salts (from either acetic acid or TFA) and ion-pair them on the quaternary amine on solid support (*see* **Note 6**). This allows a more straightforward pH control during glycan deprotection.

Following in the protocol optimized for treating a batch of 96 glycopeptides at one time.

1. Evenly distribute approximately 1.5 g of dry PL-HCO3 resin in a 96-well filter plate.
2. Place the filter plate onto the vacuum manifold fitted with the receiving 96-well plate (2 mL deep).
3. Dissolve the glycopeptides in MeOH-water solution (1:1, 400 μL) taking advantage of sonication if needed to aid solubilization.
4. Aliquot the glycopeptides inside the filter plate wells and allow 5 min incubation.
5. Apply vacuum to the manifold so as to drain the filter plate and collect the filtrate inside the receiving plate.
6. Wash the beads sequentially with 200 μL water and 200 μL MeOH, applying vacuum in between additions to collect the filtrate in the receiving plate along with the first filtrate.
7. Remove MeOH with a speed-vac and retrieve the free-base glycopeptides by lyophilization.
8. Dissolve the glycopeptides in 400 μL dry MeOH and add 200 μL of MeONa 0.015 M in MeOH, thus giving pH around 9–10 (as judged from pre-wetted pH strips), and allow incubation under gentle shaking at room temperature (*see* **Note 7**).
9. Once the reactions are completed, add 100 μL acetic acid in water (0.1 M) to quench the excess of base. The resulting pH will typically be between 6 and 7.
10. Lyophilize the glycopeptide solutions to retrieve the fully deprotected crude glycopeptides as fluffy solid.

3.5 Deprotection of Glycan Hapten for O-Glycopeptide Sequences with Carboxylic Acid Side Chains

When acidic amino acid side chains (i.e., Glu, Asp) are present, it is advisable to avoid hydrogen carbonate beads as the glycopeptides will be absorbed on the beads by ion paring with the quaternary amine (*see* **Note 8**) leading to loss of material. Accordingly, deacetylation will be undertaken directly on the lyophilized glycopeptides product obtained after cleavage from resin (Subheading 3.3). The main drawback of this approach is that the acetic acid salts present with the glycopeptides along with the acid side chains present will cause a "buffer effect," making in more laborious to tune the correct pH for proficient deacetylation. No set concentration of MeONa can be used to fit all structures; hence, the pH for each structure has to be tuned individually, hampering the throughput of the process.

1. Dissolved the lyophilized glycopeptides from Subheading 3.3 in 400 μL in dry MeOH in snap cap tubes.

2. Carefully add MeONa 0.015 M to each glycopeptides solution until pH around 9–10 is reached (as judged from pre-wetted pH strips for each structure) and seal the tube.
3. Allow incubation under gentle shaking at room temperature.
4. Control the pH on pre-wetted pH strips at least once every hour and adjust with fresh MeONa 0.015 M solution whenever it drops below pH 9.
5. Follow the reaction course with MS at regular intervals.
6. Once the reactions are completed, add 100 μL acetic acid in water (0.1 M) to quench excess of basicity. The resulting pH will typically be between 6 and 7.
7. Lyophilize the glycopeptide solutions to retrieve the fully deprotected crude glycopeptides as fluffy solids.

3.6 Covalent O-Glycopeptide Microarray Fabrication

The *O*-glycopeptides prepared in our research group are mostly applied to the fabrication of covalent *O*-glycopeptide microarrays. As previously mentioned, because of the systematic capping following each elongation step, the glycopeptides can be printed as crude mixture since the full-length sequence is the only species in the mixture retaining the free N-terminal amine to react with the NHS-ester on slide surface. Procedure for glycopeptide microarray fabrication is as follows:

1. Dissolve the lyophilized glycopeptides from glycan deacetylation step (Subheading 3.6) in 200 μL water with 0.03 % NaN_3 as preservative. This constitutes the glycopeptide stock solution.
2. Dissolve the glycopeptide stock solution in print buffer (PB) at a 1:5 dilution, thus giving the printing solution.
3. Aliquot 20 μL of printing solution per pin to be used during printing in a 384-well plate (30 μL/well) according to the pattern required by the microarray design (*see* **Note 9**).
4. Load the number of thawed Nexterion slide H that needs to be printed in the microarray printer and perform the print (refer to the printer instruction manual on how to set up and perform the print).
5. Once the print is complete, remove the slides and incubate them for 1 h at 70 % relative humidity (*see* **Note 10**).
6. Block the unprinted areas of the slides by submerging the slides in blocking buffer (BB) for 1 h at room temperature.
7. Wash the slides with water and dry them by centrifugation. The slides are now ready for staining.

4 Notes

1. Typically, all the different Fmoc-AA-OH solutions are prepared ahead of synthesis. In practical terms, this is best achieved by preparing at first a 0.5 M solution of the auxiliary nucleophiles HOBt and HOAt (9:1 mol/mol) in DMF. Then each Fmoc-AA-OH is placed in a 50 mL Falcon tube and dissolved in the above solution in order to obtain a concentration of 0.5 M in amino acid as well. Most amino acids will give cloudy solutions at first, which become homogeneous after a few minutes under sonication.
2. For Fmoc-Phe-OH NMP is used in place of DMF for solubility reasons. For Fmoc-Arg(Pbf)-OH, an improved coupling efficiency is obtained by omitting HOBt and using 0.5 M HOAt instead.
3. When incorporating the more costly and bulky glycosylated amino acid, it can be convenient to perform double coupling. It is of benefit to both the coupling efficiency and the material economy to perform two consecutive coupling steps of 90 min each with approximately 1.5 equiv. each of glyco-aa, 1.5 equiv. HOBt:HOAt (9:1), 1.4 equiv. HBTU in DMF, and 3 equiv. DIPEA in NMP.
4. TFA-scavenger cleavage cocktails are highly corrosive. Always carry on all the operation relative to peptide cleavage inside a fume hood.
5. Alternatively, the TFA can be removed under a stream of nitrogen. Perform the operation inside a fume hood.
6. It is possible to proceed to free the bases of the glycopeptides starting directly from the pellet, thus skipping lyophilizing the glycopeptides from acetic acid 2 M (Subheading 3.3, **step 8**).
7. When no more than two glycan units are present in a glycopeptide, deacetylation proceeds swiftly and usually reaches completion within 4 h. When further sugar units are present, overnight incubation is often required along with pH adjustment with fresh MeONa 0.015 M solution. It is advisable to monitor the reaction course by MS at regular intervals.
8. The amount of material absorbed will vary in a sequence-specific manner (usually, the more acid side chains present, the more efficient the ion pairing onto the beads) and in dependence of the amount of beads used.
9. The microarray design will vary depending on the amount of different glycopeptides to be arrayed. The most common slide design in our lab allows printing of either 16 or 48 identical subarrays per slide. In 16 subarray design (2 columns and 8 rows), the pitch between spots is 0.21 mm for a maximum of

28 × 28 spots hosted by each subarray. In 48 subarray design (4 columns and 12 rows), the pitch is 0.19 mm for a maximum of 12 × 12 spots hosted by each subarray. The number of pins used can vary and is not tied to a given slide design.

10. After humidification, the slides can be stored at −20 °C for several months prior use. When needed a printed slide can be taken out of the freezer and immediately exposed to blocking buffer.

Acknowledgements

This work was supported by The Benzon Foundation, The Danish Agency for Science, Technology and Innovation (FTP), EU FP7/2007-2013-EuroGlycoArrays 215536, EU FP7-GlycoBioM, and University of Copenhagen Programme of Excellence.

References

1. Varki A (2009) Essentials of glycobiology, 2nd edn. Cold Spring Harbor Laboratory Press, New York
2. Haltiwanger RS, Lowe JB (2004) Role of glycosylation in development. Annul Rev Biochem 73:491–537
3. Paulson JC, Blixt O, Collins BE (2006) Sweet spots in functional glycomics. Nat Chem Biol 2:238–248
4. Bennett EP, Mandel U, Clausen H et al (2011) Control of mucin-type O-glycosylation—a classification of the polypeptide GalNAc-transferase gene family. Glycobiology 22:736–756
5. Alvarez RA, Blixt O (2006) Identification of ligand specificities for glycan-binding proteins using glycan arrays. Methods Enzymol 415:292–310
6. Blixt O, Head S, Mondala T et al (2004) Printed covalent glycan array for ligand profiling of diverse glycan binding proteins. Proc Natl Acad Sci U S A 101:17033–17038
7. Paulson JC, Blixt O, Collins BE (2006) Sweet spots in functional glycomics. Nat Chem Biol 2:238–248
8. Blixt O, Boos I, Mandel U (2012) In: Kosma P (ed) Anticarbohydrate antibodies. New York, Springer
9. Tarp MA, Sorensen AL, Mandel U et al (2007) Identification of a novel cancer-specific immunodominant glycopeptide epitope in the MUC1 tandem repeat. Glycobiology 17:197–209
10. Ju T, Otto VI, Cummings RD (2011) The Tn antigen-structural simplicity and biological complexity. Angew Chem Int Ed Engl 50:1770–1791
11. Clo E, Kracun SK, Nudelman AS et al (2012) Characterization of the viral O-glycopeptidome: a novel tool of relevance for vaccine design and serodiagnosis. J Virol 86:6268–6278
12. Wandall HH, Blixt O, Tarp MA et al (2010) Cancer biomarkers defined by autoantibody signatures to aberrant O-glycopeptide epitopes. Cancer Res 70:1306–1313
13. Blixt O, Bueti D, Burford B et al (2011) Autoantibodies to aberrantly glycosylated MUC1 in early stage breast cancer are associated with a better prognosis. Breast Canc Res 13:R25
14. Sabbatini PJ, Ragupathi G, Hood C et al (2007) Pilot study of a heptavalent vaccine-keyhole limpet hemocyanin conjugate plus QS21 in patients with epithelial ovarian, fallopian tube, or peritoneal cancer. Clin Cancer Res 13:4170–4177
15. Davis BG (2002) Synthesis of glycoproteins. Chem Rev 102:579–602
16. Schultheiss-Reimann P, Kunz H (1983) O-glycopeptide synthesis using 9-fluorenyl methoxycarbonyl (Fmoc)-protected synthetic units. Angew Chem Int Ed Eng 22:62–63
17. Paulsen H, Schultz M, Klamann JD et al (1985) Building units of oligosaccharides. 66. Synthesis of O-glycopeptide blocks of glycophorine. Liebigs Annalen Der Chemie 2028–2048
18. Tofteng AP et al (2011) Microwave heating in solid-phase synthesis of glycopeptides. In: Dawson P, Schneider J (eds) 22nd American peptide symposium, San Diego. Pep Sci 96:512
19. Kihlberg J, Vuljanic T (1993) Piperidine is preferable to morpholine for Fmoc cleavage in

solid-phase synthesis of O-linked glycopeptides. Tetrahedron Lett 34:6135–6138
20. Brocke C, Kunz H (2004) Synthetic tumor-associated glycopeptide antigens from the tandem repeat sequence of the epithelial mucin MUC4. Synthesis 525–542
21. Vuljanic T, Bergquist KE, Clausen H et al (1996) Piperidine is preferred to morpholine for Fmoc cleavage in solid phase glycopeptide synthesis as exemplified by preparation of glycopeptides related to HIV gp120 and mucins. Tetrahedron 52:7983–8000
22. Peters S, Lowary TL, Hindsgaul O et al (1995) Solid-phase synthesis of a fucosylated glycopeptide of human factor-Ix with a fucose-alpha-(1-]O)-serine linkage. J Chem Soc Perkin Trans 1:3017–3022
23. Elofsson M, Salvador LA, Kihlberg J (1997) Preparation of Tn and sialyl Tn building blocks used in Fmoc solid-phase synthesis of glycopeptide fragments from HIV gp120. Tetrahedron 53:369–390
24. Kračun SK, Cló E, Clausen H et al (2010) Random glycopeptide bead libraries for seromic biomarker discovery. J Proteome Res 9:6705–6714
25. Hilaire P, Lowary T, Meldal M et al (1998) Oligosaccharide mimetics obtained by novel, rapid screening of carboxylic acid encoded glycopeptide libraries. J Am Chem Soc 120:13312
26. Brocke C, Kunz H (2002) Synthesis of tumor-associated glycopeptide antigens. Bioorg Med Chem 10:3085–3112
27. Blixt O, Razi N (2006) Chemoenzymatic synthesis of glycan libraries. Glycobiology 415:137–153
28. Koeller KM, Wong CH (2000) Synthesis of complex carbohydrates and glycoconjugates: enzyme-based and programmable one-pot strategies. Chem Rev 100:4465–4493
29. Tomizaki KY, Usui K, Mihara H (2005) Protein-detecting microarrays: current accomplishments and requirements. ChemBioChem 6:783–799
30. Blixt O, Clo E, Nudelman AS et al (2010) A high-throughput O-glycopeptide discovery platform for seromic profiling. J Proteome Res 9:5250–5261
31. Westerlind U, Schroder H, Hobel A et al (2009) Tumor-associated MUC1 tandem-repeat glycopeptide microarrays to evaluate serum- and monoclonal-antibody specificity. Angew Chem Int Ed Engl 121:8413–8417
32. Blixt O, Allin K, Bohorov O et al (2008) Glycan microarrays for screening sialyltransferase specificities. Glycoconj J 25:59–68
33. Tarp MA, Clausen H (2008) Mucin-type O-glycosylation and its potential use in drug and vaccine development. Biochim Biophys Acta 1780:546–563
34. Esko JD, Selleck SB (2002) Order out of chaos: assembly of ligand binding sites in heparan sulfate. Annu Rev Biochem 71:435–471
35. Hart GW, Slawson C, Ramirez-Correa G et al (2011) Cross talk between O-GlcNAcylation and Phosphorylation: roles in signaling, transcription, and chronic disease. Annu Rev Biochem 80:825–858
36. Hewitt JE (2011) Glycomarkers for muscular dystrophy. Biochem Soc Trans 39:336–339
37. Luther KB, Haltiwanger RS (2009) Role of unusual O-glycans in intercellular signaling. Int J Biochem Cell Biol 41:1011–1024

Chapter 15

Instruments for Automated Peptide Synthesis

Søren L. Pedersen and Knud J. Jensen

Abstract

This chapter provides an introduction to and an overview of current instrumentation for solid-phase peptide synthesis (SPPS). Presently, the two most common designs differ in their mode of liquid handling: the first relies on valves and valve blocks for distribution of reagents, while the second uses a robotic platform. They also differ in their mode of mixing the reactants in the reaction vessel, where the former can utilize sparging, 180° rotational shaking, or vortexing, while the latter typically uses vortexing. Valve-based instruments are often single channel (one peptide at a time), but can also be expanded to allow parallel synthesis of up to 12 and even 24 peptides, however, at the price of added complexity. Valve systems often use inert gas for their operation. The X–Y robotic platforms are ideal for parallel synthesis of large numbers of peptides up to 192 and even peptide libraries. However, although less common, the robotic platform is also very suitable for single-channel operation and can also be used for operations under inert gas. Some single-channeled synthesizers are available with UV feedback monitoring of the Fmoc removal which can be useful for some applications. Importantly, single-channel synthesizers can be equipped with fast and precise microwave heating to accelerate the synthesis and to overcome synthetic difficulties. A whole range of synthesizers with different designs are commercially available. The choice of peptide synthesizer will depend on intended application, for example on the type of chemistry, scale, and the number of peptides that are required and so on.

Key words Automated solid-phase peptide synthesis, Microwave heating, Parallel peptide synthesis

1 Introduction

Following Merrifield's pioneering invention of solid-phase peptide synthesis, his group also developed the concept of automated solid-phase peptide synthesis and constructed the first synthesizers in the mid-1960s [1–4]. Using one of the first synthesizers, Gutte and Merrifield synthesized Ribonuclease A, a 124 amino acid protein, using an astonishing 369 chemical reactions and 11,391 steps on the automated peptide synthesizer, relying on the synthetic chemistry and analytical methods available at that time [5]. Later other laboratories and companies developed a diverse range of laboratory synthesizers [6–11]. The introduction of Fmoc-based SPPS [12] and more sophisticated hardware as well as software

Knud J. Jensen et al. (eds.), *Peptide Synthesis and Applications: Second Edition*, Methods in Molecular Biology, vol. 1047, DOI 10.1007/978-1-62703-544-6_15,

allowed for simpler designs of automated peptide synthesizers. The simplified synthesizers have over the years led to comparably robust and reliable synthesizers, which have resulted in high-quality peptides and good reproducibility. However, in our view the viscous and corrosive solutions of reagents used in SPPS still pose a challenging environment for synthesizers. In the 1980s, the ABI 430–433A family of peptide synthesizers was introduced. The ABI 433A is a single-channeled peptide synthesizer equipped with UV monitoring and feedback control for Fmoc removal. The ABI family of synthesizers has been workhorse in many research laboratories around the world. However, in 2008 Applied Biosystems Inc. seized producing these instruments. While the ABI instruments, like many other instruments since then, operated in a batch mode, several instruments were developed to utilize continuous flow, where the reagent solutions are pumped continuously through the reaction vessel. However, none of these continuous flow instruments are currently commercially available. In the 1980s Cambridge Research Biochemicals introduced the semiautomated Fmoc-based synthesizer, PEPSYNthesizer. The latter also monitored the Fmoc removal reaction by UV absorbance (increased absorbance), and additionally it monitored the coupling steps (decrease in absorbance). The reaction solutions were recirculated if the UV absorbance did not sufficiently change [13]. Later, Pharmacia introduced the LKB synthesizer Biolynx 4170 using a similar concept, which was completely automated. The Milligen (Perceptive) 9050 peptide synthesizer also relied on the continuous-flow principle.

As the interest in biopharmaceuticals and peptide science has increased over the last 20–30 years, new manufactures of peptide synthesizers have emerged. The increased competition on the marked and the need for better synthesizers have led to even more sophisticated instruments. The synthesizers of today can have many special features such as parallel synthesis capability, conductive or microwave heating, UV monitoring, ability for small- or large-scale synthesis, as well as the ability for TFA cleavage (Table 1). However, more options are not always better, especially if it comes with a risk of reduced robustness of operation.

Generally, two types of liquid-handling techniques exist: volumetric dispensing using digital injection pumps and robotic arms or nitrogen pressure and valves. Volumetric dispensing allows for acurate dosing even in microliter scale that provides optimal results in terms of yield and purity while minimizing the waste and the cost of synthesis. Moving liquids with nitrogen pressure allows for an inert handling of reagents, however, flow valves are required. These valves, for example, in the ABI 433A, needed to be replaced often, otherwise, they can clog and cause termination of the synthesis. In the first synthesizers the reaction mixture was mixed by rotating the reaction vessel 180°, which is still a common type of mixing. Two other types of mixing are presently preferred, vortexing and sparging. Vortex mixing is achieved by vertical rotation of the reactor such that the fluid inside the reactor is swirled rapidly around the center, and sparging is performed by bubbling nitrogen through the reaction mixture. Nowadays, the most popular reaction vessels are cylindrical or spherical made from glass or polypropylene equipped with a polyethylene, glass filter, or teflon frit but also tips or microtiter plates with filters are used for small scale. The frit or filter allows for filtration which is the most common method for the removal of reactants and wash liquids subsequent to each washing and synthesis step. However, previously, liquids have been removed by pipetting. The first commercially available automated multiple synthesizer, the SMPS350 from Zinsser Analytic [6, 10], was capable of removing the liquid from the solid-phase slurry by immersing a needle equipped with a stainless steel mesh into the reactors. Clearly, the needle had to be thoroughly washed after each liquid removal resulting in long cycle times and the risk of cross contamination. Another technique which allows for parallel processing is tilted plate centrifugation [14]. Microtiter plates are mounted on a centrifugal plate and tilted slightly down towards the center of the centrifuge thus generating a pocket in each well, in which the solid support is collected during centrifugation while the supernatant solutions are removed from the well [15].

2 Automated Peptide Synthesizers

Peptide synthesizers have different capacities and capabilities and are targeting various applications, and many factors determine the best choice of synthesizer. First, will the instrument be used for Fmoc-SPPS or Boc-SPPS? If the synthesizer is only to be used for Fmoc chemistry, it is not necessary to acquire an instrument capable of both Fmoc and Boc chemistries. In practice, many laboratories dedicate instruments to either Fmoc or Boc chemistries and do not switch between these synthesis modes. Secondly, what is the preferred scale? Synthesizers aiming to provide peptides at the research scale, i.e., 0.05–0.5 mmol scale (Table 1), will typically provide 2–250 mg of peptide, depending on the sequences. If scale-up is an important factor in choosing a synthesizer, there are dedicated and often custom-made instruments for multi-gram and even kilogram scale. Parallel synthesizers can be used to scale-out, i.e., large-scale preparation of peptides by carrying out the same synthesis in multiple reaction vessels in parallel. In the other end of the spectrum, peptide syntheses in micromolar scale in tips, 96- or 384-well microtiter plates, are also commercially available (Table 1). Thirdly, what is the purpose of the synthesis? Is it regular synthesis of relatively simple peptides? Or are complicated syntheses of long or difficult peptide sequences needed? Is there a need for process development? Or is the need also to be able to make non-standard peptide chemistry? Routine SPPS can without problems

Table 1
List of selected automated peptide synthesizers

Applied biosystems	ABI 433A	Capacity	1
		Synthesis scale	0.1–1.0 mmol
		Amino acid vials	50 (one per coupling)
		Reagent/solvent flasks	9
		Solution transfer	Nitrogen pressure
		Type of mixing	Vortexing
		Special features	UV monitoring
Activotec	Activo-P11	Capacity	1
		Synthesis scale	0.05–2.0 mmol
		Amino acid vials	60 (one per coupling)
		Reagent/solvent flasks	9
		Type of mixing	Inversion mixing
		Solution transfer	Volumetric dispensing, robotic arm, and nitrogen pressure
		Special features	Inert gas
			UV monitoring
			Conductive heating
		Company web page	www.activotec.com
Biotage AB	Syro *Wave*™	Capacity	1 (Microwave cavity) or 24 or 48 (parallel)
		Synthesis scale	0.001–0.3 mmol
		Amino acid vials	32 (stock)
		Reagent/solvent flasks	5
		Type of mixing	Vortexing
		Solution transfer	Volumetric dispensing, robotic arm
		Special features	Microwave heating
		Company web page	www.biotage.com
CEM	Liberty 1	Capacity	1
		Synthesis scale	0.05–5.0 mmol
		Amino acid vials	20 (stock)
		Reagent/solvent flasks	4
		Type of mixing	Sparging
		Solution transfer	Nitrogen pressure
		Special features	Microwave heating
			In situ fiber optic
		Company web page	www.cem.com

(continued)

Table 1
(continued)

CEM	Liberty—12 channels	Capacity	12 Sequentially
		Synthesis scale	0.025–5.0 mmol
		Amino acid vials	25 (stock)
		Reagent/solvent flasks	7
		Type of mixing	Sparging
		Solution transfer	Nitrogen pressure
		Special features	Microwave heating TFA cleavage
		Company web page	www.cem.com
CS Bio CO	CS 136XT	Capacity	1
		Synthesis scale	0.2–5.0 mmol
		Amino acid vials	22 (stock)
		Reagent/solvent flasks	9
		Type of mixing	Inversion mixing
		Solution transfer	Nitrogen pressure
		Special features	TFA cleavage
		Company web page	www.csbio.com
Intavis	MultiPep RS	Capacity	72 column receptors 386 in microtiter plates (Parallel)
		Synthesis scale	0.025–0.1 mmol 1–10 μmol
		Amino acid vials	30 (stock)
		Reagent/solvent flasks	15
		Type of mixing	Non
		Solution transfer	Volumetric dispensing, robotic arm
		Special features	Parallel synthesis in micromolar scale
		Company web page	www.intavis.com/en/
Peptide scientific	PSI300A	Capacity	1
		Synthesis scale	0.02–0.5
		Amino acid vials	22 (Stock)
		Reagent/solvent flasks	8
		Type of mixing	Inversion mixing
		Solution transfer	Nitrogen pressure
		Special features	Inert gas
		Company web page	www.peptidescientific.com/psi/

(continued)

Table 1
(continued)

Protein Technologies Inc.	Tribute	Capacity	2 (parallel)
		Synthesis scale	0.005–2.0 mmol
		Amino acid vials	101 (one per coupling)
		Reagent/solvent flasks	5
		Type of mixing	Sparging and vortex mixing
		Solution transfer	Nitrogen pressure
		Special features	UV monitoring
		Company web page	www.peptideinstruments.com
Protein Technologies Inc.	Symphony	Capacity	12/24 (parallel)
		Synthesis scale	0.005–7.5 mmol
		Amino acid vials	20 (stock)
		Reagent/solvent flasks	6
		Type of mixing	Sparging
		Solution transfer	Nitrogen pressure
		Special features	
		Company web page	www.peptideinstruments.com
Protein Technologies Inc.	Overture	Capacity	96 or 24
		Synthesis scale	0.001–24 mmol
		Amino acid vials	49 (stock)
		Reagent/solvent flasks	6
		Type of mixing	Vortexing
		Solution transfer	Nitrogen pressure
		Special features	TFA cleavage
		Company web page	www.peptideinstruments.com
MultiSynTech/Biotage	Syro II	Capacity	48 or 96 (parallel) 3×96 (tip)
		Synthesis scale	0.001–0.300 mmol
		Amino acid vials	40 (stock)
		Reagent/solvent flasks	9
		Type of mixing	Vortexing
		Solution transfer	Volumetric dispensing, robotic arms
		Special features	Inert gas Micromolar synthesis (tip)
		Company web page	www.multisyntech.com www.biotage.com

be performed on simple and easy-to-use systems using standard protocols. Simplicity of operation can also be an advantage in laboratories with many users. Obviously, the price of instrument acquisition and the the cost of operation are also very important factors. For example, the instruments have very different solvent consumptions, which affects the cost of operation.

Protein Technologies Inc. (PTI), MultiSynTech/Biotage, CEM, and CS bio have a wide range of peptide synthesizers for research-oriented synthesis. The instruments from CS Bio are meant for scale-up (from mg to multi-kilograms), as all of these systems more or less use the same technology for small- and large-scale syntheses. However, they may require dedicated operators and may not be suitable for syntheses at a very small scale. The synthesizers from PTI range from valve-based small scale to mid-scale and from 1 to 12 parallel channels (in mmol scale) and to 96 parallel channels in μmol scale. The valve-based PTI systems generally use mixing by nitrogen bubbling, except the Overture parallel synthesizer which vortexes, and the PTI synthesizers generally use nitrogen pressure and valves to handle the solvents during synthesis. Some of the PTI systems are very flexible in their chemistry and can easily be used for peptide synthesis development and optimization. CEM currently manufactures two fully automated valve-based systems applying microwave heating during coupling and deprotection (for application protocol, *see* Chapter 17).

High flexibility is also found in the Biotage/MultiSynTech synthesizers. Due to the setup of the Biotage/MultiSynTech platforms, they are even more adaptable to different chemistries (for application protocols, *see* Chapter 7). They can perform up to 48 channels in parallel in mmol scale and 3×96 channels in μmol scale, and the solvents are handled by a volumetric dispensing using X–Y robotic arms which makes these valve-free systems very stable and reliable. The Syro II has in our own laboratory proven to be a very flexible, reliable, and high-throughput instrument for synthesis of both large libraries of, e.g., glycopeptides [16] (*see* Chapter 14) and long and difficult peptide sequences that demand very optimized protocols [17–19].

3 Automated Microwave-Assisted Solid-Phase Peptide Synthesizers

In convetional, conductive heating of chemical reactions, the energy is transferred from the heat source by convective currents or thermal conductivity to the reaction mixture. In microwave heating (dielectric heating at 2.45 GHz), the heating occurs by disposing the energy directly to the solvent and some reagents, due to interactions of the material with the alternating electric field. Materials interact with the electromagnetic field differently, i.e., materials store and convert the energy to heat to different extents, which have huge impact on their ability to be heated by microwaves. The most common solvents in solid-phase peptide synthesis, DMF and NMP, have a very good ability to absorb microwave irradiation (loss factor: tanδ 0.161 and tanδ 0.275, DMF [20] and NMP [21], respectively). The addition of amino acids and coupling reagents in the latter solvents leads to an increased absorption of microwave energy [22]. Besides the solvent, several other factors influence the properties, such as sample volume, vessel material, and the mode of stirring, i.e., vortexing or N_2 bubbling. If sufficient mixing is not achieved during microwave heating in solid-phase peptide synthesis, homogeneous temperature distributions in the reactor vessel cannot be guaranteed. DMF and NMP are strong microwave energy absorbers, which may lead to short microwave penetration. In addition DMF and NMP are viscous and the addition of a resin will increase the viscosity. Moreover, as the most common commercially available microwave synthesizers measure the temperature by an IR sensor at a fixed point, the mixing is of high importance for precise temperature measurement. The groups of Kappe and Martinez have shown huge differences between the temperature measured by IR and internal fiber-optic probes in the absence of adequate mixing [22–25]. Standard magnetic stirring is normally not recommended during SPPS due to the risk of grinding of the solid support however, it might be preferred over unmixed reactions and maybe a good option if the magnet is levitated.

Presently, two different microwave reactor designs are available for peptide synthesis, multimode and monomode (also referred to as single-mode) reactors. However, only the monomode microwave heaters are available on fully automated peptide synthesizers. Multimode instruments use the same principles as in domestic microwave oven. The microwave irradiation is reflected by the walls of the cavity, which generates pockets of high and low energy as the microwaves either reinforce or cancel out each other leading to a nonhomogeneous microwave field in the cavity. To ensure that the field is as homogeneous as possible, multimode instruments are often equipped with a mechanical mode stirrer that continuously changes the instantaneous field pattern inside the cavity. In monomode cavities a standing wave is created when the microwaves are passed by a waveguide that directs them through the reaction vessel that is set at a fixed distance from the radiation irradiation source. The major differences between the two microwave designs include are that the multimode instruments allow for manual parallel synthesis, but suffers from inhomogeneous heating inside the cavity. In contrast, the monomode reactor (manual, semiautomated, or fully automated) provides a high degree of reproducibility due to a very precise heating.

Parallel microwave-assisted peptide synthesis can be performed in 96-well polypropylene microtiter filter plates in a multimode microwave instrument like the MARS from CEM [26–28]. The microtiter filter plates are heat stable under the conditions used for microwave-assisted solid-phase peptide synthesis. Although the temperature is controlled by an internal fiber-optic sensor, the setup suffers from inhomogeneity throughout the microtiter plate, as a temperature variation of ±5 °C at different positions in the 96-wells has been observed [26]. The reagents and wash solvents are added manually, and the plate is moved by hand to a draining station in between the microwave-irradiated reaction steps [26–28].

Currently, there are two manufacturers of peptide synthesizers with microwave heating, CEM and Biotage. The manufacturers both have manual, semiautomated, and fully automated synthesizers in their portfolio—CEM has the Discover SPS, Discover SPS Plus, and Liberty, while Biotage has the Initiator 1, Initiator+ SP *Wave*, the Alstra, and the Syro *Wave*™. The Initiator 1 equipped with a peptide Initiator workstation and the Discover SPS with a wash platform (the reactor vial is moved back and forward between wash station and the microwave cavity) provides the opportunity to perform manual microwave-assisted peptide synthesis. The Biotage Initiator+ SP *Wave* allows for semi-automate synthesis. The Initiator+ SP *Wave* also has the capability of performing standard microwave-assisted organic synthesis (Table 1). The CEM Liberty, the Biotage Syro *Wave*™, and the Biotage Alstra can perform fully automated microwave-assisted synthesis. The Liberty is available in two variants, both single channeled, but one is able of synthesizing up to 12 different peptides consecutively per setup. This relies on a large number of valves, while the smaller Liberty 1 has fewer valves. The Syro *Wave*™ is equipped with both a single-channel microwave cavity and the 24 or 48 parallel channel reactor block. Thus, the instrument can either perform single peptide synthesis with microwave heating or conventional parallel synthesis at room temperature. The major differences between the CEM instruments and the Biotage/MultiSynTech instruments are the mode of mixing (nitrogen bubbling vs. vortexing), nitrogen pressured liquid handling (valves) versus volumetric X–Y robotic arm (syringes), and temperature measurement from the bottom versus the side at a fixed height, respectively.

4 Final Remark

This chapter offered an overview of the different automated peptide synthesizers for laboratory-based peptide synthesis as well as their specifications. We attempted to point out the differences between a number of available synthesizers and point out their capabilities and limitations. When chosing a synthesizer, we suggest to not just consider the possible options but also to consider, whether this allows robut and reliable operation of the instrument.

References

1. Merrifield RB (1965) Automated synthesis of peptides. Science 150:178–185
2. Merrifield RB, Stewart JM (1965) Automated peptide synthesis. Nature 207:522–523
3. Merrifield RB, Stewart JM, Jernberg N (1966) Instrument for automated synthesis of peptides. Anal Chem 38:1905–1914
4. Brunfeldt K, Halstrqm J, Roepstorff P (1969) A punched tape controlled peptide synthesizer. Acta Chem Scand 23:2830–2838
5. Gutte B, Merrifield RB (1971) The synthesis of ribonuclease A. J Biol Chem 246:1922–1941
6. Schnorrenberg G, Wiesmüller K-H, Beck-Sickinger AG et al (1990) In: Giralt E, Andreu D (eds) Peptides 1990, Proceedings of 21st EPS, ESCOM Science Publishers B.V., pp 202–203
7. Daniels SB, Hantman SF, Solé NA et al (1996) In: Ramage R, Epton R (eds) Peptides 1996, Proceeings of 24th EPS, Mayflower Scientific Ltd., pp 323–324
8. Saneii HH, Shannon JD, Miceli RM et al (1993) In: Hodges RS, Smith JA (eds) Peptides: chemistry, structure and biology. Proceedings of 13th APS, ESCOM Science Publisher B.V., pp 1018–1020.
9. Cameron LR, Holder JL, Meldal M et al (1988) Peptide synthesis. Part 13. Feedback control in solid phase synthesis. Use of fluorenylmethoxycarbonyl amino acid 3,4-dihydro-4-oxo-1,2,3-benzotriazin-3-yl esters in a fully automated system. J Chem Soc Perkin Trans 1:2895–2901
10. Schnorrenberg G, Gerhardt H (1989) Fully automatic simultaneous multiple peptide synthesis in micromolar scale—rapid synthesis of series of peptides for screening in biological assays. Tetrahedron 45:7759–7764
11. Gausepohl H, Boulin C, Kraft M et al (1992) Automated multiple peptide-synthesis. Pept Res 5:315–320
12. Carpino LA, Han GY (1970) 9-Fluorenylmethoxycarbonyl function, a new base-sensitive amino-protecting group. J Am Chem Soc 92:5748–5749
13. Dryland A, Sheppard RC (1986) Peptide synthesis. Part 8. A system for solid-phase synthesis under low pressure continuous flow conditions. J Chem Soc Perkin Trans 1:125–137
14. Lebl M (2003) Centrifugation based automated synthesis technologies. J Lab Automation 30:30–35
15. Lebl M (1999) New technique for high-throughput synthesis. Bioorg Med Chem Lett 9:1305–1310
16. Blixt O, Clo E, Nudelman AS et al (2010) A high-throughput O-glycopeptide discovery platform for seromic profiling. J Prot Res 9:5250–5261
17. Sørensen KK, Maolanon NM, Simonsen JB, Stougaard J, Jensen KJ (2013), Linear chemical synthesis of LysM domains, in preparation
18. Pedersen SL, Holst B, Vrang N et al (2009) Modifying the conserved C-terminal tyrosine of the peptide hormone PYY3-36 to improve Y2 receptor selectivity. J Pept Sci 15:753–759
19. Pedersen SL, Steentoft C, Vrang N et al (2010) Glyco-scan: varying glycosylation in the sequence of the peptide hormone PYY3-36 and its effect on receptor selectivity. Chembiochem 11:366–374
20. Gabriel C, Gabriel S, Grant EH et al (1998) Dielectric parameters relevant to microwave dielectric heating. Chem Soc Rev 27:213–224
21. Kappe CO (2004) Controlled microwave heating in modern organic synthesis. Angew Chem Int Ed 43:6250–6284
22. Bacsa B, Horváti K, Bõsze S et al (2008) Solid-phase synthesis of difficult peptide sequences at elevated temperatures: a critical comparison of microwave and conventional heating technologies. J Org Chem 73:7532–7542
23. Herrero MA, Kremsner JM, Kappe CO (2008) Nonthermal microwave effects revisited: on the importance of internal temperature monitoring and agitation in microwave chemistry. J Org Chem 73:36–47
24. Robinson J, Kingman S, Irvine D et al (2010) Understanding microwave heating effects in single mode type cavities-theory and experiment. Phys Chem Chem Phys 12:4750–4758
25. Coantic S, Subra G, Martinez J (2008) Microwave-assisted solid phase peptide synthesis on high loaded resins. Int J Pept Res Ther 14:143–147
26. Murray JK, Gellman SH (2007) Parallel synthesis of peptide libraries using microwave irradiation. Nat Protocols 2:624–631
27. Murray JK, Sadowsky JD, Scalf M et al (2008) Exploration of structure—activity relationships among foldamer ligands for a specific protein binding site via parallel and split-and-mix library synthesis. J Comb Chem 10:204–215
28. Murray JK, Gellman SH (2005) Microwave-assisted parallel synthesis of a 14-helical β-peptide library. J Comb Chem 8:58–65

Chapter 16

Microwave-Assisted Solid-Phase Peptide Synthesis Using the Biotage Syro *Wave*™

Leila Malik and Knud J. Jensen

Abstract

Fast and precise heating by microwave irradiation during solid-phase peptide synthesis (SPPS) can reduce reaction times as well as provide better purities and greater yields for the synthesis of difficult peptides. Microwave- assisted SPPS has proven to be a useful and reliable tool for the synthesis of peptides as well as small proteins. It is particularly well suited for sequences with a high propensity to form β-sheet-type structures and for sterically difficult couplings. In this protocol, conditions and detailed procedures are described for performing microwave-assisted SPPS using the Syro *Wave*™. Here we describe the synthesis of two difficult peptide sequences: the first is derived from the C-terminus of the MuLV CTL epitope, the second is a de novo designed peptide with a C-terminal alkyne.

Key words Microwave-assisted peptide synthesis, Solid-phase peptide synthesis, Difficult peptides, Microwave heating, Alkyne

1 Introduction

One of the frontiers in peptide science is the linear assembly of long peptides and small proteins, which requires that the accumulation of side reactions is minimized and that the hurdles posed by the so-called difficult sequences are overcome. In recent years precise and fast heating by microwave irradiation has emerged as a new parameter for further optimization of SPPS [1–8]. It has especially enabled the synthesis of long and difficult peptide sequences. During the synthesis of difficult sequences, the peptide chain most likely becomes partially inaccessible typically due to the formation of secondary structures, especially β-sheets [9]. In addition, steric hindrance caused by β-branched amino acids can be a problem. Several chemical methods to suppress intramolecular aggregation have been described and include pseudo-prolines [9], solvent composition [10], and chaotropic salts [11, 12]. Heating is likely to

Knud J. Jensen et al. (eds.), *Peptide Synthesis and Applications: Second Edition*, Methods in Molecular Biology, vol. 1047, DOI 10.1007/978-1-62703-544-6_16, © Springer Science+Business Media New York 2013

reduce both the inter- and intramolecular-derived self-assemblies and thereby decrease the reaction time and improve the coupling efficiency of bulky and β-branched amino acids.

Microwave irradiation is a promising tool in peptide chemistry because of the rapid and precise elevation of the temperature and the efficient temperature control during the synthesis. The possibility of a so-called microwave effect [13] has been discussed but studies have indicated that the observed enhancement was purely of a thermal nature, as similar results were obtained with conventional heating [14]. Another significant question is how to prevent the occurrence of epimerization and aspartimide formation. It has been suggested that when the *N*α-Fmoc group is removed while heating, the addition of 0.1 M HOBt to the deprotection solution with either 20 % piperidine or 5 % piperazine can suppress the aspartimide formation and epimerization of peptides containing Asp [13, 15]. It should be emphasized that aspartimide formation also is very sequence dependent [15, 16]. Furthermore, studies on the risk of epimerization in peptides containing Cys and His residues during microwave-assisted SPPS indicated that His and Cys should be coupled at 50 °C or maybe even at room temperature [13]. However, this is likely to be dependent on the side-chain protecting group.

Recently, we published the synthesis of difficult peptide sequences using an automated microwave and parallel peptide synthesis instrument from Biotage AB (Syro *Wave*™) [6]. This instrument is built around a valve-free robot originally developed for parallel peptide synthesis, where the robotic arm transports reagents instead of pumping via valves. Additionally, the reaction vessel is vortexed in the microwave cavity. This is the first example of an "X–Y" robotic microwave-assisted synthesizer developed for the assembly of long peptides. Several groups have shown that mixing is extremely important during microwave-assisted SPPS [5, 17, 18]. Martinez and coworkers reported that major differences in crude purity were observed in the presence or absence of magnetic stirring, 77 and 42 %, respectively [17]. Overall the use of microwave heating during SPPS will often aid the synthesis of difficult peptides and small proteins, provided that commonly occurring side reactions can be avoided.

The decapeptide H-WFTTLISTIM-NH_2, derived from the C-terminus of the MuLV CTL epitope [19], was reported by Jung and Redeman to be a "difficult" sequence, giving rise to major deletion products [20]. The synthesis of this peptide was used to investigate optimal coupling conditions, especially temperature and time, on the automated microwave-assisted peptide synthesizer Syro *Wave*™ [6]. Previously, the synthesis of this peptide had been evaluated on a custom-made semiautomated microwave-assisted peptide synthesizer [5]. The peptide purity with and without heating was 46 % (75 °C, 5 min, HBTU) and 35 % (RT, 45 min, HBTU), respectively. The purity was further optimized by substituting HBTU with other coupling reagents. Interestingly, for this particular peptide, the results for COMU gave a superior

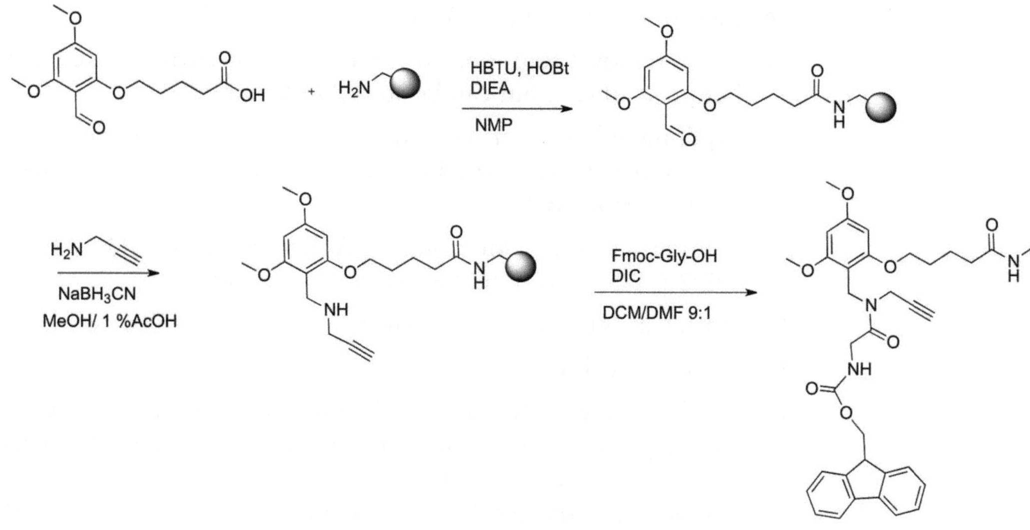

Fig. 1 Synthesis of C-terminal peptide alkynes via the BAL strategy

outcome compared to other coupling reagents. To verify the epimerization under these conditions, the crude peptide was subjected to chiral amino acid analysis, which showed only insignificant levels of epimerization [6].

This protocol outlines the synthesis of a model peptide together with a protocol for the full microwave synthesis of a C-terminal peptide alkyne, originally derived from a de novo designed three-helical coiled coil [21]. Peptide alkynes are starting materials for the Huisgen 1,3-dipolar-type cycloaddition reactions, providing the potential to assemble peptides into small proteins. They can easily be synthesized by a backbone amide linker (BAL) strategy (Fig. 1) (*see* Chapter 9 for further introduction to BAL strategy).

2 Materials

2.1 Solid-Phase Peptide Synthesis of Decapeptide (H-WFTTLISTIM-NH_2) Derived from the C-Terminus of MuLV CTL Epitope Using a Fully Automated Peptide Synthesizer Syro Wave™

1. Solid support: Fmoc-RAM-TG resin (loading 0.24 mmol/g, 0.1 mmol scale).
2. Reaction vessel: 10 mL disposable polypropylene reactor vials with a filter bottom.
3. Wash solvents: *N*-methyl-2-pyrrolidinone (NMP) in a separate 4 L bottle.
4. N^{α}-Fmoc amino acid: Weigh each of the amino acids in the WFTTLISTIM sequence (Fmoc-Phe-OH, Fmoc-Ile-OH, Fmoc-Leu-OH, Fmoc-Met-OH, Fmoc-Ser(*t*Bu)-OH, Fmoc-Thr(*t*Bu)-OH, Fmoc-Trp(Boc)-OH). All the N^{α}-Fmoc amino acids (5.2 equiv., 0.5 M) were dissolved in a freshly made 4:1 mixture of HOBt/HOAt in NMP.
5. Coupling solution: HBTU (5 equiv. 0.43 M) in NMP or COMU (5 equiv, 0.43 M) in NMP. Freshly prepared prior to start of the synthesis.

6. Base solution: DIEA (10 equiv, 2 M) in NMP.
7. Fmoc removal solution: 40 % piperidine in DMF. Make fresh before synthesis, but can be stored indefinitely.
8. Equipment: Fully automated X–Y robot using a single-mode microwave-assisted solid-phase peptide synthesizer (Syro *Wave*™, Biotage AB).
9. Peptide cleavage solution: TFA-Triethylsilane (TES)-H_2O (95:2.5:2.5). Caution: TFA is extremely corrosive.
10. Wash solvent: 15 mL cold ether.
11. Equipment: Freeze dryer.
12. Peptides were analyzed on a Phenomenex Gemini 110 Å C18 column (3 μm, 4.6 × 50 mm) or a similar Resolux column, with a flow of 1.0 mL/min with a linear gradient with increasing amount of buffer for 10–20 min (buffer A: 0.1 % formic acid in H_2O; buffer B: 0.1 % formic acid in CH_3CN).

2.2 Preparation of the Dipeptidyl Resin for the Synthesis of C-Terminal Propargylamide Peptide (Ac-YVAALESKVQAL-EKKVEALEKG-$NHCH_2CH{\equiv}CH$) Derived from CoilV$_a$L$_d$ Using a BAL Strategy

1. Solid support: Aminomethylated Tentagel resin (loading 0.26 mmol/g, 0.385 g).
2. Reaction vessel: Biotage glass vial with a cap for manual steps using the Biotage Initiator™ 10 mL disposable polypropylene syringe for peptide chain elongation.
3. Wash solvents: NMP, dichloromethane (DCM), *N*,*N*-dimethylformamide (DMF), methanol (MeOH), 1 % acetic acid (HOAc) in MeOH.
4. Coupling reagent and chemicals: HBTU, HOBt, DIEA, DIC, $NaBH_3CN$, propargylamine, and 5-(2-formyl-3,5-dimethoxy-phenoxy)pentanoic acid (o-PALaldehyde, Peptides International).
5. *N*α-Fmoc amino acid: Fmoc-Gly-OH as the secondary amino acid.
6. Equipment: A Biotage Initiator™ microwave instrument.

2.3 Solid-Phase Peptide Synthesis of C-Terminal Propargylamide Peptide (Ac-YVAALESK-VQALEKKVEALEKG-$NHCH_2C{\equiv}CH$) Derived from CoilV$_a$L$_d$ Using a Fully Automated Peptide Synthesizer Syro Wave™

1. Solid support: Dipeptidyl resin from Subheading 3.2 (444 mg, 0.18 mmol/g).
2. Reaction vessel: 10 mL disposable polypropylene reactor vials with a filter bottom.
3. Wash solvents: NMP in a separate 4 L bottle.
4. Nα-Fmoc amino acid: Weigh each of the amino acids in the Ac-YVAALESK-VQALEKKVEALEKG-$NHCH_2C{\equiv}CH$ sequence, excluding the first two preloaded C-terminal amino acids, Fmoc-Ala-OH, Fmoc-Glu(*t*Bu)-OH, Fmoc-Lys(Boc)-OH, Fmoc-Leu-OH, Fmoc-Gln(Trt)-OH, Fmoc-Ser(*t*Bu)-OH, Fmoc-Val-OH, Fmoc-Tyr(*t*Bu)-OH. All Nα-Fmoc amino acids (5.2 equiv. 0.5 M) were dissolved in a freshly made 4:1 mixture of HOBt/HOAt in NMP.

5. Coupling solution: HBTU (5 equiv. 0.43 M) in NMP. Freshly made before synthesis.
6. Base solution: DIEA (10 equiv. 2 M) in NMP.
7. Fmoc removal solution: 40 % piperidine in DMF. Make fresh before synthesis, but can be stored.
8. Equipment: Fully automated X–Y robot using a single-mode microwave-assisted solid-phase peptide synthesizer (Syro *Wave*™, Biotage AB).
9. Acetylation solution: Acetic anhydride (Ac_2O) and DCM.
10. Peptide cleavage solution: TFA-H_2O (95:5). Caution: TFA is extremely corrosive.
11. Wash solvent: 15 mL cold ether.
12. Equipment: Freeze dryer.
13. Peptides can be analyzed on a Phenomenex Gemini 110 Å C18 column (3 μm, 4.6 × 50 mm) or a similar Resolux column, with a flow of 1.0 ml/min with a linear gradient with increasing amount of buffer for 10–20 min (e.g., buffer A: 0.1 % formic acid in H_2O; buffer B: 0.1 % formic acid in CH_3CN).

3 Methods

3.1 Solid-Phase Peptide Synthesis of Decapeptide (H-WFTTLISTIM-NH_2) Derived from the C-Terminus of MuLV CTL Epitope Using a Fully Automated Peptide Synthesizer Syro Wave™

3.1.1 Method Programming

1. In the Synthesis software for the Syro, create a synthesis file under the file menu, and choose sequence list as the file type in the Library dialog box (*see* **Note 1**). Enter the sequence of the peptide from the N-terminus to the C-terminus: WFTTLISTIM.
2. The Workfile/cycle relation dialog box will appear on the monitor when you press OK. A Workfile is assigned to every cycle of the synthesis here (*see* **Note 2**). Select the appropriate workfiles. Scale factor = 1 (*see* **Note 3**). Press Generate and the Synthesis file is generated and opened in the Activities window.
3. Click the Calc button and the Calcstart dialog opens. Enter the start and end cycle for the calculation (in this case 1–10). All the required reagents for all the cycles starting from cycle 1 are then calculated.
4. Enter the loading of the resin (in this case 0.24 mmol/g) and the number of equivalents of amino acids (in this case 5.2 equiv). Recalculate and print the load list. A list of all the reagent and amino acids that needs to be weighted out and prepared will appear.
5. Save your synthesis file and select Automatic from the Activity window. Hold and drag the whole synthesis file over to the Automatic window. Press update and the program is ready to start the synthesis after setting up the instrument.

3.1.2 Instrument Setup

1. Prepare chemicals. Use at least the volumes from the load list and load the following reagent to the system: wash solvent (NMP) and *N*α-deprotection solution.
2. Weigh out the resin amount shown in the load list and load the reactor vial into the microwave cavity using the vial loading tool supplied with the system.
3. Weigh out the amino acids shown in the load list and dissolve them in the appropriate amount of HOBt/HOAt solution (*see* **Note 4**).
4. Load all the chemicals as it is in the software layout (*see* **Note 5**).
5. Be sure that all waste containers are empty and that all bottles are properly tightened.
   (a) Close the system doors.
6. To start the synthesis, click the Execute button.

3.1.3 Peptide Release and Analysis

1. Take the reactor vial out of the microwave cavity and transfer the resin to a syringe.
2. Wash with DCM (3×) and dry the resin under vacuum for 1 h.
3. Peptide release is performed using the cleavage cocktail TFA-TES-H_2O (95:2.5:2.5) for 2 h at room temperature.
4. Filter the resin and collect the peptide-cleavage solution.
5. Concentrate the solution (TFA removal) under a stream of N_2 in a well-ventilated fume hood and then precipitate the peptide with diethyl ether. The ether is finally decanted after centrifugation.
6. The peptide is dissolved in a minimal amount of H_2O/CH_3CN before being lyophilized and analyzed.
7. The crude WFTTLISTIM peptide is analyzed by via B gradient elution (e.g., 0–5 min: 5–37 %, 5–8.5 min: 37–38 %, and 8.5–9.5 min: 38–100 %).

3.2 Preparation of the Dipeptidyl Resin for the Synthesis of C-Terminal Propargylamide Peptide (Ac-YVAALESKVQA-LEKKVEALEKG-$NHCH_2C{\equiv}CH$) Derived from CoilV$_a$L$_d$ Using BAL Strategy

3.2.1 Anchoring of the Aminomethylated Resin to *o*-PALaldehyde

1. Swell the resin in DCM for 30 min, then wash with DCM (3×) and NMP (3×).
2. Prepare the coupling solution in a separate container: *o*-PALaldehyde (2 equiv.), HBTU (1.8 equiv.), HOBt (2 equiv.), and DIEA (4 equiv.) in NMP (4 mL). Pre-activate for 5 min. and then add to the resin.
3. Heat in the Biotage Initiator™ for 10 min at 60 °C. Then filtrate and wash with DCM (×3) and NMP (3×). Repeat cycles 2 and 3 (*see* **Note 6**).

3.2.2 Anchoring the Propargylamine by Reductive Amination

1. Wash with MeOH (3×) and 1 % HOAc in MeOH (3×).
2. Propargylamine (10 equiv.) and NaBH3CN (10 equiv.) are dissolved in 1 % HOAc in MeOH (4 mL) and added to the resin.
3. Heat in the microwave for 10 min at 60 °C. Then filtrate and wash with MeOH (3×), NMP (3×) and DCM (3×). Repeat cycles 1–3 (*see* **Note 7**).

3.2.3 Anchoring of the Second Amino Acid by Acylation

1. Fmoc-Gly-OH (10 equiv.) and DIC (5 equiv.) are dissolved in DCM/DMF (9:1), then pre-activate for 5 min and the whole slurry is added to the resin.
2. Heat in the microwave for 10 min at 60 °C. Then filtrate and wash with DMF (3×), NMP (3×), and DCM (3×). Repeat cycles 1 and 2 (*see* **Note 8**).
3. The resin is dried under vacuum and an Fmoc quantification is performed to determine loading.

Chain elongation was performed as described below by microwave-assisted solid-phase peptide synthesis.

3.3 Solid-Phase Peptide Synthesis of C-Terminal Propargylamide Peptide (Ac-YVAALESK-VQALEKKVEALEKG-$NHCH_2C{\equiv}CH$) Derived from CoilV$_a$L$_d$ Using the Fully Automated Peptide Synthesizer, Biotage Syro Wave™

3.3.1 Method Programming

1. In the Synthesis software for the Syro *Wave*™ instrument, create a synthesis file under the file menu, and choose sequence list as the file type in the Library dialog box (*see* **Note 1**). Enter the sequence of the peptide from the N-terminus to the C-terminus: (YVAALESKVQALEKKVEALEK) (*see* **Note 9**).
2. The Workfile/cycle relation dialog box will appear on the monitor when you press OK. A Workfile is assigned to every cycle of the synthesis here (*see* **Note 10**). Select the appropriate workfiles. Scale factor = 1. Press Generate and the Synthesis files are generated and open in the Activities window.
3. Click the Calc button and the Calcstart dialog opens. Enter the start and end cycle for the calculation (in this case 1–21). All the required reagents for all the cycles starting from cycle 1 are then calculated.
4. Enter the loading of the resin (in this case 0.18 mmol/g) and the number of equivalents of amino acids (in this case 5.2 equiv.). Recalculate and print the load list. A list of all the reagent and amino acids that needs to be weighted out and prepared will appear.
5. Save your synthesis file and select Automatic from the Activity window. Hold and drag the whole synthesis file over to the Automatic window. Press update and the program is ready to start the synthesis after setting up the instrument.

3.3.2 Instrument Setup

1. Prepare chemicals. Use at least the volumes from the load list and load the following reagents to the system: wash solvent (NMP) and Nα-deprotection solution.

2. Weigh the amount of dipeptidyl resin (prepared as described above, Subheading 3.2) shown in the load list and load the reactor vial into the microwave cavity using the vial loading tool supplied with the system.
3. Weigh the amino acids shown in the load list and dissolve them in the appropriate amount of HOBt/HOAt solution (*see* **Note 4**).
4. Load all the chemicals as it is in the software layout (*see* **Note 5**).
5. Be sure that all waste containers are empty and that all bottles are properly tightened.
6. Close the system doors.
7. To start the syntheses, click the Execute button.

3.3.3 Peptide Acetylation, Release, and Analysis

1. Take the reactor vial out of the microwave cavity and transfer the resin to a syringe.
2. Wash with DCM (3×) and add a solution of DCM/Ac_2O (3:2), 6 mL. Shake for 30 min at room temperature. Then filtrate and repeat **step 2**.
3. Wash with DCM (3×) and dry the resin under vacuum for 1 h.
4. Peptide release is performed using cleavage cocktail TFA-H_2O (95:5) for 2 h at room temperature.
5. Filter the resin and collect the peptide-cleavage solution.
6. Remove the TFA under a stream of N_2 in a well-ventilated fume hood and then precipitate the peptide with diethyl ether. The ether is removed.
7. The peptide is dissolved in a minimal amount of H_2O/CH_3CN before being lyophilized and analyzed.
8. The crude Ac-YVAALESKVQALEKKVEALEKG-$NHCH_2C{\equiv}CH$ peptide was analyzed via gradient elution (e.g., 0–8.5 min: 5–65 %, 8.5–9 min: 65–100 %, and 9–12 min: 100 %).

4 Notes

1. Use notepad or a similar text editor to create the sequence using one-letter amino acids codes.
2. Consult the operation manual for instruction on how to create a workfile, with modified and special cycles. In this case the workfile should contain a deprotection time of 2 min at 60 °C (1 × with 40 % piperidine in DMF and 1 × with 20 % piperidine in DMF), with a coupling time of 5 min at 75 °C.

3. The scale factors range from 0.1 to 2.0. This means that the volumes in the used workfiles are multiplied by this factor.
4. The HOBt/HOAt solution is (4:1) (5.2 equiv. 0.5 M)
5. The reagent bottles are handmade (not uniform) and have to go into their designed places; see labels on the bottles and the illustration on the screen.
6. If a laboratory microwave instrument is not available, then heat for 2 h at room temperature (2×), preferably on a heater-shaker. The reaction should never be performed in a domestic microwave oven.
7. If a microwave instrument is not available, then overnight at room temperature.
8. If a microwave instrument is not available, then 4 h at room temperature (2×).
9. Only the sequence YVAALESKVQALEKKVEALEK is written since the two first amino acids already are anchored to the C-terminus.
10. Consult the operation manual for instruction on how to create a workfile, with modified and special cycles. In this case the workfile should contain a deprotection at room temperature (for 2 min with 40 % piperidine in DMF and 10 min 20 % piperidine in DMF) and a couple time of 5 min at 75 °C (double couple after amino acid 16).

References

1. Bacsa B, Desai B, Dibó G, Kappe CO (2006) Rapid solid-phase peptide synthesis using thermal and controlled microwave irradiation. J Pept Sci 12:633–638
2. Brandt M, Gammeltoft S, Jensen KJ (2006) Microwave heating for solid-phase peptide synthesis: general evaluation and application to 15-mer phosphopeptides. Int J Pept Res Ther 12:349–357
3. Erdélyi M, Gogoll A (2002) Rapid microwave-assisted solid-phase peptide synthesis. Synthesis 11:1592–1596
4. Matsushita T, Hinou H, Kurogochi M, Shimizu H, Nishimura S (2005) Rapid microwave-assisted solid-phase glycopeptide synthesis. Org Lett 7:877–880
5. Pedersen SL, Sørensen KK, Jensen KJ (2010) Semi-automated microwave-assisted SPPS: optimization of protocols and synthesis of difficult sequences. Biopolymers 94(2):206–212
6. Malik L, Tofteng AP, Pedersen SL, Sørensen KK, Jensen KJ (2010) Automated "X-Y" robot for peptide synthesis with microwave heating: application to difficult peptide sequences and protein domains. J Pept Sci 16:506–512
7. Bacsa B, Kappe C (2007) Rapid solid-phase synthesis of a calmodulin-binding peptide using controlled microwave irradiation. Nat Protoc 2:2222–2227
8. Pedersen SL, Tofteng AP, Malik L, Jensen KJ (2012) Microwave heating in solid-phase peptide synthesis. Chem Soc Rev 41(5):1826–1844
9. Wohr T, Wahl F, Nefzi A, Rohwedder B, Sato T, Sun X, Mutter M (1996) Pseudo-prolines as a solubilizing, structure-disrupting protection technique in peptide synthesis. J Am Chem Soc 118:9218–9227
10. Yamashiro D, Blake J, Hao Li C (1976) The use of trifluoroethanol for improved coupling in solid-phase peptide synthesis. Tetrahedron Lett 17:1469–1472
11. Seebach D, Thaler A, Beck A (1989) Solubilization of peptides in non-polar organic solvents by the addition of inorganic salts: facts and implications. Helv Chim Acta 72:857–867
12. Thaler A, Seebach D, Cardinaux F (1991) Lithium-salt effects in peptide synthesis. Part II. Improvement of degree of resin swelling and of efficiency of coupling in solid-phase synthesis. Helv Chim Acta 74:628–643

13. Palasek S, Cox Z, Collins J (2007) Limiting racemization and aspartimide formation in microwave-enhanced Fmoc solid-phase peptide synthesis. J Pept Sci 13:143–148
14. Bacsa B, Horvati K, Bosze S, Andreae F, Kappe CO (2008) Solid-phase synthesis of difficult peptide sequences at elevated temperatures: a critical comparison of microwave and conventional heating technologies. J Org Chem 73:7532–7542
15. Dolling R, Beyermann M, Haenel J, Kernchen F, Krause E, Franke P, Brudel M, Bienert M (1994) Piperidine-mediated side product formation for Asp(OBut)-containing peptides. J Chem Soc Chem Commun 853–854.
16. Wade J, Mathieu M, Macris M, Tregear G (2000) Base-induced side reactions in Fmoc-solid phase peptide synthesis: minimization by use of piperazine as N-alpha-deprotection reagent. Lett Pep Sci 7:107–112
17. Coantic S, Subra G, Martinez J (2008) Microwave-assisted solid phase peptide synthesis on high loaded resins. Int J Pep Res Ther 14:143–147
18. Bacsa B, Horváti K, Bosze S, Andreae F, Kappe CO (2008) Solid-phase synthesis of difficult peptide sequences at elevated temperatures: a critical comparison of microwave and conventional heating technologies. J Org Chem 73:7532–7542
19. Redemann T, Jung G (1996) In situ fluoride activation allows the preparation of peptides not accessible by routine synthesis protocols. In: Ramage R, Epton R (eds) Peptides, Proceedings of the 24th European Peptide Symposium, pp 749–750
20. Carpino LA, Krause E, Sferdean CD, Schümann M, Fabian H, Bienert M, Beyermann M (2004) Synthesis of "difficult" peptide sequences: application of a depsipeptide technique to the Jung-Redemann 10- and 26-mers and the amyloid peptide Ab(1-42). Tetrahedron Lett 45:7519–7523
21. Ogihara NL, Weiss MS, Degrado WF, Eisenberg D (1997) The crystal structure of the designed trimeric coiled coil coil-V_aL_d: Implications for engineering crystals and supramolecular assemblies. Protein Sci 6:80–88

Chapter 17

Microwave-Assisted Solid-Phase Peptide Synthesis Based on the Fmoc Protecting Group Strategy (CEM)

Grace S. Vanier

Abstract

Microwave-assisted peptide synthesis has become one of the most widely used tools by peptide chemists for the synthesis of both routine and difficult peptide sequences. Microwave technology significantly reduces the synthesis time while also improving the quality of the peptides produced. Microwave energy allows most amino acid couplings to be completed in just 5 min. The Fmoc removal can also be accelerated in the microwave decreasing the reaction time from at least 15 min to only 3 min in most cases. Common side reactions such as racemization and aspartimide formation are easily controllable with optimized methods that can be applied routinely. This protocol outlines the detailed procedure for performing both manual and automated microwave-assisted peptide synthesis of two difficult peptide sequences, ACP (65–74) and β-amyloid, in high purity and yield.

Key words Microwave-assisted peptide synthesis, Microwave irradiation, Solid-phase peptide synthesis, Difficult peptides, Acyl carrier protein, Beta-amyloid

1 Introduction

The development of the solid-phase peptide synthesis method nearly 50 years ago changed the field of peptide chemistry and revolutionized the way chemists synthesize peptides [1]. The solid-phase process eliminated the need for the workup and purification required during each step of solution phase synthesis and has become the method of choice for research scale synthesis of peptides. While solid-phase peptide synthesis is a well established process and routinely generates good purity peptides, there still exist problems with aggregation and peptide sequences which are particularly difficult to synthesize. This has led to new developments in the area over the past several decades including the design of improved protection strategies, the development of new reagents for promoting the coupling reaction, and the investigation of different solid supports on which to perform the chemistry.

Knud J. Jensen et al. (eds.), *Peptide Synthesis and Applications: Second Edition*, Methods in Molecular Biology, vol. 1047, DOI 10.1007/978-1-62703-544-6_17,

The use of microwave irradiation for organic synthesis has seen extensive usage by chemists due to the ability to provide faster reaction times, cleaner products, and higher yielding reactions [2–9]. The application of microwave irradiation to peptide synthesis was first reported in 1992 [10]. Using a domestic microwave system three test peptides were synthesized with a two- to threefold rate enhancement over conventional syntheses with no observable racemization. This effect was most pronounced during the coupling of sterically hindered β-branched amino acids. While this work was a major breakthrough for peptide synthesis, an un-modified domestic microwave system has no means for accurate temperature measurement, and reproducibility with this type of system can be very difficult. It was not until nearly a decade later when commercial laboratory grade microwave systems were introduced that this research realized its full potential. These commercial grade systems feature integrated temperature control along with a homogeneous microwave field that allows much more even heating.

The mechanism by which microwave energy heats a sample is through direct interaction at the molecular level via either dipole rotation or ionic conduction resulting in very rapid heating. When exposed to microwave irradiation, ions and dipoles in a liquid solution will rotate in an attempt to follow the alternating electric field. Molecules have time to align with the electric field but not to follow the oscillating field exactly. This in turn results in energy dissipation from the dipole or ion due to random collisions that give rise to dielectric heating. During the peptide synthesis process there are many polar and ionic species present that can rapidly be heated by microwave energy. The microwave absorbers include the polar solvents, the peptide backbone with the highly polar peptide bond, the terminal amine group, bases for the coupling and deprotection reactions, cleavage acids, and polar/ionic activators. The resulting temperature increase can help break up chain aggregation due to intra- and inter-chain association and allow for easier access to the growing end of the peptide chain.

The standard conditions used for microwave-assisted peptide synthesis are fairly mild, and our typical protocol for base-labile Fmoc peptide synthesis is deprotection with 20 % piperidine at 75 °C for 3 min and coupling with HBTU and DIEA at 75 °C for 5 min. Due to the temperatures used during microwave-assisted peptide synthesis there is an increased potential for unwanted side reactions. Racemization of the activated amino acid during the coupling step is a known issue even under conventional synthesis conditions and is a problem for Cys and His residues specifically. To investigate this potential problem a model 20-mer peptide that contained each of the naturally occurring amino acids was synthesized under both conventional and microwave conditions [11]. The peptides were hydrolyzed in 6N DCl/D_2O, and the resulting free amino acids were subjected to GC-MS analysis to calculate the amount of racemization. Room temperature synthesis of the model peptide (VYWTSPFMKLIHEQCNRADG-NH_2) resulted in 68 % crude purity with a maximum of 1.5 % racemization for all amino acids. The first microwave-assisted synthesis was performed with two stage N^{α}-deprotection (30 s at 50 W followed by 3 min at 50 W (T_{max} = 80 °C)) and couplings of 5 min at 40 W (T_{max} = 80 °C). The initial synthesis resulted in an increase of the crude purity to 76 % but also a significant amount of racemization of Cys and His. The use of HOBt as an additive in the coupling, (benzotriazol-1-yloxy) tripyrrolidinophosphonium hexafluorophosphate (PyBOP) as an alternative activator, and the use of a more hindered base during coupling (2,4,6-trimethylpyridine (TMP)) and 4-methylmorphine (NMM) [12]) were also explored, but all attempts to reduce the racemization under microwave conditions failed. Lowering the maximum coupling temperature of the Cys and His residues from 80 to 50 °C and using a two stage heating method (120 s at 0 W followed by 240 s at 40 W) successfully reduced the racemization. The racemization can be reduced even further by performing the coupling of Cys and His at room temperature. Using modified protocols for the incorporation of Cys and His, the advantage of microwave heating can be used for the incorporation of the other amino acids, while still maintaining a low level of racemization. This study also demonstrated that once the Cys and His are incorporated into the peptide, subsequent couplings at 80 °C do not result in increased racemization of these residues indicating that the racemization is limited to only the activated ester state.

Another well documented potential side reaction when performing peptide synthesis is aspartimide formation. This side reaction occurs during the Fmoc deprotection with piperidine with Asp containing sequences where the carboxy adjacent residue is Gly, Asn, Ser, or Ala. Each subsequent deprotection after the incorporation of an Asp, which is susceptible to aspartimide formation, will result in a decrease in the peptide purity due to an accumulation of the side reaction. The 20-mer model peptide described above to explore racemization with microwave irradiation also contains the Asp-Gly sequence that potentially will suffer from aspartimide formation [11]. In the first microwave synthesis of the test peptide, deprotection with 20 % piperidine at 80 °C led to considerable amounts of aspartimide formation. Addition of 0.1 M HOBt to the deprotection with piperidine reduced the aspartimide formation and resulted in a 10 % increase in crude purity. The use of piperazine in place of piperidine resulted in an even further decrease in the aspartimide formation and no deletion products were observed despite the decreased pK_a of the deprotection reagent.

Microwave-assisted peptide synthesis has become the method of choice for the synthesis of difficult peptides, but it has also been used extensively in academia and industry for the routine synthesis of a variety of different types of peptides [13–16]. Peptides synthesized using microwave irradiation maintain full biological activity and have been used for numerous applications including to investigate human relaxin peptides [17, 18], nucleophosmin mutations in acute myeloid leukemia [19], Alzheimer's disease amyloid β-peptide [20, 21], control interfacial properties of a solid [22], GPCR-peptide interactions [23], human melanocortin receptors [24], human urotension-II [25], insulin [26–28], nicotinic acetylcholine receptors [29], proline-rich antibacterial peptides [30, 31], human apolipoprotein C-I [32], and glucagon-like peptide-1 [33]. Microwave irradiation can also be used to promote the synthesis of modified peptides and these modifications include labeling with dyes, fluorophores, and metal spectroscopic labels, lipidation, cyclization, phosphorylation, and glycosylation. In addition, microwave energy has been used to accelerate the synthesis of a variety of peptidomimetics including β-peptides, peptoids, and peptide nucleic acid (PNA)–peptide hybrids, a new class of peptides that selectively target DNA/RNA sequences with high binding specificity and affinity [16].

This protocol outlines the general procedure for both manual and automated microwave-assisted peptide synthesis. For the manual synthesis the acyl carrier protein, ACP (65–74) sequence was selected as a representative difficult peptide. The automated synthesis features the preparation of β-amyloid (1–42), another peptide that is known to be difficult to synthesize. Both methods provide the peptides in excellent purity with good yields in a fraction of the time it would take to perform these syntheses using conventional, room temperature synthesis conditions.

2 Materials

For optimal results, all reagents should be prepared fresh with high quality solvents (ACS grade and above) unless otherwise indicated. Protective gear should be worn as outlined in the user's safety program for hazardous materials and the reagent manufacturer's material safety data sheet. Refer to these guidelines for proper decontamination, handling and disposal of reagents or any hazardous materials.

2.1 Manual Microwave-Assisted Peptide Synthesis of ACP (65–74)

1. Equipment: Single mode microwave synthesizer configured for manual peptide synthesis equipped with a fiber optic temperature probe (Discover SPS microwave-assisted peptide synthesizer, CEM Corporation) (*see* **Note 1**).
2. Vessel: 25 mL polypropylene fritted vessel equipped with Teflon seal ball and capped luer fitting (CEM Corporation).
3. Resin: 0.1 mmol Fmoc-Gly-Wang (0.164 g, 0.61 mmol/g substitution, 100–200 mesh) (*see* **Note 2**).

Table 1
Amino acid amounts for the manual microwave synthesis of ACP (65–74)

Amino acid	Number of vials	Molecular weight	Mass (g)
Fmoc-Ala-OH	2	311.3	0.156
Fmoc-Asn(Trt)-OH	1	596.7	0.298
Fmoc-Asp(OtBu)-OH	1	411.5	0.206
Fmoc-Gln(Trt)-OH	1	610.7	0.305
Fmoc-Ile-OH	2	353.4	0.177
Fmoc-Tyr(tBu)-OH	1	459.6	0.230
Fmoc-Val-OH	1	339.4	0.170

4. Wash solvents: *N,N*-dimethylformamide (DMF) and dichloromethane (DCM) in suitable wash bottles (*see* **Note 3**).
5. Solution for Fmoc removal: 100 mL of 20 % piperidine in DMF with 0.1 M 1-hydroxybenzotriazole (HOBt). This solution can be stored for up to a month (*see* **Notes 4** and **5**).
6. Fmoc-AA-OH: Weigh out 0.5 mmol (5 equiv.) of each of the Fmoc-amino acids comprising the ACP (65–74) sequence (VQAAIDYING) into separate vials (Table 1) (*see* **Notes 6** and **7**).
7. Activator: Add 0.190 g of *N*-[(1*H*-benzotriazol-1-yl)(dimethylamino)methylene]-*N*-methylmethanaminium hexafluorophosphate *N*-oxide (HBTU) to each of the amino acid vials (*see* **Notes 8** and **9**).
8. Activator base solution: 50 mL of 0.25 M *N,N-diisopropylethylamine* (DIEA) in DMF (*see* **Note 10**).

2.2 Automated Microwave-Assisted Peptide Synthesis of β-Amyloid (1–42)

1. Equipment: Automated microwave-assisted peptide synthesizer (Liberty microwave-assisted peptide synthesizer, CEM Corporation).
2. Resin: 0.1 mmol Fmoc-PAL-PEG-PS (0.5 g, 0.20 mmol/g substitution, 100–200 mesh) (*see* **Note 2**).
3. Wash solvents: 8 L of DMF and 100 mL of DCM in 4 L bottles (*see* **Note 3**).
4. Solution for Fmoc removal: 800 mL of 20 % piperidine in DMF with 0.1 M HOBt in a 1 L bottle (*see* **Notes 4** and **5**). This solution can be stored for up a month.
5. Fmoc-AA-OH: 0.2 M amino acid in DMF solutions of each of the Fmoc-amino acids comprising the β-amyloid (1–42) sequence(DAEFRHDSGYEVHHQKLVFFAEDVGSNKGAI IGLMVGGVVIA) in separate 125 mL bottles (Table 2)

Table 2
Volume of 0.2 M amino acids for the automated microwave synthesis of β-amyloid (1–42)

Amino acid	Molecular weight	Mass (g)	Volume (mL)
Fmoc-Ala-OH	311.3	0.685	11
Fmoc-Arg(Pbf)-OH	648.8	0.778	6
Fmoc-Asn(Trt)-OH	596.7	0.597	5
Fmoc-Asp(OtBu)-OH	411.5	0.658	8
Fmoc-Gln(Trt)-OH	610.7	0.611	5
Fmoc-Glu(OtBu)	425.5	0.681	8
Fmoc-Gly-OH	297.3	0.951	16
Fmoc-His(Trt)	619.7	1.983	16
Fmoc-Ile-OH	353.4	0.565	8
Fmoc-Leu-OH	353.4	0.424	6
Fmoc-Lys(Boc)-OH	468.5	0.562	6
Fmoc-Met-OH	371.5	0.372	5
Fmoc-Phe-OH	387.4	0.620	8
Fmoc-Ser(tBu)-OH	383.4	0.460	6
Fmoc-Tyr(tBu)-OH	459.6	0.460	5
Fmoc-Val-OH	339.4	1.086	16

(*see* **Notes 6**, 7, and **11**). For best results these solutions should be fresh, but most amino acids can be stored in solution for up to 2 weeks.

6. Activator: 55 mL of 0.5 M HBTU in DMF in a 250 mL amber bottle. Cap the bottle and shake to ensure complete dissolution of the activator. It may take up 10 min to fully go into solution (*see* **Notes 8** and **9**).
7. Activator base solution: 30 mL of 2.0 M DIEA in NMP in a 250 mL bottle (*see* **Notes 10** and **12**).
8. Resin tube: 50 mL polypropylene centrifuge tube.

2.3 Peptide Release and Side Chain Deprotection

1. Equipment: Single mode microwave synthesizer configured for manual peptide cleavage equipped with a fiber optic temperature probe (Accent microwave-assisted peptide cleavage system, CEM Corporation).
2. Equipment: centrifuge.
3. Equipment: lyophilizer.

4. Vessel: 25 mL polypropylene fritted vessel equipped with Teflon seal ball and capped luer fitting (CEM Corporation).
5. Product Tube: 50 mL polypropylene centrifuge tube.
6. Cleavage Cocktail: To 9.25 mL of trifluoroacetic acid (TFA) add 0.25 mL each of water, triisopropylsilane (TIS), and 3,6-dioxa-1,8-octane-dithiol (DODT) to give 10 mL total volume of cleavage cocktail. For best results this mixture should be prepared just prior to performing the peptide cleavage (*see* **Note 13**).
7. Solvent: 100 mL cold diethyl ether.

2.4 Peptide Analysis

1. Equipment: analytical high performance liquid chromatography (HPLC) system equipped with a mass spectrometer (Thermo Fisher Scientific Surveyor HPLC with LCQ Advantage ion trap mass spectrometer).
2. Analytical HPLC column: ACP (65–74)—Waters Atlantis C18 column, 3 μm, 2.1 × 100 mm and β-amyloid—Waters Symmetry300 C4 Column, 5 μm, 2.1 × 150 mm.
3. Solvent A: HPLC grade water with 0.1 % formic acid.
4. Solvent B: HPLC grade acetonitrile with 0.1 % formic acid.

3 Methods

3.1 Manual Microwave-Assisted Peptide Synthesis of ACP (65–74)

3.1.1 Resin Preparation

1. Place a luer plug on the bottom of the reaction vessel. Attach the fiber optic clip and insert the thermowell.
2. Weigh 0.164 g of the Fmoc-Gly-Wang resin into the reaction vessel.
3. Add 7 mL of DMF to the resin, and swirl for 10 s.
4. Wait for 15 min to swell the resin.
5. Remove the luer plug and immediately place the reaction vessel in to the vacuum manifold. Remove the solvent by vacuum filtration.
6. Add 5 mL of DMF to the resin to wash and swirl for 10 s.
7. Remove the solvent by vacuum filtration and replace the luer plug.

3.1.2 Fmoc Removal

1. Add 7 mL of the 20 % piperidine in DMF with 0.1 M HOBt solution to the resin. Swirl for 10 s (*see* **Note 14**).
2. Place the vessel in the holder, and insert the fiber optic probe (*see* **Note 15**).
3. Place the vessel assembly in the microwave cavity. Program the Deprotection method (Table 3) and press play (*see* **Note 16**).

Table 3
Microwave methods for manual microwave-assisted peptide synthesis

Method name	Power (W)	Temperature (°C)	Delta temperature (°C)	Time (min:s)
Deprotection	35	75	5	3:00
Coupling	20	75	5	5:00
Cleavage	15	38	5	30:00

4. Remove the vessel assembly from the cavity, remove the luer plug, and immediately place the reaction vessel in to the vacuum manifold. Remove the deprotection mixture by vacuum filtration.
5. Add 5 mL of DMF to the resin to wash, and swirl for 10 s. Remove the solvent by vacuum filtration.
6. Repeat **step 5** four times. Replace the luer plug.

3.1.3 Amino Acid Coupling

1. To the vial containing 0.298 g of Fmoc-Asn(Trt)-OH and 0.190 g of HBTU add 4.0 mL of the 0.25 M DIEA in DMF. Cap the vial and shake to ensure complete dissolution of the amino acid and activator. If necessary, vortex the mixture until the mixture is fully dissolved (*see* **Note 17**).
2. Add the preactivated amino acid solution to resin. Swirl for 10 s.
3. Place the vessel in the holder, and insert the fiber optic probe (*see* **Note 15**).
4. Place the vessel assembly in the microwave cavity. Program the Coupling method (Table 3) and press play (*see* **Note 16**).
5. Remove the vessel assembly from the cavity, remove the luer plug, and immediately place the reaction vessel in to the vacuum manifold. Remove the coupling mixture by vacuum filtration.
6. Add 5 mL of DMF to the resin to wash, and swirl for 10 s. Remove the solvent by vacuum filtration.
7. Repeat **step 6** four times. Replace the luer plug (*see* **Note 18**).

3.1.4 Peptide Chain Elongation

1. Repeat Subheadings 3.1.2 and 3.1.3 for each successive amino acid in the sequence in the following order: Fmoc-Ile-OH, Fmoc-Tyr(tBu)-OH, Fmoc-Asp(OtBu)-OH, Fmoc-Ile-OH, Fmoc-Ala-OH, Fmoc-Ala-OH, Fmoc-Gln(Trt)-OH, and Fmoc-Val-OH (*see* **Note 19**).
2. Repeat **steps 1–4** of Subheading 3.1.2 to remove the Fmoc from the N-terminal Val.
3. Add 5 mL of DCM to the resin to wash, and swirl for 10 s. Remove the solvent by vacuum filtration.
4. Repeat **step 3** four times. Replace the luer plug.

3.2 Automated Microwave-Assisted Peptide Synthesis of β-Amyloid (1–42)

3.2.1 Method Programming

1. Use the Sequence Editor in the PepDriver™ controller software to enter the sequence of the peptide from the N-terminus to the C-terminus: DAEFRHDSGYEVHHQKLVFFAED-VGSNKGAIIGLMVGGVVIA.
2. In the Method Editor highlight the sequence created in **step 1** and select the following options: Skip Resin Load = No, C-terminus = amide (*see* **Note 20**), Final Deprotection = Yes, Scale = 0.10, Resin = Standard-Resin, Cleaving = Non-cleavage, Final Depr. = 0.10. Also, enter the resin substitution under the resin information.
3. While still in the Method Editor verify the "0.10-Single" method is selected for each of the amino acids with the exception of the histidine residues that require the "0.10-Single 50C Temp" method and the arginine residue that requires the "0.10-Double Arg" method (*see* **Notes 21** and **22**).

3.2.2 Instrument Set-Up

1. Prior to beginning the synthesis verify all the regularly scheduled maintenance procedures have been performed on the system.
2. Place a dip tube filter on each of the positions necessary for the synthesis and load the following reagents on to the system: wash solvents (DMF and DCM), solution for Fmoc removal, Fmoc-AA-OH, activator, and activator base solution. Use the Change Bottle function when necessary (*see* **Note 23**).
3. Verify the following: waste container has sufficient capacity, ample nitrogen supply, all necessary reagents are loaded on the system, and all bottles are properly tightened.
4. Weigh 0.5 g of Fmoc-PAL-PEG-PS resin into a 50 mL polypropylene centrifuge tube, and place the tube on the system in position 1 (*see* **Note 24**).
5. Load the method created in Subheading 3.2.1 into position 1 of the PepDriver software.
6. Press Play to start the method.
7. The system will go through an initialization to prepare for the synthesis, and then the system will perform all the steps necessary for the synthesis without user intervention including the resin transfer to the reaction vessel, resin swelling, Fmoc removal, amino acid coupling, peptide chain elongation, and finally return of the resin to the tube with the completed peptide as per the method programmed.
8. When the synthesis is complete, the peptide resin will be in the position 1 resin tube along with a mixture of DMF and DCM.
9. Transfer the peptide resin to a 25 mL polypropylene fritted vessel using DCM to wash the centrifuge tube. Remove the solvent by vacuum filtration.

10. Add 5 mL of DCM to the resin to wash, and swirl for 10 s. Remove the solvent by vacuum filtration.
11. Repeat **step 10** four times. Place a luer plug on the vessel.

3.3 Peptide Release and Side Chain Deprotection

1. Add 10 mL of freshly prepared cleavage cocktail to the reaction vessel, and swirl for 10 s (*see* **Notes 25** and **26**).
2. Place the vessel in the holder, and insert the fiber optic probe into the thermowell (*see* **Note 15**).
3. Place the vessel assembly in the microwave cavity. Program the Cleavage method (Table 3) and press play (*see* **Note 16**).
4. Place a clean 50 mL centrifuge tube on the product collection position. Remove the vessel assembly from the cavity, remove the luer plug, and immediately place the reaction vessel in to the vacuum manifold. Transfer the cleaved peptide by vacuum filtration into the 50 mL centrifuge tube. Wash any residual peptide from the resin with a minimal amount of TFA (up to 1 mL).
5. Add 35 mL of ice-cold diethyl ether to the peptide to precipitate.
6. Allow the peptide to precipitate for at least 30 min on ice or in a freezer.
7. Centrifuge the precipitated peptide to isolate the crude product 4 min at 1,300 × *g*. Carefully decant the supernatant. Rinse the peptide with an additional 10 mL of cold diethyl ether, and then repeat the centrifugation to isolate the precipitate.
8. Dissolve the peptide in a minimal amount (<5 mL) of 10 % aqueous acetic acid. Freeze the sample with liquid nitrogen and lyophilize overnight: ACP (65–74) 102 mg (theoretical yield 106 mg, isolated crude yield 96 %) and β-amyloid (1–42) 325 mg (theoretical yield 451 mg, isolated crude yield 72 %).

3.4 Peptide Analysis

1. Dissolve a minimal amount of peptide in water (less than 1 mL), and add methanol dropwise and sparingly to get the peptide into solution. Sonicate the peptide to assist solubilization.
2. Analyze the peptide by LC-MS using the following conditions:

   ACP (65–74): Waters Atlantis C18 column with a gradient of 5–95 % Solvent B over 20 min.

   β-Amyloid (1–42): Waters Symmetry300 C4 column at 40 °C with a gradient of 5–95 % Solvent B over 25 min.
3. *See* Figs. 1 and 2 for the LC-MS chromatogram and mass spectra for ACP (65–74) and beta-amyloid (1–42), respectively: ACP (65–74) crude purity 87 % and β-amyloid (1–42) crude purity 90 %.

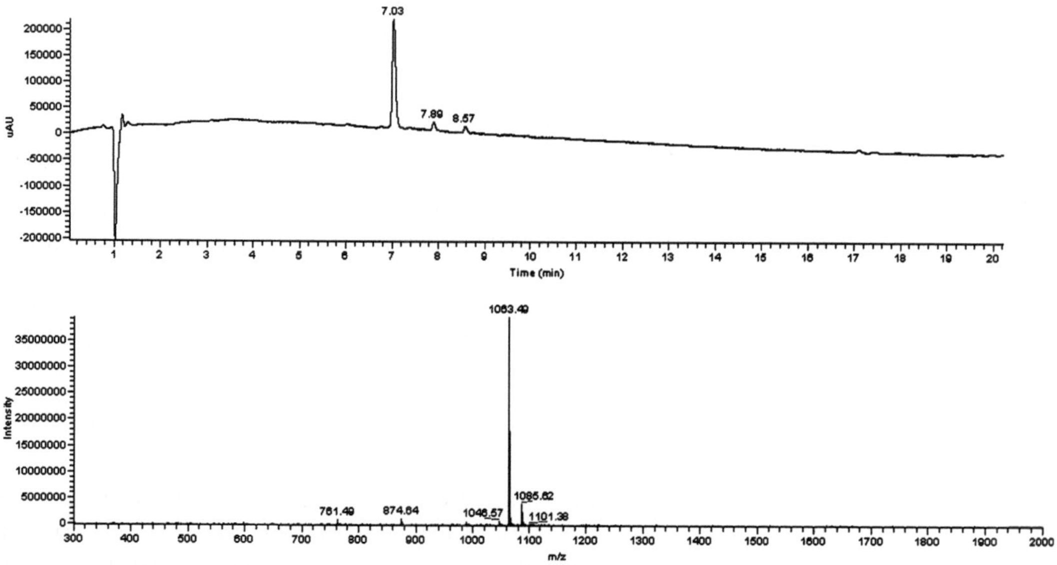

Fig. 1 Crude HPLC chromatogram and mass spectra for the microwave synthesis of ACP (65–74). The major peak eluting at 7.03 min comprises a mass of 1,063.49, the $[M + H]^+$ of ACP (65–74)

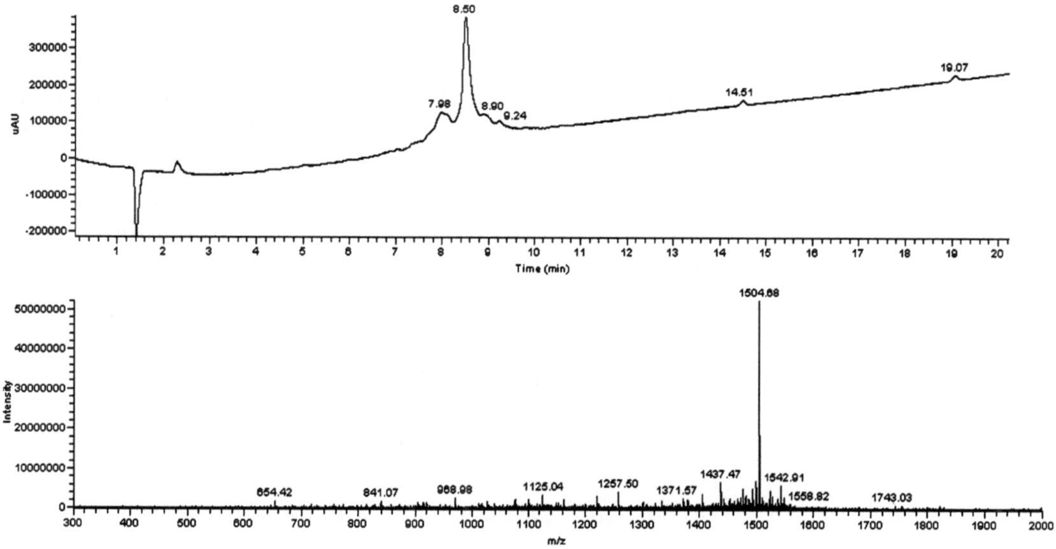

Fig. 2 Crude HPLC chromatogram and mass spectra for the microwave synthesis of β-amyloid (1–42). The major peak eluting at 8.50 min comprises a mass of 1,504.68, the $[M + 3H]^{3+}$ of beta-amyloid (1–42)

4 Notes

1. Manual microwave-assisted peptide synthesis can also be performed in a multimode microwave system [34].
2. Any standard resin used for solid-phase peptide synthesis can be used with microwave irradiation including, but not limited to, polystyrene (PS) resin, polyethylene glycol-PS (PEG-PS) resin, PEG, and cross-linked ethoxylate acrylate (CLEAR) resin.

3. Alternatively, *N*-methylpyrrolidinone (NMP) can be used as the wash solvent as well as the solvent for making up the reagents.
4. Alternatively, 5 % piperazine in DMF can be used for the solution for Fmoc removal.
5. For peptide sequences that are susceptible to aspartimide formation, add 0.1 M HOBt to the Fmoc deprotection solution to prevent this side reaction [11]. HOBt hydrate can be used rather than the anhydrous form. Oxyma Pure is a non-explosive alternative to HOBt that will also suppress aspartimide formation during the Fmoc removal.
6. When using a pre-loaded resin, it is not necessary to weigh out the C-terminal amino acid.
7. Any Fmoc-amino acids can be used, but selection of the specific derivatives should be based on the requirements for the synthesis. Typically amino acids with orthogonal side chain protecting groups are used, but other derivatives may be necessary depending on the sequence desired. For example, Fmoc-Lys(Fmoc)-OH can be used to generate a branched sequence while or Fmoc-Cys(Acm)-OH may be used when disulfide bridge formation is desired.
8. It is not necessary to add HOBt as an additive to the coupling reaction when using HBTU activation. Microwave irradiation provides ample enhancement of the coupling efficiency to render the HOBt unnecessary. Also, it has been demonstrated that HOBt does not reduce racemization during the coupling reaction [11].
9. Alternative activation strategies can be used with microwave irradiation including, but not limited to, *N,N'*-diisopropylcarbodiimide (DIC)-HOBt, *N*-[(1*H*-6-chlorobenzotriazol-1-yl)(dimethylamino)methylene]-*N*-methylmethanaminium hexafluorophosphate *N*-oxide (HCTU), *N*-[(dimethylamino)-1*H*-1,2,3-triazolo[4,5-b]pyridino-1ylmethylene]-*N*-methylmethanaminium hexafluorophosphate *N*-oxide (HATU), (Benzotriazol-1-yloxy) tripyrrolidinophosphonium hexafluorophosphate (PyBOP), and(1-Cyano-2-ethoxy-2-oxoethylidenaminooxy)dimethylamino-morpholino-carbenium hexafluorophosphate (COMU).
10. Alternative bases such as 4-methylmorpholine (NMM) or 2,4,6-trimethylpyridine (TMP) may be used, but the microwave coupling is not as efficient and can lead to greater racemization of Cys and His.
11. The actual amount used by the automated microwave-assisted peptide synthesizer will be lower than the volumes listed. These volumes include a 10 % overage to ensure smooth operation of the synthesizer.

12. At a concentration of 2.0 M, DIEA is not miscible with DMF. NMP is required for a homogeneous solution.
13. Appropriate scavengers should be selected for the peptide sequence. The TFA/water/TIS/DODT mixture is a good general cocktail for the majority of peptide sequences.
14. The microwave methods are rapid and typically do not require additional agitation. If in situ agitation is desired, a glass or Teflon tube can be inserted into the reaction mixture to introduce nitrogen or argon bubbling. While the microwave system is equipped with electromagnetic stirring, magnetic stirring is generally not recommended for solid-phase peptide synthesis.
15. The fiber optic probe is used to accurately measure the temperature of the reaction mixture. It is important to remember to insert the probe into the thermowell prior to microwave irradiation. Failure to do so can result in overheating of the reaction mixture and potential decomposition of the peptide.
16. The power settings are only suggestions as individual microwave performance will vary. Select a power setting that will heat the reaction mixture to 75 °C in about 70 s while limiting the temperature overshoot to less than 85 °C for the deprotection and coupling methods. The cleavage method should be limited to a maximum temperature of 42 °C.
17. This step can be performed during the previous deprotection step to maximize time management and decrease the overall synthesis time.
18. Optional: Test a small portion of the resin to ensure the coupling has gone to completion using a qualitative protocol such as the Kaiser test. If the coupling has successfully gone to completion, proceed to the next step. Otherwise, repeat the amino acid coupling steps in Subheading 3.1.3.
19. If necessary, the synthesis can be paused at the end of every coupling cycle, when the N-terminal amino group is still Fmoc protected. To avoid undesired removal of the Fmoc group due to storage in DMF, wash the resin five times with DCM and let it dry at room temperature. Cover the vessel, replace the luer plug, and store in the freezer until the synthesis can be resumed. Allow the sample reach room temperature before resuming the synthesis, and swell the dry resin as described in Subheading 3.1.1.
20. The C-terminus selection signals to the software if the coupling of the first amino acid must be performed by the instrument. The software will assume with the selection of "amide" the first amino acid is not loaded on the resin and therefore will couple the first amino acid in the sequence. Conversely, the software will assume with the selection of "acid" the first amino acid is pre-loaded on the resin and will skip the coupling of the first amino acid.

21. These instructions utilize the default cycles in the software. Consult the operation manual for instructions on how to create modified or specialized cycles, and how to incorporate these cycles into the method.
22. To limit racemization of the cysteine and histidine residues, the coupling should be performed at 50 °C rather than the standard 75 °C used for the other amino acids [11]. In addition, due to the potential for γ-lactam formation, the coupling of arginine residues may require a slow ramp to temperature method to ensure complete coupling. The "Double Arg" method features a double coupling method where the first coupling is performed for 30 min with microwave irradiation at 75 °C for only the final 5 min. The second coupling is a standard microwave method at 75 °C for 5 min.
23. The Usage Calculator within the PepDriver software can be used to verify the reagent amounts needed for the synthesis.
24. The resin may cling to the sides of the tube. If necessary, wash the resin from the sides of the tube using a small amount (<5 mL) of DCM.
25. Due to the corrosive and toxic nature of TFA, all peptide cleavage and deprotection steps should be performed in a fume hood.
26. To avoid an exothermic reaction, ensure that all DMF has been washed from the resin with DCM before adding the TFA.

References

1. Merrifield RB (1963) Solid phase peptide synthesis. I. The synthesis of a tetrapeptide. J Am Chem Soc 85:2149–2154
2. Lew A, Krutzik PO, Hart ME, Chamberlin AR (2002) Increasing rates of reaction: microwave-assisted organic synthesis for combinatorial chemistry. J Comb Chem 4:95–105
3. Loupy A (ed) (2006) Microwaves in organic synthesis. Wiley-VCH, Weinheim
4. Hayes BL (2002) Microwave synthesis: chemistry at the speed of light. CEM, Matthews
5. Loupy A (ed) (2002) Microwaves in organic synthesis. Wiley-VCH, Weinheim
6. Kappe CO, Stadler A (2005) Microwaves in organic and medicinal chemistry. Wiley-VCH, Weinheim
7. Lidström P, Tierney JP (eds) (2005) Microwave-assisted organic synthesis. Blackwell, Oxford
8. Kappe CO, Dallinger D, Murphee SS (2009) Practical microwave synthesis for organic chemists: strategies, instruments, and protocols. Wiley-VCH, Weinheim
9. Leadbeater NE (ed) (2010) Microwave heating as a tool for sustainable chemistry. CRC Press, Boca Raton
10. Yu HM, Chen ST, Wang KT (1992) Enhanced coupling efficiency in solid-phase peptide-synthesis by microwave irradiation. J Org Chem 57:4781–4784
11. Palasek SA, Cox ZJ, Collins JM (2007) Limiting racemization and aspartimide formation in microwave-enhanced Fmoc solid phase peptide synthesis. J Pept Sci 13:143–148
12. Collins JM, Unpublished results
13. Collins JM (2006) Microwave-enhanced solid-phase peptide synthesis. In: Loupy A (ed) Microwaves in organic synthesis. Wiley-VCH, Weinheim
14. Vanier GS (2010) Microwave heating as a tool for the biosciences. In: Leadbeater NE (ed) Microwave heating as a tool for sustainable chemistry. CRC Press, Boca Raton, pp 231–269
15. Pedersen SL, Tofteng AP, Malik L, Jensen KJ (2012) Microwave heating in solid-phase peptide synthesis. Chem Soc Rev 41:1826–1844
16. Collins JM, Leadbeater NE (2007) Microwave energy: a versatile tool for the biosciences. Org Biomol Chem 5:1141–1150
17. Hossain MA, Rosengren KJ, Haugaard-Jonsson LM, Zhang S, Layfield S, Ferraro T,

Daly NL, Tregear GW, Wade JD, Bathgate RAD (2008) The A-chain of human relaxin family peptides has distinct roles in the binding and activation of the different relaxin family peptide receptors. J Biol Chem 283:17287–17297

18. Hossain MA, Rosengren KJ, Zhang S, Bathgate RA, Tregear GW, van Lierop BJ, Robinson AJ, Wade JD (2009) Solid phase synthesis and structural analysis of novel A-chain dicarba analogs of human relaxin-3 (INSL7) that exhibit full biological activity. Org Biomol Chem 7:1547–1553
19. Grummitt CG, Townsley FM, Johnson CM, Warren AJ, Bycroft M (2008) Structural consequences of nucleophosmin mutations in acute myeloid leukemia. J Biol Chem 283:23326–23332
20. Drew SC, Masters CL, Barnham KJ (2009) Alanine-2 carbonyl is an oxygen ligand in Cu2+ coordination of Alzheimer's disease amyloid-beta peptide—relevance to N-terminally truncated forms. J Am Chem Soc 131:8760–8761
21. Drew SC, Noble CJ, Masters CL, Hanson GR, Barnham KJ (2009) Pleomorphic copper coordination by Alzheimer's disease amyloid-beta peptide. J Am Chem Soc 131:1195–1207
22. Mosse WK, Koppens ML, Gengenbach TR, Scanlon DB, Gras SL, Ducker WA (2009) Peptides grafted from solids for the control of interfacial properties. Langmuir 25:1488–1494
23. Harterich S, Koschatzky S, Einsiedel J, Gmeiner P (2008) Novel insights into GPCR-peptide interactions: mutations in extracellular loop 1, ligand backbone methylations and molecular modeling of neurotensin receptor 1. Bioorg Med Chem 16:9359–9368
24. Grieco P, Cai M, Liu L, Mayorov A, Chandler K, Trivedi D, Lin G, Campiglia P, Novellino E, Hruby VJ (2008) Design and microwave-assisted synthesis of novel macrocyclic peptides active at melanocortin receptors: discovery of potent and selective hMC5R receptor antagonists. J Med Chem 51:2701–2707
25. Bin Zhang H, Chi YS, Huang WL, Ni SJ (2007) Total synthesis of human urotension-II by microwave-assisted solid phase method. Chin Chem Lett 18:902–904
26. Tofteng AP, Jensen KJ, Schaffer L, Hoeg-Jensen T (2008) Total synthesis of desB30 insulin analogues by biomimetic folding of single-chain precursors. Chembiochem 9:2989–2996
27. Schäffer L, Brand CL, Hansen BF, Ribel U, Shaw AC, Slaaby R, Sturis J (2008) A novel high-affinity peptide antagonist to the insulin receptor. Biochem Biophys Res Commun 376:380–383
28. Shabanpoor F, Hughes RA, Bathgate RA, Zhang S, Scanlon DB, Lin F, Hossain MA, Separovic F, Wade JD (2008) Solid-phase synthesis of europium-labeled human INSL3 as a novel probe for the study of ligand-receptor interactions. Bioconjug Chem 19:1456–1463
29. Armishaw C, Jensen AA, Balle T, Clark RJ, Harpsoe K, Skonberg C, Liljefors T, Stromgaard K (2009) Rational design of alpha-conotoxin analogues targeting alpha7 nicotinic acetylcholine receptors: improved antagonistic activity by incorporation of proline derivatives. J Biol Chem 284:9498–9512
30. Cassone M, Vogiatzi P, La Montagna R, De Olivier Inacio V, Cudic P, Wade JD, Otvos L Jr (2008) Scope and limitations of the designer proline-rich antibacterial peptide dimer, A3-APO, alone or in synergy with conventional antibiotics. Peptides 29:1878–1886
31. Noto PB, Abbadessa G, Cassone M, Mateo GD, Agelan A, Wade JD, Szabo D, Kocsis B, Nagy K, Rozgonyi F, Otvos L Jr (2008) Alternative stabilities of a proline-rich antibacterial peptide in vitro and in vivo. Protein Sci 17:1249–1255
32. James PF, Dogovski C, Dobson RCJ, Bailey MF, Goldie KN, Karas JA, Scanlon DB, O'Hair RAJ, Perugini MA (2009) Aromatic residues in the C-terminal helix of human apoC-I mediate phospholipid interactions and particle morphology. J Lipid Res 50:1384–1394
33. Chi Y, Zhang H, Huang W, Zhou J, Zhou Y, Qian H, Ni S (2008) Microwave-assisted solid phase synthesis, PEGylation, and biological activity studies of glucagon-like peptide-1(7-36) amide. Bioorg Med Chem 16:7607–7614
34. Murray JK, Aral J, Miranda LP (2011) Solid-phase peptide synthesis using microwave irradiation. In: Satyanarayanajois SD (ed) Drug design and discovery: methods and protocols. Springer Science+Business Media, LLC, New York

INDEX

Knud J. Jensen et al. (eds.), *Peptide Synthesis and Applications: Second Edition*, Methods in Molecular Biology, vol. 1047, DOI 10.1007/978-1-62703-544-6,

T

W

MIX
Papier aus verantwortungsvollen Quellen
Paper from responsible sources
FSC® C105338

If you have any concerns about our products,
you can contact us on
ProductSafety@springernature.com

In case Publisher is established outside the EU,
the EU authorized representative is:
Springer Nature Customer Service Center GmbH
Europaplatz 3, 69115 Heidelberg, Germany

Printed by Libri Plureos GmbH
in Hamburg, Germany